Concrete Technology

Concrete Technology

A. M. Neville

Civil Engineering Consultant
Formerly Principal and Vice-Chancellor
University of Dundee

J. J. Brooks

Senior Lecturer in Civil Engineering Materials
University of Leeds

Longman
Scientific &
Technical

Copublished in the United States with
John Wiley & Sons, Inc., New York

Longman Scientific & Technical,
Longman Group UK Limited,
Longman House, Burnt Mill, Harlow,
Essex CM20 2JE, England
and Associated Companies throughout the world.

Copublished in the United States with
John Wiley & Sons, Inc., 605 Third Avenue, New York, NY 10158

First published 1987
Revised reprint 1990
Reprinted 1991, 1993

British Library Cataloguing in Publication Data
Neville, A. M.
  Concrete technology.
  1. Concrete
  I. Title   II. Brooks, J. J.
  666'.893   TA439
ISBN 0-582-98859-4

Library of Congress Cataloging-in-Publication Data
Neville, Adam M.
  Concrete technology.
  Bibliography: p.
  Includes index.
  1. Concrete. 2. Cement I. Brooks, J. J.
  II. Title.
  TA439.N46 1987   666'.893   86-27231
  ISBN 0-470-20716-7 (USA only)

Produced by Longman Singapore Publishers Pte Ltd
Printed in Singapore

Contents

Preface

This book is aimed at British and American university, college, and polytechnic students who wish to understand concrete for the purpose of using it in professional practice.

The large incidence of material (as distinct from structural) failure of concrete structures in recent years – bridges, buildings, pavements and runways – is a clear indication that the professional engineer does not *know* enough about concrete. Perhaps, in consequence of his ignorance, he does not take sufficient care to ensure the selection of correct ingredients for concrete making, to achieve a suitable mix, and to obtain a technically sound execution of concrete works. The effects of climate and temperature, and of exposure conditions, do not seem always to be taken into account so as to ensure lasting and durable concrete structures.

The remedy lies in acquiring appropriate knowledge at the same time as structural design is learned, because the purpose of understanding concrete and its behaviour is to support the structural design so that its objectives are fully achieved and not vitiated by the passage of time and by environmental agencies. Indeed, the structural designer should be adequately familiar with his concrete so that structural detailing is predicated on a sound understanding of how concrete behaves under load, under temperature and humidity changes, and under the relevant conditions of environmental and industrial exposure. This book sets out to meet these needs.

Since construction is governed by contractual documents and specifications, the various properties of concrete have to be described in terms of national standards and recognized testing methods. The book refers to the important British and American standards[1] and shows how they link to the essential features of concrete behaviour.

An engineer involved in construction of a concrete structure, from a dam to a runway, from a bridge to a high-rise building, must design his concrete mix: unlike steel, this cannot be bought by reference to a

---

[1] The main authorities which produce standards, standard specifications and codes of practice are the British Standards Institution (BSI), the American Society for Testing and Materials (ASTM), and the American Concrete Institute (ACI). Extracts from British Standards are reproduced by permission of the British Standards Institution. Complete copies can be obtained from BSI at Linford Wood, Milton Keynes, MK14 6LE. The address of the American Society for Testing and Materials is 1619 Race Street, Philadelphia, Pennsylvania 19103. The address of the American Concrete Institute is Box 19150, 22400 West Seven Mile Road, Detroit, Michigan 48219.

xi

supplier's catalogue. The book discusses, with full examples, two of the most widespread methods of mix design, one American, the other British.

Finally, it should be pointed out that, since the success of a concrete structure is the concern both of the structural designer and of the contractor, no graduate engineer, whatever his career plans, can be ignorant of concrete technology. And even if his specialization is not in concrete, he will still need the material for retaining walls and foundations, for fireproofing and finishing, and for a multitude of ancillary works. He is therefore well advised to become thoroughly familiar with the contents of this book.

Acknowledgements

We are indebted to Miss Angela C. Ranby for her most valuable help in preparing the text for printing and to Miss Sarah J. Griffiths for her assistance in checking and handling the proofs.

1

Concrete as a structural material

The reader of this book is presumably someone interested in the use of concrete in structures, be they bridges or buildings, highways or dams. Our view is that, in order to use concrete satisfactorily, both the designer and the contractor need to be familiar with concrete technology. *Concrete Technology* is indeed the title of this book, and we ought to give reasons for this need.

These days, there are two commonly used structural materials: concrete and steel. They sometimes complement one another, and sometimes compete with one another, so that many structures of a similar type and function can be built in either of these materials. And yet, universities, polytechnics and colleges teach much less about concrete than about steel. This in itself would not matter were it not for the fact that, in actual practice, the man on the job needs to know more about concrete than about steel. This assertion will now be demonstrated.

Steel is manufactured under carefully controlled conditions, always in a highly sophisticated plant; the properties of every type of steel are determined in a laboratory and described in a manufacturer's certificate. Thus the designer of a steel structure need only specify the steel complying with a relevant standard, and the constructor need only ensure that correct steel is used and that connections between the individual steel members are properly executed.

On a concrete building site, the situation is totally different. It is true that the quality of cement is guaranteed by the manufacturer in a manner similar to that of steel, and, provided a suitable cement is chosen, its quality is hardly ever a cause of faults in a concrete structure. But cement is not the building material: concrete is. Cement is to concrete what flour is to a fruit cake, and the quality of the cake depends on the cook.

It is possible to obtain concrete of specified quality from a ready-mix supplier but, even in this case, it is only the raw material that is bought. Transporting, placing and, above all, compacting greatly influence the final product. Moreover, unlike the case of steel, the choice of mixes is virtually infinite and therefore the selection cannot be made without a sound knowledge of the properties and behaviour of concrete. It is thus the competence of the designer and of the specifier that determines the *potential* qualities of concrete, and the competence of the contractor and

1

the supplier that controls the *actual* quality of concrete in the finished structure. It follows that they must be thoroughly conversant with the properties of concrete and with concrete making and placing.

What is concrete?

An overview of concrete as a material is difficult at this stage because we must refrain from discussing specialized knowledge not yet presented, so that we have to limit ourselves to some selected features of concrete.

Concrete, in the broadest sense, is any product or mass made by the use of a cementing medium. Generally, this medium is the product of reaction between hydraulic cement and water. But, these days, even such a definition would cover a wide range of products: concrete is made with several types of cement and also containing pozzolan, fly ash, blast-furnace slag, a 'regulated set' additive, sulphur, admixtures, polymers, fibres, and so on; and these concretes can be heated, steam-cured, autoclaved, vacuum-treated, hydraulically pressured, shock-vibrated, extruded, and sprayed. This book is restricted to considering no more than a mixture of cement, water, aggregate (fine and coarse) and admixtures.

This immediately begs the question: what is the relation between the constituents of this mixture? There are three possibilities. First, one can view the cementing medium, i.e. the products of hydration of cement, as *the* essential building material, with the aggregate fulfilling the role of a cheap, or cheaper, dilutant. Second, one can view the coarse aggregate as a sort of mini-masonry which is joined together by mortar, i.e. by a mixture of hydrated cement and fine aggregate. The third possibility is to recognize that, as a first approximation, concrete consists of two phases: hydrated cement paste and aggregate, and, as a result, the properties of concrete are governed by the properties of the two phases and also by the presence of interfaces between them.

The second and third view each have some merit and can be used to explain the behaviour of concrete. The first view, that of cement paste diluted by aggregate, we should dispose of. Suppose you could buy cement more cheaply than aggregate – should you use a mixture of cement and water alone as a building material? The answer is emphatically *no* because the so-called volume changes[1] of hydrated cement paste are far too large: shrinkage[2] of neat cement paste is almost ten times larger than shrinkage of concrete with 250 kg of cement per cubic metre. Roughly the same applies to creep.[3] Furthermore, the heat generated by a large amount of hydrating cement,[4] especially in a hot climate,[5] may lead to cracking.[6] One can also observe that most aggregates are less

---

[1] Chapter 12   [4] Chapter 2
[2] Chapter 13   [5] Chapter 9
[3] Chapter 12   [6] Chapter 13

prone to chemical attack[7] than cement paste, even though the latter is, itself, fairly resistant. So, quite independently of cost, the use of aggregate[8] in concrete is beneficial.

Good concrete

Beneficial means that the influence is good and we could, indeed we should, ask the question: what is good concrete? It is easier to precede the answer by noting that *bad* concrete is, alas, a most common building material. By bad concrete we mean a substance with the consistence[9] of soup, hardening into a honeycombed,[10] non-homogeneous and weak mass, and this material is made simply by mixing cement, aggregate and water. Surprisingly, the ingredients of good concrete are exactly the same, and the difference is due entirely to 'know-how'.

With this 'know-how' we can make good concrete, and there are two overall criteria by which it can be so defined: it has to be satisfactory in its hardened state[11] and also in its fresh state[12] while being transported from the mixer and placed in the formwork. Very generally, the requirements in the fresh state are that the consistence of the mix is such that the concrete can be compacted[13] by the means which are actually available on the job, and also that the mix is cohesive[14] enough to be transported[15] and placed without segregation[16] by the means available. Clearly, these requirements are not absolute but depend on whether transport is by a skip with a bottom discharge or by a flat-tray lorry, the latter, of course, not being a very good practice.

As far as the hardened state[17] is considered, the usual requirement is a satisfactory compressive strength.[18] We invariably specify strength because it is easy to measure, although the 'number' that comes out of the test is certainly *not* a measure of the intrinsic strength of concrete in the structure but only of its quality. Thus, strength is an easy way of ascertaining compliance with the specification[19] and sorts out contractual obligations. However, there are also other reasons for the preoccupation with compressive strength, namely, that many properties of concrete are related to its compressive strength. These are: density,[20] impermeability,[21] durability,[22] resistance to abrasion,[23] resistance to impact,[24] tensile strength,[25] resistance to sulphates,[26] and some others, but not shrinkage[27] and not necessarily creep.[28] We are not saying that these properties are a single and unique function of compressive strength,

---

[7] Chapter 14
[8] Chapter 3
[9] Chapter 5
[10] Chapter 6
[11] Chapter 6
[12] Chapter 5

[13] Chapter 7
[14] Chapter 5
[15] Chapter 7
[16] Chapter 5
[17] Chapter 6
[18] Chapter 6

[19] Chapter 17
[20] Chapter 6
[21] Chapter 14
[22] Chapter 14
[23] Chapter 11

[24] Chapter 11
[25] Chapter 11
[26] Chapter 14
[27] Chapter 13
[28] Chapter 12

3

and we are aware of the issue of whether durability[29] is best ensured by specifying strength,[30] water/cement ratio,[31] or cement content.[32] But the point is that, in a very *general* way, concrete of higher strength has more desirable properties. A detailed study of all this is of course what concrete technology is all about.

Composite materials

We have referred to concrete as a two-phase material and we should now consider this topic further, with special reference to the modulus of elasticity[33] of the composite product. In general terms, a composite material consisting of two phases can have two fundamentally different forms. The first of these is an ideal composite *hard* material, which has a continuous matrix of an elastic phase with a high modulus of elasticity, and embedded particles of a lower modulus. The second type of structure is that of an ideal composite *soft* material, which consists of elastic particles with a high modulus of elasticity, embedded in a continuous matrix phase with a lower modulus.

The difference between the two cases can be large when it comes to the calculation of the modulus of elasticity of the composite. In the case of a composite hard material, it is assumed that the strain is constant over any cross-section, while the stresses in the phases are proportional to their respective moduli. This is the case on the left-hand side of Fig. 1.1. On the other hand, for composite soft material, the modulus of elasticity is calculated from the assumption that the stress is constant over any cross-section, while the strain in the phases is inversely proportional to their respective moduli; this is the picture on the right-hand side of Fig. 1.1. the corresponding equations are:

for a composite hard material

$$E = (1 - g)E_m + gE_p$$

and for a composite soft material

$$E = \left[\frac{1-g}{E_m} + \frac{g}{E_p} \right]^{-1}$$

where E = modulus of elasticity of the composite material,
E_m = modulus of elasticity of the matrix phase,
E_p = modulus of elasticity of the particle phase, and
g = fractional volume of the particles.

---

[29] Chapter 14   [32] Chapter 19
[30] Chapter 6   [33] Chapter 12
[31] Chapter 6

4

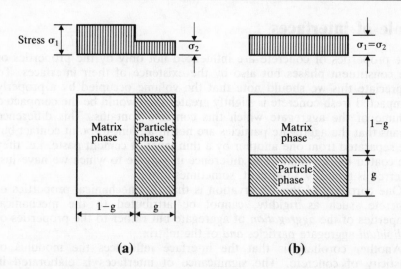

Fig. 1.1: Models for: (a) composite hard, and (b) composite soft materials

We must not be deceived by the simplicity of these equations and jump to the conclusion that all we need to know is whether the modulus of elasticity of aggregate is higher or lower than that of the paste. The fact is that these equations represent boundaries for the modulus of elasticity of the composite. With the practical random distribution of aggregate in concrete, neither boundary can be reached as neither satisfies the requirements of both equilibrium and compatibility. For practical purposes, a fairly good approximation is given by the expression for the composite soft material for mixes made with normal aggregates;[34] for lightweight aggregate mixes,[35] the expression for the composite hard material is more appropriate.

From the scientific point of view, there is something more that should be said on the subject of the two-phase approach, and that is that we can apply it to the cement phase alone as a sort of second step. Cement paste[36] can be viewed as consisting of hard grains of unhydrated cement in a soft matrix of products of hydration.[37] The products of hydration, in turn, consist of 'soft' capillary pores[38] in a hard matrix of cement gel.[39] Appropriate equations can be readily written down but, for the present purpose, it is sufficient to note that hard and soft are relative, and not absolute terms.

---

[34] Chapter 3   [37] Chapter 2
[35] Chapter 18   [38] Chapter 2
[36] Chapter 2   [39] Chapter 2

Role of interfaces

The properties of concrete are influenced not only by the properties of the constituent phases but also by the existence of their interfaces. To appreciate this we should note that the volume occupied by a properly compacted fresh concrete is slightly greater than would be the compacted volume of the aggregate which this concrete contains. This difference means that the aggregate particles are not in a point-to-point contact but are separated from one another by a thin layer of cement paste, i.e. they are coated by the paste. The difference in volume to which we have just referred is typically 3 per cent, sometimes more.

One corollary of this observation is that the mechanical properties of concrete, such as rigidity, cannot be attributed to the mechanical properties of the *aggregation* of aggregate but rather to the properties of *individual* aggregate particles *and* of the matrix.

Another corollary is that the interface influences the modulus of elasticity of concrete. The significance of interfaces is elaborated in Chapter 6, and a figure in that chapter (Fig. 6.11) shows the stress–strain relations[40] for aggregate, neat cement paste, and concrete. Here we have what at first blush is a paradox: aggregate alone exhibits a linear stress–strain relation and so does hydrated neat cement paste. But the composite material consisting of the two, i.e. concrete, has a curved relation. The explanation (which will be elaborated in Chapter 6) lies in the presence of the interfaces and in the development of microcracking[41] at these interfaces under load. These microcracks develop progressively at interfaces, making varying angles with the applied stress, and therefore there is a progressive increase in local stress intensity and in the magnitude of strain. Thus, strain increases at a faster rate than the applied stress, and so the stress–strain curve continues to bend over, with an apparently pseudo-plastic behaviour.

Approach to study of concrete

The preceding *mis en scène* introduces perforce many terms and concepts which may not be entirely clear to the reader. The best approach is to study the following chapters and then to return to this one.

The order of presentation is as follows. First, the ingredients of concrete: cement,[42] normal aggregate,[43] and mixing water.[44] Then, the concrete in its fresh state.[45] The following chapter[46] discusses the strength

---

[40] Chapter 12
[41] Chapter 6
[42] Chapter 2
[43] Chapter 3
[44] Chapter 4
[45] Chapter 5
[46] Chapter 6

of concrete because, as already mentioned, this is one of the most important properties of concrete and one that is always prominent in the specification.

Having established how we make concrete and what we fundamentally require, we turn to some techniques: mixing and handling,[47] use of admixtures to modify the properties at this stage,[48] and methods of dealing with temperature problems.[49]

In the following chapters, we consider the development of strength,[50] strength properties other than compressive and tensile strengths,[51] and behaviour under stress.[52] Next come the behaviour in normal environment,[53] durability,[54] and, in a separate chapter, resistance to freezing and thawing.[55]

Having studied the various properties of concrete, we turn to testing[56] and compliance with specifications,[57] and finally to mix design;[58] after all, this is what we must be able to do in order to choose the right mix for the right job. Two chapters extend our knowledge to less common materials: lightweight concrete[59] and special concretes.[60] As a *finale,* we review the advantages and disadvantages of concrete as a structural material.[61]

2

Cement

Ancient Romans were probably the first to use concrete – a word of Latin origin – based on *hydraulic cement,* that is a material which hardens under water. This property and the related property of not undergoing chemical change by water in later life are most important and have contributed to the widespread use of concrete as a building material. Roman cement fell into disuse, and it was only in 1824 that the modern cement, known as Portland cement, was patented by Joseph Aspdin, a Leeds builder.

Portland cement is the name given to a cement obtained by intimately mixing together calcareous and argillaceous, or other silica-, alumina-, and iron oxide-bearing materials, burning them at a clinkering temperature, and grinding the resulting clinker. The definitions of the British Standard (BS 12: 1978) and of the American Standard (ASTM C 150–84) are on those lines; no material, other than gypsum, water, and grinding aids may be added after burning.

Manufacture of Portland cement

From the definition of Portland cement given above, it can be seen that it is made primarily from a combination of a calcareous material, such as limestone or chalk, and of silica and alumina found as clay or shale. The process of manufacture consists essentially of grinding the raw materials into a very fine powder, mixing them intimately in predetermined proportions and burning in a large rotary kiln at a temperature of about 1400 °C (2550 °F) when the material sinters and partially fuses into clinker. The clinker is cooled and ground to a fine powder, with some gypsum added, and the resulting product is the commercial Portland cement used throughout the world.

The mixing and grinding of the raw materials can be done either in water or in a dry condition; hence, the names wet and dry process. The mixture is fed into a rotary kiln, sometimes (in the wet process) as large as 7 m (23 ft) in diameter and 230 m (750 ft) long. The kiln is slightly inclined. The mixture is fed at the upper end while pulverized coal (or other source of heat) is blown in by an air blast at the lower end of the kiln, where the temperature may reach about 1500 °C (2750 °F). The amount of coal required to manufacture one tonne (2200 lb) of cement is between 100 kg (220 lb) and about 350 kg (770 lb), depending on the process used.

As the mixture of raw materials moves down the kiln, it encounters a progressively higher temperature so that various chemical changes take place along the kiln: First, any water is driven off and CO_2 is liberated from the calcium carbonate. Further on, the dry material undergoes a series of chemical reactions until, finally, in the hottest part of the kiln, some 20 to 30 per cent of the material becomes liquid, and lime, silica and alumina recombine. The mass then fuses into balls, 3 to 25 mm ($\frac{1}{8}$ to 1 in.) in diameter, known as clinker.

Afterwards, the clinker drops into coolers, which provide means for an exchange of heat with the air subsequently used for the combustion of the pulverized coal. The cool clinker, which is very hard, is interground with gypsum in order to prevent flash-setting of the cement. The ground material, that is cement, has as many as 1.1×10^{12} particles per kilogramme (0.5×10^{12} per lb).

A single kiln of modern design (using the dry process) can produce as much as 6200 tonnes of clinker a day. To put this figure into perspective we can quote recent annual cement production figures: 70 million tonnes in the US and 14.5 million tonnes in the UK. Expressing the cement consumption (which is not the same as production because of imports and exports) in another way, we can note that the quantity of cement per capita was 342 kg (754 lb) in US and 265 kg (584 lb) in UK; the highest consumption in a large industrialized country was 678 kg (1494 lb) in Italy. Another figure of interest is the consumption of about 2000 kg (4400 lb) per capita in Saudi Arabia, Qatar and United Arab Emirates.

Basic chemistry of cement

We have seen that the raw materials used in the manufacture of Portland cement consist mainly of lime, silica, alumina and iron oxide. These compounds interact with one another in the kiln to form a series of more complex products, and, apart from a small residue of uncombined lime which has not had sufficient time to react, a state of chemical equilibrium is reached. However, equilibrium is not maintained during cooling, and the rate of cooling will affect the degree of crystallization and the amount of amorphous material present in the cooled clinker. The properties of this amorphous material, known as glass, differ considerably from those of crystalline compounds of a nominally similar chemical composition. Another complication arises from the interaction of the liquid part of the clinker with the crystalline compounds already present.

Nevertheless, cement can be considered as being in frozen equilibrium, i.e. the cooled products are assumed to reproduce the equilibrium existing at the clinkering temperature. This assumption is, in fact, made in the calculation of the compound composition of commercial cements: the 'potential' composition is calculated from the measured quantities of oxides present in the clinker as if full crystallization of equilibrium products had taken place.

Four compounds are regarded as the major constituents of cement: they are listed in Table 2.1 together with their abbreviated symbols. This shortened notation, used by cement chemists, describes each oxide by one letter, viz.: $CaO = C$; $SiO_2 = S$; $Al_2O_3 = A$; and $Fe_2O_3 = F$. Likewise, H_2O in hydrated cement is denoted by H.

Table 2.1: **Main compounds in Portland cement**

Name of compound	Oxide composition	Abbreviation
Tricalcium silicate	$3CaO.SiO_2$	C_3S
Dicalcium silicate	$2CaO.SiO_2$	C_2S
Tricalcium aluminate	$3CaO.Al_2O_3$	C_3A
Tetracalcium aluminoferrite	$4CaO.Al_2O_3.Fe_2O_3$	C_4AF

The calculation of the potential composition of Portland cement is based on the work of R. H. Bogue and others, and is often referred to as 'Bogue composition'. Bogue's equations for the percentages of main compounds in cement are given below. The terms in brackets represent the percentage of the given oxide in the total mass of cement.

$$C_3S = 4.07(CaO) - 7.60(SiO_2) - 6.72(Al_2O_3) - 1.43(Fe_2O_3) - 2.85(SO_3)$$

$$C_2S = 2.87(SiO_2) - 0.754(3CaO.SiO_2)$$

$$C_3A = 2.65(Al_2O_3) - 1.69(Fe_2O_3)$$

$$C_4AF = 3.04(Fe_2O_3).$$

The silicates, C_3S and C_2S, are the most important compounds, which are responsible for the strength of hydrated cement paste. In reality, the silicates in cement are not pure compounds, but contain minor oxides in solid solution. These oxides have significant effects on the atomic arrangements, crystal form, and hydraulic properties of the silicates.

The presence of C_3A in cement is undesirable: it contributes little or nothing to the strength of cement except at early ages, and when hardened cement paste is attacked by sulphates, the formation of calcium sulphoaluminate (ettringite) may cause disruption. However, C_3A is beneficial in the manufacture of cement in that it facilitates the combination of lime and silica.

C_4AF is also present in cement in small quantities, and, compared with the other three compounds, it does not affect the behaviour significantly;

10

however, it reacts with gypsum to form calcium sulphoferrite and its presence may accelerate the hydration of the silicates.

The amount of gypsum added to the clinker is crucial, and depends upon the C_3A content and the alkali content of cement. Increasing the fineness of cement has the effect of increasing the quantity of C_3A available at early ages, and this raises the gypsum requirement. An excess of gypsum leads to expansion and consequent disruption of the set cement paste. The optimum gypsum content is determined on the basis of the generation of the heat of hydration (see page 13) so that a desirable rate of early reaction occurs, which ensures that there is little C_3A available for reaction after all the gypsum has combined. ASTM C 150–84 and BS 12: 1989 specify the amount of gypsum as the mass of sulphur trioxide (SO_3) present.

In addition to the main compounds listed in Table 2.1, there exist minor compounds, such as MgO, TiO_2, Mn_2O_3, K_2O, and Na_2O; they usually amount to not more than a few per cent of the mass of cement. Two of the minor compounds are of interest: the oxides of sodium and potassium, Na_2O and K_2O, known as *the alkalis* (although other alkalis also exist in cement). They have been found to react with some aggregates, the products of the *alkali–aggregate* reaction causing disintegration of the concrete (see page 273), and have also been observed to affect the rate of the gain of strength of cement. It should, therefore, be pointed out that the term 'minor compounds' refers primarily to their quantity and not necessarily to their importance.

A general idea of the composition of cement can be obtained from Table 2.2, which gives the oxide composition limits of Portland cements. Table 2.3 gives the oxide composition of a typical cement and the calculated compound composition, obtained by means of Bogue's equations given on page 10.

Table 2.2: **Approximate composition limits of Portland cement**

Oxide	Content, per cent
CaO	60–67
SiO_2	17–25
Al_2O_3	3–8
Fe_2O_3	0.5–6.0
MgO	0.1–4.0
Alkalis	0.2–1.3
SO_3	1–3

Two terms used in Table 2.3 require explanation. The *insoluble residue*, determined by treating with hydrochloric acid, is a measure of adulteration of cement, largely arising from impurities in gypsum. BS 12: 1989 limits the insoluble residue to 1.5 per cent of the mass of cement; the corresponding limit of ASTM C 150–84 is 0.75 per cent. The *loss on ignition* shows the extent of carbonation and hydration of free lime and free magnesia due to the exposure of cement to the atmosphere. The specified limit both of ASTM C 150–84 and of BS 12: 1989 is 3 cent but the latter allows a loss on ignition of 4 per cent for cements in the tropics. Since hydrated free lime is innocuous, for a given free lime content of cement, a greater loss on ignition is really advantageous.

Table 2.3: **Oxide and compound compositions of a typical Portland cement**

Typical oxide composition per cent		Hence, calculated compound composition (using formulae of page 10), per cent	
CaO	63	C_3A	10.8
SiO_2	20	C_3S	54.1
Al_2O_3	6	C_2S	16.6
Fe_2O_3	3	C_4AF	9.1
MgO	$1\frac{1}{2}$	Minor compounds	–
SO_3	2		
K_2O			
Na_2O	1		
Others	1		
Loss on ignition	2		
Insoluble residue	$\frac{1}{2}$		

Hydration of cement

So far, we have discussed cement in powder form but the material of interest in practice is the set cement paste. This is the product of reaction of cement with water. What happens is that, in the presence of water, the silicates and aluminates (Table 2.1) of Portland cement form products of hydration or hydrates, which in time produce a firm and hard mass – the hardened cement paste. As stated earlier, the two calcium silicates (C_3S and C_2S) are the main cementitious compounds in cement, the former hydrating much more rapidly than the latter. In commercial cements, the calcium silicates contain small impurities from some of the oxides present in the clinker. These impurities have a strong effect on the properties of the hydrated silicates. The 'impure' C_3S is known as alite and the 'impure' C_2S as belite.

The product of hydration of C_3S is the microcrystalline hydrate $C_3S_2H_3$ with some lime separating out as crystalline $Ca(OH)_2$; C_2S behaves similarly but clearly contains less lime. Nowadays, the calcium silicate hydrates are described as C-S-H (previously referred to as tobermorite gel), the approximate hydration reactions being written as follows:

For C_3S:

$$2C_3S + 6H \longrightarrow C_3S_2H_3 + 3Ca(OH)_2.$$
$$[100] \quad\, [24] \qquad\; [75] \qquad\;\; [49]$$

For C_2S:

$$2C_2S + 4H \longrightarrow C_3S_2H_3 + Ca(OH)_2.$$
$$[100] \quad\, [21] \qquad\; [99] \qquad\; [22]$$

The numbers in the square brackets are the corresponding masses, and on this basis both silicates require approximately the same amount of water for hydration, but C_3S produces more than twice as much $Ca(OH)_2$ as is formed by the hydration of C_2S.

The amount of C_3A in most cements is comparatively small; its hydrate structure is of a cubic crystalline form which is surrounded by the calcium silicate hydrates. The reaction of pure C_3A with water is very rapid and would lead to a *flash set,* which is prevented by the addition of gypsum to the cement clinker. Even so, the rate of reaction of C_3A is quicker than that of the calcium silicates, the approximate reaction being

$$C_3A + 6H \longrightarrow C_3AH_6.$$
$$[100] \quad\; [40] \qquad\;\; [140]$$

The bracketed masses show that a higher proportion of water is required than for the hydration of silicates.

It may be convenient at this stage to summarize the pattern of formation and hydration of cement: this is shown schematically in Fig. 2.1.

Heat of hydration and strength

In common with many chemical reactions, the hydration of cement compounds is exothermic, and the quantity of heat (in joules) per gram of unhydrated cement, evolved upon complete hydration at a given temperature, is defined as the heat of hydration. Methods of determining its value are described in BS 4550: Part 3: Section 3.8: 1978, and in ASTM C 186–82.

The temperature at which hydration occurs greatly affects the rate of heat development, which for practical purposes is more important than the total heat of hydration; the same total heat produced over a longer period can be dissipated to a greater degree with a consequent smaller rise in temperature. This problem is discussed on page 169.

For the usual range of Portland cements, about one-half of the total heat is liberated between 1 and 3 days, about three-quarters in 7 days,

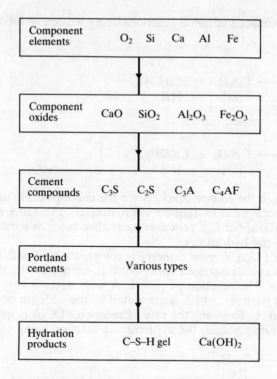

Fig. 2.1: Schematic representation of the formation and hydration of Portland cement

and nearly 90 per cent in 6 months. In fact, the heat of hydration depends on the chemical composition of the cement, and is approximately equal to the sum of the heats of hydration of the individual pure compounds when their respective proportions by mass are hydrated separately; typical values are given in Table 2.4.

It follows that by reducing the proportions of C_3A and C_3S, the heat of hydration (and its rate) of cement can be reduced. Fineness of cement affects the rate of heat development but not the total amount of heat

Table 2.4: Heat of hydration of pure compounds

Compound	Heat of hydration	
	J/g	Cal/g
C_3S	502	120
C_2S	260	62
C_3A	867	207
C_4AF	419	100

14

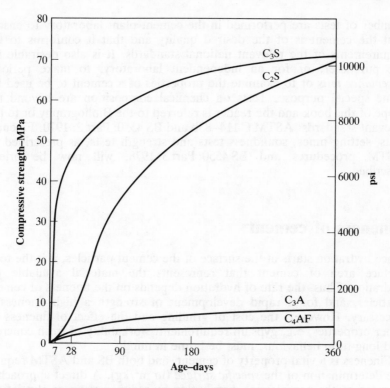

Fig. 2.2: Development of strength of pure compounds
(From: R. H. BOGUE, *Chemistry of Portland Cement* (New York, Reinhold, 1955).)

liberated, which can be controlled in concrete by the quantity of cement in the mix (richness).

It may be noted that there is no relation between the heat of hydration and the cementing properties of the individual compounds. As we have said, the two compounds primarily responsible for the strength of hydrated cement are C_3S and C_2S, and a convenient rule assumes that C_3S contributes most to the strength development during the first four weeks and C_2S influences the later gain in strength. At the age of about one year, the two compounds, mass for mass, contribute approximately equally to the strength of hydrated cement. Figure 2.2 shows the development of strength of the four pure compounds of cement. However, in contrast to the prediction of heat of hydration of cement from its constituent compounds, it has not been found possible to predict the strength of hydrated cement on the basis of compound composition.

Tests on cement

Because the quality of cement is vital for the production of good concrete, the manufacture of cement requires stringent control. A number of tests are performed in the cement plant laboratory to ensure that the cement is of the desired quality and that it conforms to the requirements of the relevant national standards. It is also desirable for the purchaser, or for an independent laboratory, to make periodic *acceptance tests* or to examine the properties of a cement to be used for some special purpose. Tests on chemical composition are beyond the scope of this book and the reader is referred to the Bibliography or to the relevant standards: ASTM C 114–83b and BS 4550: Part 2: 1970. Fineness tests, setting times, soundness tests and strength tests, as prescribed by ASTM procedures and BS 4550: Part 3: 1978, will now be briefly described.

Fineness of cement

Since hydration starts at the surface of the cement particles, it is the total surface area of cement that represents the material available for hydration. Thus, the rate of hydration depends on the fineness of cement particles, and for a rapid development of strength a high fineness is necessary. However, the cost of grinding and the effect of fineness on other properties, e.g. gypsum requirement, workability of fresh concrete and long-term behaviour, must be borne in mind.

Fineness is a vital property of cement, and both BS and ASTM require the determination of the *specific surface* (in m^2/kg). A direct approach is to measure the particle size distribution by sedimentation or elutriation; these methods are based on Stoke's law, giving the terminal velocity of fall under gravity of a spherical particle in a fluid medium. A development is the Wagner turbidimeter, as specified by ASTM C 115–79b. Here, the concentration of particles in suspension at a given level in kerosene is determined using a beam of light, the percentage of light transmitted (and hence the area of particles) being measured by a photocell. A typical curve of particle size distribution is shown in Fig. 2.3, which also gives the corresponding contribution of these particles to the total surface area of the sample.

The specific surface of cement can be determined by the air permeability (Lea and Nurse) method (BS 4550: Part 3: Section 3.3: 1978) which measures the pressure drop when dry air flows at a constant velocity through a bed of cement of known porosity and thickness. From this, the surface area per unit mass of the bed can be related to the permeability of the bed. A modification of this method is that of Blaine (ASTM C 204–84), in which the air does not pass through the bed at a constant rate, but a known volume of air passes at a prescribed average pressure, the rate of flow diminishing steadily; the time taken for the flow to take place is measured, and for a given apparatus and standard porosity, the specific surface can be calculated.

Both of the above air permeability methods give similar values of specific surface but very much higher than the Wagner turbidimeter method (see Table 2.5). This is due to Wagner's assumption about the

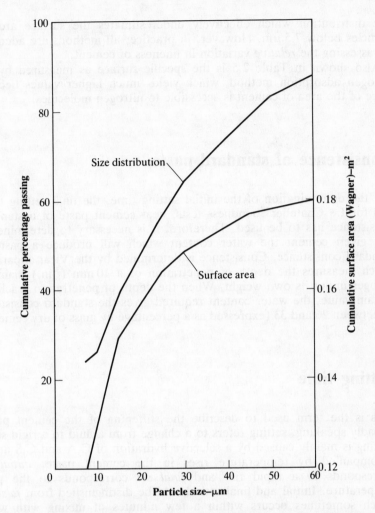

Fig. 2.3: Example of particle size distribution and cumulative surface area contributed by particles up to any size for 1 gram of cement

Table 2.5: **Examples of specific surface of cement measured by different methods**

Cement	Specific surface (m²/kg) measured by:		
	Wagner method	**Lea and Nurse method**	**Nitrogen adsorption method**
A	180	260	790
B	230	415	1000

17

size distribution which effectively underestimates the surface area of particles below 7.5 μm. However, in practice, all methods are adequate for assessing the *relative* variation in fineness of cement.

Also shown in Table 2.5 is the specific surface as measured by the nitrogen adsorption method, which yields much higher values because more of the area of cement is accessible to nitrogen molecules.

Consistence of standard paste

For the determination of the initial setting time, the final setting time, and for Le Chatelier soundness tests, neat cement paste of a standard consistence has to be used. Therefore, it is necessary to determine for any given cement the water content which will produce a paste of standard consistence. Consistence is determined by the Vicat apparatus, which measures the depth of penetration of a 10 mm ($\frac{3}{8}$ in.) diameter plunger under its own weight. When the depth of penetration reaches a certain value, the water content required gives the standard consistence of between 26 and 33 (expressed as a percentage by mass of dry cement).

Setting time

This is the term used to describe the stiffening of the cement paste. Broadly speaking, setting refers to a change from a fluid to a rigid state. Setting is mainly caused by a selective hydration of C_3A and C_3S and is accompanied by temperature rises in the cement paste; *initial set* corresponds to a rapid rise and *final set* corresponds to the peak temperature. Initial and final sets should be distinguished from *false set* which sometimes occurs within a few minutes of mixing with water (ASTM C 451–83). No heat is evolved in a false set and the concrete can be re-mixed without adding water. *Flash set* has previously been mentioned and is characterized by the liberation of heat.

For the determination of initial set, the Vicat apparatus is again used, this time with a 1 mm (0.04 in.) diameter needle, acting under a prescribed weight on a paste of standard consistence. When the needle penetrates to a point 5 mm (0.2 in.) from the bottom of a special mould, initial set is said to occur (time being measured from adding the mixing water to the cement). A minimum time of 45 minutes is prescribed by BS 12: 1978 for ordinary and rapid-hardening Portland (Types I and III) cements, and also for Portland blast-furnace (Type IS) cement; for low heat Portland cement (BS 1370: 1979), the minimum time is 60 minutes.

A similar procedure is specified by ASTM C 191–82 except that a smaller depth of penetration is required; a minimum setting time of 60 minutes is prescribed for Portland cements (ASTM C 150–84).

Final set is determined by a needle with a metal attachment hollowed out so as to leave a circular cutting edge 5 mm (0.2 in.) in diameter and set 0.5 mm (0.02 in.) behind the tip of the needle. Final set is said to have occurred when the needle makes an impression on the paste surface but the cutting edge fails to do so. British Standards prescribe the final setting time as a maximum of 10 hours for Portland cements, which is the same as that of the American Standards. An alternative method is that of the Gillmore test, as prescribed by ASTM C 266–81.

The initial and final setting times are approximately related:

final time (min.) = 90 + 1.2 [initial time (min.)]

(except for high alumina cement). Since temperature affects the setting times, BS 4550: Part 3: Section 3.6: 1978 specifies that the mixing has to be undertaken at a temperature of $20 \pm 2\,°C$ ($68 \pm 4\,°F$) and minimum relative humidity of 65 per cent, and the cement paste stored at $20 \pm 1\,°C$ ($68 \pm 2\,°F$) and maximum relative humidity of 90 per cent.

Soundness

It is essential that the cement paste, once it has set, does not undergo a large change in volume. One restriction is that there must be no appreciable expansion, which under conditions of restraint could result in disruption of the hardened cement paste. Such expansion may occur due to reactions of free lime, magnesia and calcium sulphate, and cements exhibiting this type of expansion are classified as unsound.

Free lime is present in the clinker and is intercrystallized with other compounds; consequently, it hydrates very slowly occupying a larger volume than the original free calcium oxide. Free lime cannot be determined by chemical analysis of cement because it is not possible to distinguish between unreacted CaO and $Ca(OH)_2$ produced by a partial hydration of the silicates when the cement is exposed to the atmosphere.

Magnesia reacts with water in a manner similar to CaO, but only the crystalline form is deleteriously reactive so that unsoundness occurs. Calcium sulphate is the third compound liable to cause expansion through the formation of calcium sulphoaluminate (ettringite) from excess gypsum (not used up by C_3A during setting).

Le Chatelier's accelerated test is prescribed by BS 4550: Part 3: Section 3.7: 1978 for detecting unsoundness due to free lime only. Essentially, the test is as follows. Cement paste of standard consistence is stored in water for 24 hours. The expansion is determined after increasing the temperature and boiling for 1 hour, followed by cooling to the original temperature. If the expansion exceeds a specified value, a further test is made after the cement has been spread and aerated for 7 days. At the end of this period, lime may have hydrated or carbonated, so that a second expansion test should fall within 50 per cent of the original specified value. A cement which fails to satisfy at least one of these tests should not be used. In practice, unsoundness due to free lime is very rare.

Magnesia is rarely present in large quantities in the raw materials used for making cement in the UK, but in the US this is not the case. For this reason, ASTM C 151–84 specifies the autoclave test which is sensitive to both free magnesia and free lime. Here, a neat cement paste specimen of known length is cured in humid air for 24 h and then heated by high-pressure steam (2 MPa (295 psi)) for about 1 h so that a temperature of 216 °C (420 °F) is attained. After maintaining that temperature and pressure for a further 3 h, the autoclave is cooled so that the pressure falls within 1.5 h and the specimen is cooled in water to 23 °C (73 °F) in 15 min. After a further 15 min, the length of the specimen is measured: the expansion due to autoclaving must not exceed 0.8 per cent of the original length. This accelerated test gives no more than a broad indication of the risk of long-term expansion in practice.

No test is available for the detection of unsoundness due to an excess of calcium sulphate, but its content can be easily determined by chemical analysis.

Strength

Strength tests are not made on neat cement paste because of difficulties in obtaining good specimens and in testing with a consequent large variability of test results. Cement–sand mortar and, in some cases, concrete of prescribed proportions, made with specified materials under strictly controlled conditions, are used for the purpose of determining the strength of cement.

There are several forms of strength tests: direct tension, compression, and flexure. In recent years, the tension test has been gradually superseded by the compression test and therefore will not be discussed here.

There are two British Standard methods for testing the compressive strength of cement: one uses mortar, the other concrete (BS 4550: Part 3: Section 3.4: 1978). In the mortar test, a 1:3 cement–sand mortar is used, with the mass of water in the mix being 10 per cent of the mass of the dry materials. The sand is a standard material of nearly one size. Using standard mixing and casting procedures, 71 mm (2.78 in.) cubes are demoulded after 24 hours and cured in water until they are tested in a wet-surface condition. Since the strength of mortar is not directly representative of that of concrete, and for other reasons, the concrete test was introduced in 1958 and is now covered by BS 4550: Part 3: 1978.

ASTM C 109–80 prescribes a cement–sand mix with proportions of 1:2.75 and a water/cement ratio of 0.485, using Ottawa (Illinois) sand for making 51 mm (2 in.) cubes. The mixing and casting procedure is similar to that of BS 4550:1978 but the cubes are cured in saturated lime water at 23 °C (73 °F) until they are tested.

An alternative compression test is the *modified cube* method (ASTM C 349–82) which utilizes the sections of failed flexural prisms (see below).

BS 4550: Part 3: Section 3.4: 1978 prescribes a concrete cube test with one of three water/cement ratios: 0.60, 0.55 and 0.45. The amount of aggregate, which has to come from particular quarries, is specified in BS 4550: Parts 4 and 5: 1978. Batches of 100 mm (4 in.) cubes are made by hand in a prescribed manner under specified conditions of mixing and curing, and BS 12: 1978 requires certain minimum values of strength at each age (see Table 2.6).

Table 2.6: **BS 12: 1989 and ASTM C 150–86 requirements for strength of cement**

Age (days)	Minimum compressive strength															
	BS 12: 1989 (Mortar)						BS 12: 1989 (Concrete)						ASTM C 150–86 (Concrete)			
	Controlled fineness Portland		Ordinary Portland		Rapid hardening		Controlled fineness Portland		Ordinary Portland		Rapid hardening		Type I		Type III	
	MPa	psi	MPa	psi	MPa	psi	MPa	psi	MPa	psi	MPa	psi	MPa	psi	MPa	psi
1	—	—	—	—	—	—	—	—	—	—	—	—	—	—	12	1740
2	—	—	—	—	25	3630	—	—	—	—	15	2180	—	—	—	—
3	23	3340	25	3630	—	—	13	1890	15	2180	—	—	12	1740	24	3480
7	—	—	—	—	—	—	—	—	—	—	—	—	19	2760	—	—
28	41	5950	47*	6820*	52	7540	29	4210	34**	4930**	38	5510	28+	4060	—	—

* and not more than 67 MPa (9720 psi)
** and not more than 52 MPa (7540 psi)
+ not normally specified

The flexural test, prescribed in ASTM C 348–80, uses simply-supported 40 × 40 × 160 mm mortar prisms loaded at mid-span; the mix proportions, storage, and curing procedures are the same as for the compression test. As stated earlier, an advantage of this test is that the modified cube test can be undertaken as well.

Types of Portland cement

So far, we have considered Portland cement as a generic material. However, when hydrated, cements differing in chemical composition may

Table 2.7: Main types of Portland cement

British classification		American classification	
Description	BS	Description	ASTM
Controlled fineness Portland	12: 1989	—	—
Ordinary Portland	12: 1989	Type I	C 150–86
Rapid-hardening Portland	12: 1989	Type III	C 150–86
Ultra-high early strength Portland	—	—	—
Low-heat Portland	1370: 1979	Type IV	C 150–86
Modified cement	—	Type II	C 150–86
Sulphate-resisting Portland	4027: 1980	Type V	C 150–86
Portland blast-furnace (slag cement)	146: Part 2: 1973	Type IS Type IS(MS)	C 595–86
Low-heat Portland blast-furnace	4246: Part 2: 1974	—	—
White Portland	12: 1989	—	C 150–86
	4627: 1970	Type IP	
Portland-pozzolan	6588: 1985	Type P	C 595–86
	3892: Part 1: 1982	Type I(PM)	

NB Cements Type I, IS, IP, P, I(PM), II and III are also available with an interground air-entraining agent, and are then denoted by letter A, e.g. Type IA (see page 291).

Table 2.8: **Typical average values of compound composition of Portland cements of different types**

Cement	Compound composition, per cent							
	C_3S	C_2S	C_3A	C_4AF	$CaSO_4$	Free CaO	MgO	Loss on ignition
Type I	59	15	12	8	2.9	0.8	2.4	1.2
Type II	46	29	6 (8 max.)	12	2.8	0.6	3.0	1.0
Type III	60	12	12 (15 max.)	8	3.9	1.3	2.6	1.9
Type IV	30 (35 max.)	46 (40 min.)	5 (7 max.)	13	2.9	0.3	2.7	1.0
Type V	43	36	4 (5 max.)	12	2.7	0.4	1.6	1.0

The numbers in parentheses are the maximum or minimum values specified by ASTM C 150–84.

exhibit different properties. It should thus be possible to select mixtures of raw materials for the production of cements with various desired properties. In fact, several types of Portland cement are available commercially, and additional special cements can be produced for special uses. Table 2.7 lists the main types of Portland cement together with the appropriate BS and ASTM Standards, while Table 2.8 gives the average values of compound composition.

Many of the cements have been developed to ensure good durability of concrete under a variety of conditions. It has not been possible, however,

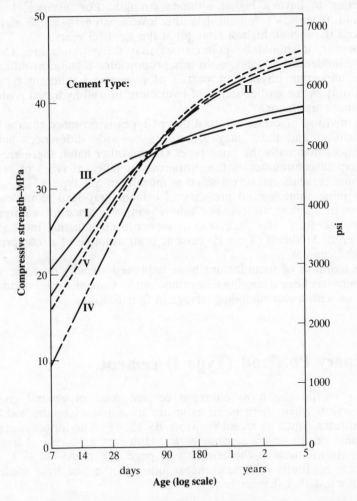

Fig. 2.4: Strength development of concretes containing 335 kg of cement per cubic metre (565 lb/yd³) and made with Portland cements of different types: ordinary (Type I), modified (Type II), rapid-hardening (Type III), low-heat (Type IV), and sulphate-resisting (Type V)
(From: US BUREAU OF RECLAMATION, *Concrete Manual*, 8th Edn (Denver, Colorado, 1975).)

23

to find in the composition of cement a complete answer to the problem of durability of concrete: the principal mechanical and physical properties of hardened concrete, such as strength, shrinkage, permeability, resistance to weathering, and creep, are affected also by factors other than cement composition, although this determines to a large degree the rate of gain of strength. Figure 2.4 shows the general rate of development of strength of concretes made with cements of different types: while the rates vary considerably, there is little difference in the 90-day strength of cements of all types. The general tendency is for the cements with a slow rate of hardening to have a higher ultimate strength. For instance, low-heat Portland (Type IV) cement has the lowest strength at 28 days but develops the second highest strength at the age of 5 years.

However, it should be pointed out that these trends are, to some extent, influenced by changes in mix proportions. Significant differences in the important physical properties of cements of different types are found only in the earlier stages of hydration; in well-hydrated pastes the differences are minor.

The division of cements into different types is no more than a broad classification and there may sometimes be wide differences between cements of nominally the same type. On the other hand, there are often no sharp discontinuities in the properties of different types of cement, and some cements can be classified as more than one type.

Obtaining some special property of cement may lead to undesirable features in another respect. For this reason, a balance of requirements may be necessary, and the economic aspect of manufacture must also be considered. Modified (Type II) cement is an example of a 'compromise' all-round cement.

The methods of manufacture have improved steadily over the years, and there has been a continual development of cements to serve different purposes with a corresponding change in specifications.

Ordinary Portland (Type I) cement

This is by far the most common cement used in general concrete construction when there is no exposure to sulphates in the soil or in groundwater. Since its recent revision, BS 12: 1989 no longer contains a limitation on the *lime saturation factor*. However, an excess of free lime causes unsoundness of the cement (see page 19). In ASTM C 150–84, there are no limits of lime content, although the free lime content is generally less that 0.5 per cent.

Further standard requirements are as follows:

	BS 12 : 1989	ASTM C 150–86
magnesium oxide	≯ 4 per cent	≯ 6 per cent
insoluble residue	≯ 1.5 per cent	≯ 0.75 per cent
loss on ignition	≯ 3 per cent	≯ 3 per cent
gypsum content (expressed as SO_3) when C_3A content is:		
unspecified	3[1] or 3.5[2]	—
3.5 per cent	2.5 per cent	—
≯ 8 per cent	—	3 per cent
> 8 per cent	—	3.5 per cent

Over the years, there have been changes in the characteristics of ordinary Portland cement: modern cements have a higher C_3S content and a greater fineness than 40 years ago. BS 12: 1989 specifies a minimum of 275 m^2/kg for ordinary Portland cement and 225 kg/m^2 for controlled fineness Portland cement. In consequence, modern cements have a higher 28-day strength than in the past, but the later gain in strength is smaller. A practical consequence of this is that we can no longer expect 'improvement with age'. This is an important point to remember since construction specifications are usually related to the 28-day strength of concrete. Moreover, using a high early strength cement for a given specified 28-day strength of concrete, it is possible to use a leaner mix with a higher water/cement ratio. Some of these mixes have an inadequate durability.

Ordinary Portland (Type I) cement is an excellent general cement and is the cement most widely used.

Rapid-hardening Portland (Type III) cement

This cement is similar to Type I cement and is covered by the same standards. As the name implies, the strength of this cement develops rapidly because (as can be seen from Table 2.8) of a higher C_3S content (up to 70 per cent) and a higher fineness (minimum 325 m^2/kg); these days, it is the fineness that is the distinguishing factor between the ordinary and the rapid-hardening Portland cements, and there is generally little difference in chemical composition.

The principal reason for the use of Type III cement is when formwork is to be removed early for re-use or where sufficient strength for further construction is required quickly. Rapid-hardening Portland cement

---

[1]For controlled fineness cement.
[2]For ordinary Portland cement.

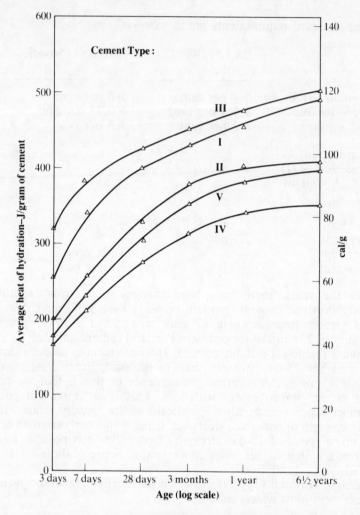

Fig. 2.5: Development of heat of hydration of different Portland cements
cured at 21 °C (70 °F) (water/cement ratio of 0.40): ordinary (Type
I), modified (Type II), rapid-hardening (Type III), low-heat (Type
IV), and sulphate-resisting (Type V)
(From: G. J. VERBECK and C. W. FOSTER, Long-time study of
cement performance in concrete, Chapter 6: The heats of hydration
of the cements, *Proc. ASTM,* 50, pp. 1235–57 (1950).)

should not be used in mass concrete construction or in large structural
sections because of its higher rate of heat development (see Fig. 2.5). On
the other hand, for construction at low temperatures, the use of this
cement may provide a satisfactory safeguard against early frost damage
(see Ch. 15).

The setting time of Type III and Type I cements is the same. The cost
of Type III cement is only marginally greater than that of ordinary
Portland cement.

Special rapid-hardening Portland cements

These are specially manufactured cements which are highly rapid-hardening. For instance, ultra-high early strength Portland cement is permitted for structural use in the UK. The high early strength is achieved by a higher fineness (700 to $900 \, m^2/kg$) and a higher gypsum content, but this does not affect long-term soundness. Typical uses are early prestressing and urgent repairs.

In some countries, a regulated-set cement (or jet cement) is made from a mixture of Portland cement and calcium fluoraluminate with an appropriate retarder (usually citric acid). The setting time (1 to 30 min.) can be controlled in the manufacture of the cement as the raw materials are interground and burnt together, while the early strength development is controlled by the content of the calcium fluoraluminate. This cement is expensive but valuable when an extremely early high strength is needed.

Low-heat Portland (Type IV) cement

Developed in the US for use in large gravity dams, this cement has a low heat of hydration. Both ASTM C 150–84 and BS 1370: 1979 limit the heat of hydration to $250 \, J/g$ ($60 \, cal/g$) at the age of 7 days, and $290 \, J/g$ ($70 \, cal/g$) at 28 days.

According to BS 1370: 1979, the limits of the lime saturation factor (see page 24) are 0.66 to 1.08, and, because of the lower content of C_3S and C_3A, there is a slower development of strength than with ordinary Portland cement, but the ultimate strength is unaffected. The fineness must not be less than $320 \, m^2/kg$ to ensure a sufficient rate of gain of strength.

In the US, Portland-pozzolan (Type P) cement can be specified to be of the low-heat variety, while Type IP cement can be required to have a moderate heat of hydration. ASTM C 595–83a deals with these cements.

Modified (Type II) cement

In some applications, a very low early strength may be a disadvantage, and for this reason a modified cement was developed in the US. This cement has a higher rate of heat development than that of Type IV cement, and a rate of gain of strength similar to that of Type I cement. Type II cement is recommended for structures where a moderately low heat generation is desirable or where moderate sulphate attack may occur (see page 267). This cement is not available in the United Kingdom.

Sulphate-resisting (Type V) cement

This cement has a low C_3A content so as to avoid sulphate attack from outside the concrete; otherwise the formation of calcium sulphoaluminate and gypsum would cause disruption of the concrete due to an increased volume of the resultant compounds. The salts particularly active are magnesium and sodium sulphate, and sulphate attack is greatly accelerated if accompanied by alternate wetting and drying, e.g. in marine structures subject to tide or splash.

To achieve sulphate resistance, the C_3A content in sulphate-resisting cement is limited to 3.5 per cent (BS 4027: 1980) with a minimum fineness of $250 \, m^2/kg$; otherwise this cement conforms to the specification for ordinary Portland cement. In the US, when the limit of sulphate expansion is not specified, the C_3A content is limited to 5 per cent (ASTM C 150–84), and the total content of C_4AF plus twice the C_3A content is limited to 20 per cent; also, the gypsum content is limited to 2.3 per cent when the maximum C_3A content is 8 per cent or less.

In the US, there exist also cements with moderate sulphate-resisting properties. These are produced by blending Portland cement with slag (Type IS(MS)) or with pozzolan (Type IP(MS)). They have the C_3A content limited to 8 per cent and are covered by ASTM C 595–83a.

Provision for a low-alkali sulphate-resisting cement is made in BS 4027: 1980.

The heat developed by sulphate-resisting cement is not much higher than that of low-heat cement, which is an advantage, but the cost of the former is higher due to the special composition of the raw materials. Thus, in practice, sulphate-resisting cement should be specified only when necessary; it is not a cement for general use.

Portland blast-furnace (Type IS) cement

This type of cement is made by intergrinding or blending Portland cement clinker with granulated blast-furnace slag, which is a waste product in the manufacture of pig iron; thus, there is a lower energy consumption in the manufacture of cement. Slag contains lime, silica and alumina, but not in the same proportions as in Portland cement, and its composition can vary a great deal. Sometimes, Portland blast-furnace cement is referred to as *slag* cement.

The hydration of slag is initiated when lime liberated in the hydration of Portland cement provides the correct alkalinity; subsequent hydration does not depend on lime.

The amount of slag should be between 25 and 70 per cent of the mass of the mixture, according to ASTM C 595–83a, and less than 65 per cent according to BS 146: Part 2: 1973. However, BS 4246: Part 2: 1974 allows a slag content of 50 to 90 per cent for the manufacture of low-heat Portland blast-furnace cement.

In the UK, it is proposed to issue a standard specification for ground granulated blast-furnace slag for use in concrete, mortar or grout, in combination with Portland cement. The tests required are similar to those for Portland cement, the limiting values being: minimum fineness of $275 \, m^2/kg$; maximum insoluble residue of 1.5 per cent; maximum magnesia content of 14 per cent; maximum sulphur content of 2 per cent; maximum loss on ignition of 3 per cent; and maximum manganese content of 2 per cent. In addition, the maximum lime/silica ratio is 1.4 and the minimum *chemical modulus* is 1, the latter being given as

$$\frac{(CaO) + (MgO) + (Al_2O_3)}{(SiO_2)}$$

where each symbol in brackets refers to the percentage by mass in the total slag of the particular oxide as determined according to BS 4550: Part 2: 1970. Other proposed requirements are that the glass content should be a specified minimum and the moisture content at delivery should not exceed 1 per cent of the mass of the dry slag. Limits of compressive strength, setting times and soundness are specified.

Portland blast-furnace cement is similar to ordinary Portland (Type I) cement as regards fineness, setting times and soundness. However, early strengths are generally lower than in Type I cement; later strengths are similar. BS 146: Part 2: 1973 requires a 28-day cube strength of 34 MPa (4900 psi) for mortar or 22 MPa (3200 psi) for concrete.

Typical uses are in mass concrete because of a lower heat of hydration and in sea-water construction because of a better sulphate resistance (due to a lower C_3A content) than with ordinary Portland cement. Slag with a low alkali content can also be used with an aggregate suspected of alkali reactivity (see page 273).

A variant used in the UK is part-replacement, at the mixer, of cement by dry-ground granulated slag of the same fineness.

Portland blast-furnace cement is in common use in countries where slag is widely available and can be considered to be a cement for general use.

Supersulphated (slag) cement

Because it is made from granulated blast-furnace slag, supersulphated cement will be considered at this stage, even though it is not a Portland cement.

Supersulphated cement is made by intergrinding a mixture of 80 to 85 per cent of granulated slag with 10 to 15 per cent of calcium sulphate (in the form of dead-burnt gypsum or anhydrite) and about 5 per cent of Portland cement clinker. A fineness of 400 to 500 m^2/kg is usual. Supersulphated cement has a low heat of hydration (about 200 J/g (48 cal/g) at 28 days). Although not available in the UK, the cement is covered by BS 4248: 1974.

The advantages of supersulphated cement lie in a high resistance to sea water and to sulphate attack, as well as to peaty acids and oils. The use of this cement requires particular attention as its rate of strength development is strongly affected at low and high temperatures, and it should not be mixed with Portland cements; also, the range of mix proportions is limited so as not to affect the strength development. The cement has to be stored under very dry condtions as otherwise it deteriorates rapidly.

White and coloured Portland cements

For architectural purposes, white concrete, or, particularly in tropical countries, a pastel colour paint finish is sometimes required. For these purposes, white cement is used. It is also used because of its low content of soluble alkalis so that staining is avoided. White cement is made from china clay, which contains little iron oxide and manganese oxide, together with chalk or limestone free from specified impurities. In addition, special precautions are required during the grinding of the clinker so as to avoid contamination. For these reasons, the cost of white cement is high (twice that of ordinary Portland cement). Because of this, white concrete is often used in the form of a well-bonded facing against normal concrete backing.

Pastel colours can be obtained by painting or by adding pigments to the mixer, provided there is no adverse effect on strength. Air-entraining pigments are available in the US, and an improved uniformity of colour is achieved by using a superplasticizer (see page 156). Alternatively, it is possible to obtain white cement interground with a pigment (BS 1014: 1975 (1986)). White high-alumina cement is also manufactured but is expensive (see page 34).

Portland-pozzolan (Types IP, P and I(PM)) cements

These cements are made by intergrinding or blending pozzolans (see page 33) with Portland cement. ASTM C 618–84 describes a *pozzolan* as a siliceous or siliceous and aluminous material which in itself possesses little or no cementitious value but will, in finely divided form and in the presence of moisture, chemically react with lime (liberated by hydrating Portland cement) at ordinary temperatures to form compounds possessing cementitious properties.

As a rule, Portland-pozzolan cements gain strength slowly and therefore require curing over a comparatively long period, but the long-term strength is high (see Fig. 2.6). Figure 2.6 shows that similar behaviour occurs where the pozzolan replaces part of cement, but the long-term strength depends on the level of replacement.

ASTM C 595–83a describes Type IP for general construction and Type P for use when high strengths at early ages are not required; Type I(PM) is a pozzolan-modified Portland cement for use in general construction. The pozzolan content is limited to between 15 and 40 per cent of the total mass of the cementitious material for Types IP and P while Type I(PM) requires less than 15 per cent pozzolan. According to BS 6588: 1985, the pozzolan content in Portland-pozzolan cement is limited to 35 per cent by

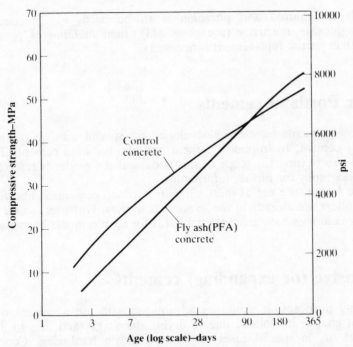

Fig. 2.6: Typical relative rates of strength development of Portland cement (control) concrete and concrete with fly ash (PFA) replacement

mass.

BS 3892: Part 1: 1982 specifies the content of pulverized fuel ash (PFA – see page 33), which is a common type of pozzolan, as between 25 and 40 per cent by mass. The use of PFA improves sulphate resistance. PFA can be used with low-heat, Portland blast-furnace or low-heat Portland blast-furnace cements, provided none of the relevant properties is detrimentally affected. There exists also a British standard (BS 6610: 1985) for *pozzolanic cement,* defined as a blend of ordinary Portland cement and pozzolan, usually PFA, with a pozzolan content of between 35 and 50 per cent by mass. BS 6610: 1985 describes a test for determining the properties of PFA in pozzolanic cement, and the latter must satisfy the test for pozzolanicity (see page 33). The uses of this cement are in rolled concrete (see page 415), in concrete with low-heat characteristics, and in concrete requiring good chemical resistance.

Pozzolans may often be cheaper than the Portland cement that they replace but their chief advantage lies in slow hydration and therefore low rate of heat development. Hence, Portland-pozzolan cement or a partial replacement of Portland cement by the pozzolan is used in mass concrete construction.

Partial replacement of Portland cement by pozzolan has to be carefully defined, as its specific gravity (or relative density) (1.9 to 2.4) is much lower than that of cement (3.15). Thus, replacement by mass results in a considerably greater volume of cementitious material. If equal early

31

strength is required and pozzolan is to be used, e.g. because of alkali–aggregate reactivity (see page 273), then *addition* of pozzolan rather than partial replacement is necessary.

Other Portland cements

Numerous cements have been developed for special uses, in particular masonry cement, hydrophobic cement and anti-bacterial cement. These cements are beyond the scope of this book and the reader is referred to the Bibliography for further information.

In the future, the use of inert fillers in Portland cements is likely, and indeed fillers are already in use in some countries. However, so far only in a few countries have national standards for such cements been issued.

Expansive (or expanding) cements

For many purposes, it would be advantageous to use a cement which does not change its volume due to drying shrinkage (and thus to avoid cracking) or, in special cases, even expands on hardening. Concrete containing such a cement expands in the first few days of its life, and a form of prestress is obtained by restraining this expansion with steel reinforcement: steel is put in tension and concrete in compression. Restraint by external means is also possible. It should be noted that the use of expanding cement cannot produce 'shrinkless' concrete, as shrinkage occurs after moist curing has ceased, but the magnitude of expansion can be adjusted so that the expansion and subsequent shrinkage are equal and opposite.

All types of expansive cements produce calcium sulphoaluminate hydrate (ettringite) which causes expansion of the paste.

Type M cement of ASTM C 845–80 is manufactured by intergrinding Portland cement clinker, high-alumina cement clinker and gypsum, which produce expansion within 2 or 3 days of casting. This cement is also known as a high-energy expanding cement, being quick-setting and rapid-hardening (7 MPa (1000 psi) in 6 hours, and 50 MPa (7000 psi) in 28 days). The cement has a high resistance to sulphate attack.

The same ASTM classification includes the expanding cement *Type K*. The ingredients, gypsum, bauxite and chalk, are burnt to form the expanding agent of calcium sulphate and calcium aluminate (mainly C_5A_3). When the expanding agent is mixed with water, Portland cement and a stabilizer (blast-furnace slag), the excess calcium sulphate is taken up by the slag so that the expansion is controlled. By very careful proportioning of the materials and control of the clinkering conditions, anhydrous calcium sulphoaluminate ($C_4A_3.SO_3$) is formed. With this cement, the rate and magnitude of expansion are more reliable than in the earlier versions of expanding cements.

Type S cement is also specified by ASTM C 845–80; it has a high C_3A content and slightly more interground calcium sulphate than is usual in Portland cement.

Two types of concretes based on expanding cements are recognized by the ACI Committee 223 report 'Expansive Cement Concretes – Present State of Knowledge'. Shrinkage-compensating concrete is one in which expansion, if restrained, induces compressive stresses which approximately offset the induced tensile stresses (about 0.2 to 0.7 MPa (30 to 100 psi)). Self-stressing concrete is one in which the induced compressive stresses are large enough to result in a significant compressive stress after drying shrinkage has occurred (about 1 to 3.5 MPa (150 to 500 psi)). Both types of concrete require restraint of the expansions, usually by reinforcement, preferably triaxial.

Many of the properties of these concretes are similar to those of the corresponding Portland cement concretes, but the slump loss (see page 80) occurs faster; also, the resistance to sulphate attack may be impaired, especially with Type M and S cements. Expansive cements are used in special applications, such as the prevention of water leakage.

Pozzolans

The use of pozzolans in Portland-pozzolan cements has already been mentioned on page 30, together with the definition of a pozzolan. Typical materials of this type are volcanic ash (the original pozzolan), pumicite, opaline shales and cherts, calcined diatomaceous earth, burnt clay, and fly ash (PFA).

For an assessment of pozzolanic activity with cement, the *pozzolanic activity index* should be measured; according to ASTM C 618–84, this is the ratio of compressive strength of the mixture with a specified replacement of cement by pozzolan to the strength of a mix without replacement; BS 3892: Part 1: 1982 specifies a similar method for pulverized fuel ash. There is also a *pozzolanic activity index with lime* (total activity). BS 4550: Part 2: 1970 compares the quantity of $Ca(OH)_2$ present in a liquid phase in contact with the hydrated pozzolanic cement with the quantity of $Ca(OH)_2$ capable of saturating a medium of the same alkalinity. If the concentration of $Ca(OH)_2$ in the solution is lower than that of the saturated medium, the cement satisfies the test for *pozzolanicity*.

The most common artificial pozzolan is fly ash, or pulverized fuel ash (PFA), which is obtained by electrostatic or mechanical means from the flue gases of furnaces in coal-fired power stations. The fly ash particles are spherical and of at least the same fineness as cement so that silica is readily available for reaction. Uniformity of properties is important, and BS 3892: Part 1: 1982 specifies the fineness, expressed as the mass proportion of the ash retained on a 45 μm mesh test sieve, to be at most 12.5 per cent. Also, the loss on ignition must not exceed 7 per cent, the MgO content 4 per cent, the SO_3 content 2.5 per cent, the moisture content 0.5 per cent, and the total water requirement of the mixture of the PFA and ordinary Portland cement should not exceed 95 per cent of that for the Portland cement alone. ASTM C 618–84 requires a minimum content of 70 per cent of silica, alumina and ferric oxide all together, a maximum SO_3 content of 5 per cent, a maximum loss on ignition of 12 per cent and a maximum alkali content (expressed as Na_2O) of 1.5 per cent. The latter value is applicable only when fly ash is to be used with alkali reactive aggregate.

Recently, high-lime ashes originating from lignite coal (with up to 24 per cent lime and thus possessing hydraulic properties) have entered the pozzolan market. Compared with other ashes, they are lighter in colour and can have a higher content of MgO, which, together with some of the lime, can cause deleterious expansion; also, their strength behaviour at higher temperatures is suspect.

High-alumina cement (HAC)

High-alumina cement was developed at the beginning of this century to resist sulphate attack but it soon became used as a very rapid-hardening cement.

HAC is manufactured from limestone or chalk and bauxite, the latter consisting of hydrated alumina, oxides of iron and titanium, with small amounts of silica. After crushing, the raw materials are heated to the point of fusion at about 1600 °C (2900 °F), and the product is cooled and fragmented before being ground to a fineness of 250 to 320 m²/kg. The high hardness of the clinker, together with the high prime cost of bauxite and the high firing temperature, result in HAC being more expensive than, say, rapid-hardening Portland (Type III) cement.

Table 2.9 gives typical values of oxide composition of HAC. A minimum alumina content of 32 per cent is prescribed by BS 915: 1972 (1983), which also requires the alumina/lime ratio to be between 0.85 and 1.3.

Table 2.9: **Typical oxide composition of high-alumina cement**

Oxide	Content, per cent
SiO_2	3 to 8
Al_2O_3	37 to 41
CaO	36 to 40
Fe_2O_3	9 to 10
FeO	5 to 6
TiO_2	1.5 to 2
MgO	1
Insoluble residue	1

Considerably less is known about the compound composition of HAC than of Portland cement, and no simple method of calculation is available. The main cementitious compounds are calcium aluminates: CA and C_5A_3 (or $C_{12}A_7$). Other phases present are: $C_6A_4.FeO.S$ and an isomorphous $C_6A_4MgO.S$, while C_2S (or C_2AS) does not account for more than a few per cent. There are other minor compounds but no free lime exists and thus unsoundness is never a problem.

The hydration of CA results in the formation of CAH_{10}, a small quantity of C_2AH_8 and of alumina gel ($Al_2O_3.aq$). With time, these hexagonal CAH_{10} crystals become transformed into cubic crystals of C_3AH_6 and alumina gel. This transformation is known as *conversion*. Conversion is encouraged by a higher temperature and a higher concentration of lime or a rise in alkalinity. The product of hydration of C_5A_3 is believed to be C_2AH_8.

As mentioned earlier, HAC is highly satisfactory in resisting sulphate attack, which is mainly due to the absence of $Ca(OH)_2$ in the products of hydration. However, lean mixes are much less resistant to sulphates, and also the chemical resistance decreases drastically after conversion.

It was also mentioned that HAC exhibits a very high rate of strength development. About 80 per cent of its ultimate unconverted strength is reached at the age of 24 hours, and even at 6 to 8 hours, sufficient strength is achieved for the removal of formwork. The rapid hydration produces a high rate of heat development, which is about $2\frac{1}{2}$ times that of rapid-hardening Portland (Type III) cement, although the total heat of hydration is of the same order for both types of cement.

It should be noted that the rapidity of hardening of HAC is not accompanied by rapid setting. In fact, HAC is slow setting but the final set follows the inital set more rapidly than in Portland cement. The setting time is greatly affected by the addition of plaster, lime, Portland cement and organic matter. In the case of mixtures of Portland cement and HAC, flash set may occur when either cement constitutes between 20 and 80 per cent of the mixture. This quick-setting property is advantageous for stopping the ingress of water and the like, but the long-term strength of such a mixture is quite low.

The conversion of HAC is of particular practical interest because it leads to a loss of strength in consequence of the fact that the converted cubic C_3AH_6 hydrate has a higher density than the unconverted hexagonal CAH_{10} hydrate. Thus, if the overall volume of the body is constant, conversion results in an increase in the porosity of the paste, which has a fundamental influence on the strength of concrete (see page 101). Figure 2.7 shows a typical loss of strength due to conversion, which is a function of both temperature and water/cement ratio; at moderate and high water/cement ratios, the residual strength may be so low as to be unacceptable for most structural purposes. However, even with low water/cement ratios, conversion increases the porosity so that chemical attack may occur. In view of the effects of conversion, HAC is no longer used in structural concrete above or below ground level, but it is a valuable material for repair work of limited life and in temporary works.

HAC concrete is one of the foremost refractory materials, especially

35

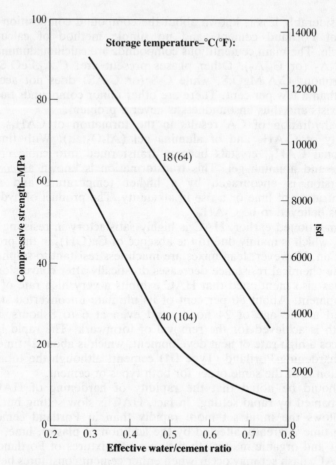

Storage temperature–°C(°F):

18(64)

40(104)

Compressive strength–MPa

psi

Effective water/cement ratio

Fig. 2.7: Influence of the effective water/cement ratio (see page 55) on the
strength of high-alumina cement concrete cubes cured in water at
18 °C (64 °F) and 40 °C (104 °F) for 100 days

above about 1000 °C (1800 °F). Depending on the type of aggregate, the
minimum strength at these temperatures varies between 5 and 25 per cent
of the inital strength, and temperatures as high as 1600–1800 °C
(2400–3300 °F) can be withstood with special aggregates. Refractory
concrete of this type has a good chemical resistance and has other
advantages in that it resists thermal movement and shock.

Bibliography

2.1 ACI COMMITTEE 223, Expansive cement concretes – present
state of knowledge, *J. Amer. Concr. Inst.*, **67**, pp. 583–610 (Aug.
1970).

2.2 ACI COMMITTEE 223–83, Standard practice for the use of shrinkage-compensating concrete, Part 1: Materials and General Properties of Concrete, pp. 36, *ACI Manual of Concrete Practice,* 1990.

2.3 E. E. BERRY and V. M. MALHOTRA, *Fly Ash in Concrete,* pp. 247 (Ottawa, Canada Centre for Mineral and Energy Technology, Nov. 1984).

2.4 R. H. BOGUE, *Chemistry of Portland Cement* (New York, Reinhold, 1955).

2.5 F. M. LEA, *The Chemistry of Cement and Concrete* (London, Arnold, 1970).

2.6 A. M. NEVILLE in collaboration with P. J. Wainwright, *High-alumina Cement Concrete,* pp. 201 (Lancaster/New York, Construction Press, 1975).

Problems

2.1 How can the heat of hydration of cement be reduced?

2.2 What are the main products of hydration of HAC?

2.3 Is there any relation between the cementing properties and heat of hydration of cement?

2.4 Why are tests on cement necessary in a cement plant?

2.5 What are the causes of unsoundness of cement?

2.6 Describe the important effects of C_3A on the properties of concrete.

2.7 Why is the C_3A content in cement of interest?

2.8 Describe the effects of C_3S on the properties of concrete.

2.9 How does gypsum influence the hydration of C_3A?

2.10 Compare the contribution of the various compounds in cement to its heat of hydration.

2.11 How is fineness of cement measured?

2.12 What is meant by the water of hydration?

2.13 How is consistence of cement paste measured?

2.14 What is the difference between false set and flash set?

2.15 What are the main stages in the manufacture of Portland cement?

2.16 What are the main stages in the manufacture of high-alumina cement?

2.17 What are the reactions of hydration of the main compounds in Portland cement?

2.18 What is the method of calculating the compound composition of Portland cement from its oxide composition?

2.19 What are the major compounds in Portland cement?

2.20 What are the minor compounds in Portland cement? What is their role?

2.21 What is meant by loss on ignition?

2.22 What is the difference between false set, initial set and final set?

2.23 How are strength tests of cement performed?

2.24 What is the difference between ordinary Portland (Type I) cement and rapid-hardening Portland (Type III) cement? Which of these cements would you use for mass concrete?

2.25 Describe the chemical reactions which take place during the first 24 hours of hydration of ordinary Portland (Type I) cement at normal temperature.

2.26 Compare the contributions of C_3S and C_2S to the 7-day strength of concrete.

2.27 What is meant by the total heat of hydration of cement?

2.28 What is meant by conversion of HAC?

2.29 What are the consequences of conversion of HAC?

2.30 Under what conditions would you recommend the use of HAC?

2.31 Describe the consequences of mixing Portland cement and HAC.

2.32 Would you recommend HAC for structural use?

2.33 Why is gypsum added in the manufacture of Portland cement?

2.34 Why is sulphate-resisting (Type V) cement suitable for concrete exposed to sulphate attack?

2.35 Why is C_3A undesirable in cement?

2.36 How is the gypsum content of Portland cement specified?

2.37 What are the alkalis in cement?

2.38 What is insoluble residue in cement?

2.39 What cement would you use for refractory purposes?

2.40 Why is the amount of gypsum added to clinker carefully controlled?

2.41 What is meant by chemical modulus?

2.42 What cement would you use to reduce alkali-aggregate reaction?

2.43 What is the pozzolanic activity index?

2.44 What produces the expansive property of expansive cements?

2.45 What is the most common artificial pozzolan and how is it used in cement?

2.46 What are the advantages of using PFA or slag?

2.47 What is a blended cement?

2.48 Under what conditions should PFA and slag not be used?

2.49 Calculate the Bogue composition of the cements with the oxide composition given below.

Oxide	Content, per cent		
	Cement A	Cement B	Cement C
SiO_2	22.4	25.0	20.7
CaO	68.2	61.0	64.2
Fe_2O_3	0.3	3.0	5.3
Al_2O_3	4.6	4.0	3.9
SO_3	2.4	2.5	2.0
Free lime	3.3	1.0	1.5

Answer:

Cement	Compound, per cent			
	C_3S	C_2S	C_3A	C_4AF
A	69.3	12.0	11.7	0.9
B	20.0	56.6	5.5	9.1
C	64.5	10.8	1.3	16.1

3

Normal aggregate

Since approximately three-quarters of the volume of concrete is occupied by aggregate, it is not surprising that its quality is of considerable importance. Not only may the aggregate limit the strength of concrete but the aggregate properties greatly affect the durability and structural performance of concrete.

Aggregate was originally viewed as an inert, inexpensive material dispersed throughout the cement paste so as to produce a large volume of concrete. In fact, aggregate is not truly inert because its physical, thermal and, sometimes, chemical properties influence the performance of concrete, for example, by improving its volume stability and durability over that of the cement paste. From the economic viewpoint, it is advantageous to use a mix with as much aggregate and as little cement as possible, but the cost benefit has to be balanced against the desired properties of concrete in its fresh and hardened state.

Natural aggregates are formed by the process of weathering and abrasion, or by artifically crushing a larger parent mass. Thus, many properties of the aggregate depend on the properties of the parent rock, e.g. chemical and mineral composition, petrographic classification, specific gravity, hardness, strength, physical and chemical stability, pore structure, colour, etc. In addition, there are other properties of the aggregate which are absent in the parent rock: particle shape and size, surface texture and absorption. All these properties may have a considerable influence on the quality of fresh or hardened concrete.

Even when all these properties are known, it is difficult to define a good aggregate for concrete. Whilst aggregate whose properties are all satisfactory will always make good concrete, aggregates appearing to have some inferior property may also make good concrete, and this is why the criterion of performance in concrete has to be used. For instance, a rock sample may disrupt on freezing but need not do so when embedded in concrete. However, in general, aggregate considered poor in more than one respect is unlikely to make a satisfactory concrete, so that aggregate testing is of value in assessing its suitability for use in concrete.

Size classification

Concrete is made with aggregate particles covering a range of sizes up to a maximum size which usually lies between 10 mm ($\frac{3}{8}$ in.) and 50 mm (2 in.); 20 mm ($\frac{3}{4}$ in.) is typical. The particle size distribution is called *grading*. Low-grade concrete may be made with aggregate from deposits containing a whole range of sizes, from the largest to the smallest, known as *all-in* or *pit-run* aggregate. The alternative, very much more common, and always used in the manufacture of good quality concrete, is to obtain the aggregate in at least two separate lots, the main division being at a size of 5 mm ($\frac{3}{16}$ in.) or No. 4 ASTM sieve. This divides *fine* aggregate (sand), from *coarse* aggregate (see Table 3.6). It should be noted that the term aggregate is sometimes used to mean coarse aggregate in contradistinction to sand, a practice which is not correct.

Sand is generally considered to have a lower size limit of about 0.07 mm (0.003 in.) or a little less. Material between 0.06 mm (0.002 in.) and 0.02 mm (0.0008 in.) is classified as *silt,* and smaller particles are termed *clay*. *Loam* is a soft deposit consisting of sand, silt and clay in about equal proportions.

Petrographic classification

From the petrological standpoint, aggregates can be divided into several groups of rocks having common characteristics, as classified by BS 812: Part 1: 1975 (see Table 3.1). The group classification does not imply suitability of any aggregate for concrete-making; unsuitable material can be found in any group, although some groups tend to have a better record than others. It should also be remembered that many trade and customary names of aggregates are in use, and these often do not correspond to the correct petrographic classification.

ASTM Standard C 294–69 (reapproved 1981) gives a description of the more common or important minerals found in aggregates, viz.:

Silica minerals – (quartz, opal, chalcedony, tridymite, cristobalite)
Feldspars
Micaceous minerals
Carbonate minerals
Sulphate minerals
Iron sulphide minerals
Ferromagnesian minerals
Zeolites
Iron oxide minerals
Clay minerals.

The details of petrological and mineralogical methods are outside the scope of this book, but it is important to realize that geological examination of aggregate is a useful aid in assessing its quality and

41

Table 3.1: Rock type classification of natural aggregates according to
BS 812: Part 1: 1975

Basalt Group
Andesite
Basalt
Basic porphyrites
Diabase
Dolerites of all kinds
 including theralite
 and teschenite
Epidiorite
Lamprophyre
Quartz-dolerite
Spilite

Flint Group
Chert
Flint

Gabbro Group
Basic diorite
Basic gneiss
Gabbro
Hornblende-rock
Norite
Peridotite
Picrite
Serpentinite

Granite Group
Gneiss
Granite
Granodiorite
Granulite
Pegnatite
Quartz-diorite
Syenite

**Gritstone Group
(including fragmental
volcanic rocks)**
Arkose
Greywacke
Grit
Sandstone
Tuff

Hornfels Group
Contact-altered rocks
 of all kinds except
 marble

Limestone Group
Dolomite
Limestone
Marble

Porphyry Group
Aplite
Dacite
Felsite
Granophyre
Keratophyre
Microgranite
Porphyry
Quartz-porphyrite
Rhyolite
Trachyte

Quartzite Group
Ganister
Quartzitic sandstones
Re-crystallized
 quartzite

Schist Group
Phyllite
Schist
Slate
All severely sheared
 rocks

especially in comparing a new aggregate with one for which service records are available. Furthermore, adverse properties, such as the presence of some unstable forms of silica, can be detected. In the case of artificial aggregates (see Chapter 18) the influence of manufacturing methods and of processing can also be studied.

Shape and texture classification

The external characteristics of the aggregate, in particular the particle shape and surface texture, are of importance with regard to the properties of fresh and hardened concrete. The shape of three-dimensional bodies is difficult to describe, and it is convenient to define certain geometrical characteristics of such bodies.

Roundness measures the relative sharpness or angularity of the edges and corners of a particle. The actual roundness is the consequence of the strength and abrasion resistance of the parent rock and of the amount of wear to which the particle has been subjected. In the case of crushed aggregate, the shape depends on the nature of the parent material and on the type of crusher and its reduction ratio, i.e. the ratio of initial size to that of the crushed product. A convenient broad classification of particle shape is that of BS 812: Part 1: 1975, given in Table 3.2.

There is no ASTM equivalent, but a classification sometimes used in

Table 3.2: **Particle shape classification of aggregates according to BS 812: Part 1: 1975 with examples**

Classification	Description	Examples
Rounded	Fully water-worn or completely shaped by attrition	River or seashore gravel; desert, seashore and wind-blown sand
Irregular	Naturally irregular, or partly shaped by attrition and having rounded edges	Other gravels; land or dug flint
Flaky	Material of which the thickness is small relative to the other two dimensions	Laminated rock
Angular	Possessing well-defined edges formed at the intersection of roughly planar faces	Crushed rocks of all types; talus; crushed slag
Elongated	Material, usually angular, in which the length is considerably larger than the other two dimensions	—
Flaky and Elongated	Material having the length considerably larger than the width, and the width considerably larger than the thickness	—

the US is as follows:

Well rounded – no original faces left
Rounded   – faces almost gone
Subrounded – considerable wear, faces reduced in area
Subangular – some wear but faces untouched
Angular   – little evidence of wear.

Since the degree of packing of particles all of one size depends on their shape, the angularity of aggregate can be estimated from the proportion of voids among particles compacted in a prescribed manner. BS 812: Part 1: 1975 defines the concept of *angularity number*; this can be taken as 67 minus the percentage of solid volume in a vessel filled with aggregate in a standard manner. The size of particles used in the test must be controlled within narrow limits, and should preferably lie within any of the following four ranges: 20.0 and 14.0 mm ($\frac{3}{4}$ and $\frac{1}{2}$ in.); 14.0 and 10.0 mm ($\frac{1}{2}$ and $\frac{3}{8}$ in.); 10.0 and 6.3 mm ($\frac{3}{8}$ and $\frac{1}{4}$ in.); 6.3 and 5.0 mm ($\frac{1}{4}$ and $\frac{3}{16}$ in.).

The number 67 in the expression for the angularity number represents the solid volume of the most rounded gravel, so that the angularity number measures the percentage of voids in excess of that in the rounded gravel (i.e. 33). The higher the number, the more angular the aggregate, the range for practical aggregates being between 0 and 11.

Another aspect of the shape of coarse aggregate is its *sphericity,* defined as a function of the ratio of the surface area of the particle to its volume (specific surface). Sphericity is related to the bedding and cleavage of the parent rock, and is also influenced by the type of crushing equipment when the size of particles has been artificially reduced. Particles with a high ratio of surface area to volume are of particular interest as they lower the workability of the mix (see page 80). Elongated and flaky particles are of this type. The latter can also adversely affect the durability of concrete as they tend to be oriented in one plane, with water and air voids forming underneath. The presence of elongated or flaky particles in excess of 10 to 15 per cent of the mass of coarse aggregate is generally considered undesirable, although no recognized limits are laid down.

The classification of such particles is made by means of simple gauges described in BS 812: Part 1: 1975. The method is based on the assumption that a particle is flaky if its thickness (least dimension) is less than 0.6 times the mean sieve size of the size fraction to which the particle belongs. Similarly, a particle whose length (largest dimension) is more than 1.8 times the mean sieve size of the size fraction is said to be elongated. The mean size is defined as the arithmetic mean of the sieve size on which the particle is just retained and the sieve size through which the particle just passes. As closer size control is necessary, the sieves considered are not those of the standard concrete aggregate series but: 75.0, 63.0, 50.0, 37.5, 28.0, 20.0, 14.0, 10.0, 6.30 and 5.00 mm (or about 3, $2\frac{1}{2}$, 2, $1\frac{1}{2}$, 1, $\frac{3}{4}$, $\frac{1}{2}$, $\frac{3}{8}$, $\frac{1}{4}$ and $\frac{3}{16}$ in.) sieves. The flakiness and elongation tests are useful for general assessment of aggregate, but they do not adequately describe the particle shape.

The mass of flaky particles, expressed as a percentage of the mass of the sample, is called the *flakiness index. Elongation index* is similarly defined. Some particles are both flaky and elongated, and are therefore counted in both categories.

BS 882: 1983 (as amended in 1984) requires the flakiness index of coarse aggregate to be:

less than 50 for uncrushed gravel and less than 40 for crushed rock or crushed gravel when these aggregates are used to make concrete of grade[1] of 20 to 35 MPa (2900 to 5100 psi); and

less than 35 for aggregates used to make concrete of grade[1] higher than 35 MPa (5100 psi).

Sea aggregates may contain shells whose content needs to be controlled because they are brittle and they also reduce the workability of the mix. The shell content is determined by weighing hand-picked shells and shell fragments from a sample of aggregate greater than 5 mm ($\frac{3}{16}$ in.); the details of the test are prescribed by BS 812: Part 1: 1975 (as amended in

Table 3.3: **Surface texture classification of aggregates according to BS 812: Part 1: 1975 with examples**

Group	Surface Texture	Characteristics	Examples
1	Glassy	Conchoidal fracture	Black flint, vitreous slag
2	Smooth	Water-worn, or smooth due to fracture of laminated or fine-grained rock	Gravels, chert, slate, marble, some rhyolites
3	Granular	Fracture showing more or less uniform rounded grains	Sandstone, oolite
4	Rough	Rough fracture of fine- or medium-grained rock containing no easily visible crystalline constituents	Basalt, felsite, porphyry, limestone
5	Crystalline	Containing easily visible crystalline constituents	Granite, gabbro, gneiss
6	Honeycombed	With visible pores and cavities	Brick, pumice, foamed slag, clinker, expanded clay

[1] Grade strength is the 28-day characteristic strength (see Chapter 6).

1976). BS 882: 1983 limits the *shell content* of aggregate to:

20 per cent for aggregate of nominal 10 mm ($\frac{3}{8}$ in.) single size, and for graded or all-in aggregate finer than 10 mm ($\frac{3}{8}$ in.) and coarser than 5 mm ($\frac{3}{16}$ in.); and

8 per cent for single sizes and for graded or all-in aggregate coarser than 10 mm ($\frac{3}{8}$ in.).

The classification of the *surface texture* is based on the degree to which the particle surfaces are polished or dull, smooth or rough; the type of roughness has also to be described. Surface texture depends on the hardness, grain size and pore characteristics of the parent material (hard, dense and fine-grained rocks generally having smooth fracture surfaces) as well as on the degree to which forces acting on the particle surface have smoothed or roughened it. Visual estimate of roughness is quite reliable, but in order to reduce misunderstanding the classification of BS 812: Part 1: 1975 should be followed (see Table 3.3).

The shape and surface texture of aggregate, especially of fine aggregate, have a strong influence on the water requirement of the mix (see page 80). In practical terms, more water is required when there is a greater void content of the loosely-packed aggregate. Generally, flakiness and shape of the coarse aggregate have an appreciable effect on the workability of concrete, the workability decreasing with an increase in the angularity number.

Mechanical properties

While the various tests described in the following sections give an indication of the quality of the aggregate, it is not possible to relate the potential strength development of concrete to the properties of the aggregate, and indeed it is not possible to translate the aggregate properties into its concrete-making properties.

Bond

Both the shape and the surface texture of aggregate influence considerably the strength of concrete, especially so for high strength concretes; flexural strength is more affected than compressive strength. A rougher texture results in a greater adhesion or bond between the particles and the cement matrix. Likewise, the larger surface area of a more angular aggregate provides a greater bond. Generally, texture characteristics which permit no penetration of the surface of the particles by the paste are not conducive to good bond, and hence softer, porous and mineralogically heterogeneous particles result in a better bond.

The determination of the quality of bond is rather difficult and no accepted test exists. Generally, when bond is good, a crushed concrete specimen should contain some aggregate particles broken right through, in addition to the more numerous ones separated from the paste matrix. However, an excess of fractured particles suggests that the aggregate is too weak.

Strength

Clearly, the compressive strength of concrete cannot significantly exceed that of the *major* part of the aggregate contained therein, although it is not easy to determine the crushing strength of the aggregate itself. A few weak particles can certainly be tolerated; after all, air voids can be viewed as aggregate particles of zero strength.

The required information about the aggregate particles has to be obtained from indirect tests: crushing strength of prepared rock samples, crushing value of bulk aggregate, and performance of aggregate in concrete. The latter simply means either previous experience with the given aggregate or a trial use of the aggregate in a concrete mix known to have a certain strength with previously proven aggregates.

Tests on prepared rock samples are little used, but we may note that a good average value of crushing strength of such samples is about 200 MPa (30 000 psi), although many excellent aggregates range in strength down to 80 MPa (12 000 psi). It should be observed that the required aggregate strength is considerably higher than the normal range of concrete strength because the actual stresses at the points of contact of individual particles may be far in excess of the nominal applied compressive stress. On the other hand, aggregate of moderate or low strength and modulus of elasticity can be valuable in preserving the integrity of concrete, because volume changes, for hygral or thermal reasons, lead to a lower stress in the cement paste when the aggregate is compressible whereas a rigid aggregate might lead to cracking of the surrounding cement paste.

The *aggregate crushing value* (ACV) test is prescribed by BS 812: Part 110: 1990, and is a useful guide when dealing with aggregates of unknown performance.

The material to be tested should pass a 14.0 mm ($\frac{1}{2}$ in.) test sieve and be retained on a 10.0 mm ($\frac{3}{8}$ in.) sieve. When, however, this size is not available, particles of other sizes may be used, but those larger than standard will in general give a higher crushing value, and the smaller ones a lower value than would be obtained with the same rock of standard size. The sample should be dried in an oven at 100 to 110 °C (212 to 230 °F) for four hours, and then placed in a cylindrical mould and tamped in a prescribed manner. A plunger is put on top of the aggregate and the whole assembly is placed in a compression testing machine and subjected to a load of 400 kN (40 tons) (pressure of 22.1 MPa (3200 psi)) over the gross area of the plunger, the load being increased gradually over a period of 10 minutes. After releasing the load, the aggregate is removed and sieved on a 2.36 mm (No. 8 ASTM) test sieve[3] in the case of a sample of the 14.0 to 10.0 mm ($\frac{1}{2}$ to $\frac{3}{8}$ in.) standard size; for samples of other sizes, the sieve size is prescribed in BS.812: Part 110: 1990. The ratio of the mass of material passing this sieve to the total mass of the sample is called the aggregate crushing value.

There is no explicit relation between the aggregate crushing value and its compressive strength but, in general, the crushing value is greater for a lower compressive strength. For crushing values of over 25 to 30, the test is rather insensitive to the variation in strength of weaker aggregates. This is so because, having been crushed before the full load of 400 kN (40 tons) has been applied, these weaker materials become compacted so that the amount of crushing during later stages of the test is reduced.

For this reason, a *ten per cent fines value* test is included in BS 812: Part 111: 1990. In this test, the apparatus of the standard crushing test is used to determine the load required to produce 10 per cent fines from the 14.0 to 10.0 mm ($\frac{1}{2}$ to $\frac{3}{8}$ in.) particles. This is achieved by applying a progressively increasing load on the plunger so as to cause its penetration in 10 minutes of about:

15 mm (0.6 in.) for rounded or partially rounded aggregate,
20 mm (0.8 in.) for crushed aggregate, and
24 mm (0.95 in.) for honeycombed aggregate (such as expanded shale or foamed slag – see Chapter 18).

These penetrations should result in a percentage of fines passing a 2.36 mm (No. 8 ASTM) sieve of between 7.5 and 12.5. If y is the actual percentage of fines due to a maximum load of x tons, then the load required to give 10 per cent fines is given by:

$$\frac{14x}{y+4}.$$

Because some aggregates have a significantly lower resistance to crushing in a saturated and surface dry condition (see page 54) BS 812: Part 111: 1990 includes that moisture state, which is more representative of the practice situation than the oven-dry state. However, after crushing, the fines have to be dried to a constant mass or for 12 h at 105 °C (221 °F).

It should be noted that in this test, unlike the standard crushing value test, a higher numerical result denotes a higher strength of the aggregate.

---

[3] For sieve sizes see Table 3.6.

BS 882: 1983 prescribes a minimum value of 150 kN (15 tons) for aggregate to be used in heavy-duty concrete floor finishes, 100 kN (10 tons) for aggregate to be used in concrete pavement wearing surfaces, and 50 kN (5 tons) when used in other concretes.

Toughness

Toughness can be defined as the resistance of aggregate to failure by impact, and it is usual to determine the *aggregate impact value* of bulk aggregate. Full details of the prescribed tests are given in BS 812: Part 112: 1990. Toughness determined in this manner is related to the crushing value, and can, in fact, be used as an alternative test. In BS 812: Part 112: 1990, the aggregate may also be tested in a saturated and surface dry condition for the reasons given on page 51. The size of the particles tested is the same as in the crushing value test, and the permissible values of the crushed fraction smaller than a 2.36 mm (No.8 ASTM) test sieve are also the same. The impact is provided by a standard hammer falling 15 times under its own weight upon the aggregate in a cylindrical container. This results in fragmentation similar to that produced by the plunger in the crushing value test. BS 882: 1983 prescribes the following maximum values of the average of duplicate samples:

- 25 per cent when the aggregate is to be used in heavy-duty concrete floor finishes,
- 30 per cent when the aggregate is to be used in concrete pavement wearing surfaces, and
- 45 per cent when to be used in other concrete.

Hardness

Hardness, or resistance to wear, is an important property of concrete used in roads and in floor surfaces subjected to heavy traffic. The *aggregate abrasion value* of the bulk aggregate is assessed using BS 812: Part 113: 1990: aggregate particles between 14.0 and 20.0 mm ($\frac{1}{2}$ and $\frac{3}{4}$ in.) are made up in a tray in a single layer, using a setting compound. The sample is subjected to abrasion in a standard machine, the grinding lap being turned 500 revolutions with single-size sand fed continuously at a prescribed rate; the sand is not re-usable. The aggregate abrasion value is defined as the percentage loss in mass on abrasion, so that a high value denotes a low resistance to abrasion.

The *Los Angeles test* combines the processes of attrition and abrasion, and gives results which show a good correlation not only with the actual wear of the aggregate in concrete, but also with the compressive and flexural strengths of concrete when made with the same aggregate. In this test, aggregate of specified grading is placed in a cylindrical drum, mounted horizontally, with a shelf inside. A charge of steel balls is added, and the drum is rotated a specified number of revolutions. The tumbling and dropping of the aggregate and of the balls results in abrasion and attrition of the aggregate, the proportion of broken material, expressed as a percentage, being measured.

The Los Angeles test can be performed on aggregates of different sizes, the same wear being obtained by an appropriate mass of the sample and of the charge of steel balls, and by a suitable number of revolutions. The various quantities are prescribed by ASTM C 131–81.

To assess any possibility of degradation of an unknown fine aggregate on prolonged mixing of fresh concrete, a wet attrition test is desirable to see how much material smaller than 75 μm (No. 200 sieve) is produced. However, the Los Angeles test is not very suitable for this latter requirement and, in fact, no standard apparatus is available.

Physical properties

Several common physical properties of aggregate, of the kind familiar from the study of elementary physics, are relevant to the behaviour of aggregate in concrete and to the properties of concrete made with the given aggregate. These physical properties of aggregate and their measurement will now be considered.

Specific gravity

Since aggregate generally contains pores, both permeable and impermeable (see page 53), the meaning of the term specific gravity (or *relative density*) has to be carefully defined, and there are indeed several types of this measure. According to ASTM C 127–84, specific gravity is defined as the ratio of mass (or weight in air) of a unit volume of material to the mass of the same volume of water at the stated temperature. BS 812: Part 107: (Draft) uses the term *particle density*, expressed in kilogrammes per cubic metre. Thus particle density is numerically 1000 times greater than specific gravity.

The *absolute* specific gravity and the particle density refer to the volume of the solid material excluding all pores, whilst the *apparent* specific gravity and the apparent particle density refer to the volume of solid material including the impermeable pores, but not the capillary ones. It is the apparent specific gravity or apparent particle density which is normally required in concrete technology, the actual definition being the ratio of the mass of the aggregate dried in an oven at 100 to 110 °C (212 to 230 °F) for 24 hours to the mass of water occupying a volume equal to that of the solid including the impermeable pores. The latter mass is determined using a vessel which can be accurately filled with water to a specified volume. This method is prescribed by ASTM C 128–84 for fine aggregate. Thus, if the mass of the oven-dried sample is D, the mass of the vessel full of water is C, and the mass of the vessel with the sample and topped up with water is B, then the mass of the water occupying the same volume as the solid is $D - (B - C)$. The apparent specific gravity is then

$$\frac{D}{D - (B - C)}.$$

The vessel referred to earlier, and known as a *pycnometer,* is usually a one-litre jar with a watertight metal conical screwtop having a small hole at the apex. The pycnometer can thus be filled with water so as to contain precisely the same volume every time.

For the apparent specific gravity of coarse aggregate, ASTM C 127–84 prescribes the *wire-basket* method. Because of difficulties in the pycnometer method and because different particles in a given aggregate may have different values of particle density, BS 812: Part 107: (Draft) also prescribes the wire-basket method for aggregate between 63 mm ($2\frac{1}{2}$ in.) and 5 mm ($\frac{3}{16}$ in.) in size, and it specifies a *gas-jar* method for aggregate not larger than 20 mm ($\frac{3}{4}$ in.). The wire basket, which has apertures 1 to 3 mm (0.04 to 0.12 in.) in size, is suspended from a balance by wire hangers into a watertight tank. A gas jar is a wide-mouthed vessel of a 1 to 1.5 litre capacity and has a flat-ground lip to ensure that it can be made watertight by a disc of plate glass. The apparent particle density (in kg/m^3) is given by

$$\frac{1000\, D}{D - (B - C)}.$$

where the symbols have the same meaning as before, except that B is the apparent mass in water of the basket (or mass of the gas-jar vessel) containing the sample of saturated aggregate, and C is the apparent mass in water of the empty basket (or the mass of the gas-jar vessel filled with water only); all the values of mass are in grammes.

Calculations with reference to concrete are generally based on the saturated and surface-dry (SSD) condition of the aggregate (see page 54) because the water contained in *all* the pores does not participate in the chemical reactions of cement and can, therefore, be considered as part of the aggregate. Thus, if the mass of a sample of the saturated and surface-dry aggregate is A, the term *bulk* specific gravity (SSD) is used, viz.

$$\frac{A}{A - (B - C)}.$$

51

Alternatively, the bulk particle density (in kg/m^3) is given by

$$\frac{1000\,A}{A - (B - C)}.$$

The bulk specific gravity (SSD) and the bulk particle density (SSD) are most frequently and easily determined, and are necessary for calculations of yield of concrete or of the quantity of aggregate required for a given volume of concrete. BS 812: Part 107: (Draft) prescribes the procedure for the determination of the bulk particle density (SSD) while ASTM C 127–84 and C 128–84 prescribe the procedure for the measurement of the bulk specific gravity (SSD).

The majority of natural aggregates have an apparent specific gravity of between 2.6 and 2.7, whilst the values for lightweight and artificial aggregates extend considerably from below to very much above this range (see Chapter 18). Since the actual value of specific gravity or particle density is not a measure of the quality of the aggregate, it should not be specified unless we are dealing with a material of a given petrological character when a variation in specific gravity or particle density would reflect a change in the porosity of the particles. An exception to this is the case of construction such as a gravity dam, where a minimum density of concrete is essential for the stability of the structure.

Bulk density

It is well known that in the metric system the density (or unit weight in air, or unit mass) of a material is numerically equal to the specific gravity although, of course, the latter is a ratio while density is expressed in kilogrammes per litre, e.g. for water, 1.00 kg per litre. However, in concrete practice, expressing the density in kilogrammes per cubic metre is more common. In the American system, the absolute specific gravity has to be multiplied by the unit mass of water (62.4 lb/ft^3) in order to be converted into absolute density expressed in pounds per cubic foot.

This absolute density, it must be remembered, refers to the volume of individual particles only, and of course it is not physically possible to pack these particles so that there are no voids between them. Thus, when aggregate is to be batched by volume it is necessary to know the *bulk density* which is the actual mass that would fill a container of unit volume, and this density is used to convert quantities by mass to quantities by volume.

The bulk density depends on how densely the aggregate is packed and, consequently, on the size distribution and shape of the particles. Thus, for test purposes, the degree of compaction has to be specified. BS 812: Part 2: 1975 recognizes two degrees: *loose* and *compacted*. The test is performed using a metal cylinder of prescribed diameter and depth, depending on the maximum size of the aggregate and also on whether compacted or loose bulk density is being determined. For the latter, the dried aggregate is gently placed in the container to overflowing and then levelled by rolling a rod across the top. In order to find the compacted bulk density, the container is filled in three stages, each one-third of the volume being tamped a prescribed number of times with a 16 mm ($\frac{5}{8}$ in.) diameter round-nosed rod. Again the overflow is removed. The nett mass of the aggregate in the container divided by its volume then represents the bulk density for either degree of compaction. The ratio of the loose bulk density to the compacted bulk density lies usually between 0.87 and 0.96. ASTM C 29–78 prescribes a similar procedure.

Knowing the bulk specific gravity (SSD) for the saturated and surface-dry condition, ρ the *voids ratio* can be calculated from the expression:

$$\text{voids ratio} = 1 - \frac{\text{bulk density}}{\rho \times \text{unit mass of water}}.$$

Thus, the voids ratio indicates the volume of mortar required to fill the space between the coarse aggregate particles. However, if the aggregate contains surface water it will pack less densely owing to the bulking effect (see page 56). Moreover, the bulk density as determined in the laboratory may not represent that on site and may, therefore, not be suitable for the purposes of converting mass to volume in the batching of concrete.

As mentioned earlier, the bulk density depends on the size distribution of the aggregate particles; particles all of one size can be packed to a limited extent but smaller particles can be added in the voids between the larger ones, thus increasing the bulk density. In fact, the maximum bulk density of a mixture of fine and coarse aggregates is achieved when the mass of the fine aggregate is approximately 35 to 40 per cent of the total mass of aggregate. Consequently, the minimum remaining volume of voids determines the minimum cement paste content and, therefore, the minimum cement (powder) content; this latter is, of course, of economic importance.

Porosity and absorption

The porosity, permeability and absorption of aggregate influence the bond between it and the cement paste, the resistance of concrete to freezing and thawing, as well as chemical stability, resistance to abrasion, and specific gravity.

The pores in aggregate vary in size over a wide range, but even the smallest pores are larger than the gel pores in the cement paste. Some of the aggregate pores are wholly within the solid whilst others open onto the surface of the particle so that water can penetrate the pores, the amount and rate of penetration depending on their size, continuity and total volume. The range of porosity of common rocks varies from 0 to 50 per cent, and since aggregate represents some three-quarters of the volume of concrete it is clear that the porosity of the aggregate materially contributes to the overall porosity of concrete (see page 109).

When all the pores in the aggregate are full, it is said to be *saturated and surface-dry*. If this aggregate is allowed to stand free in dry air, some water will evaporate so that the aggregate is *air-dry*. Prolonged drying in an oven would eventually remove the moisture completely and, at this stage, the aggregate is *bone-dry* (or *oven-dry*). These various stages, including an initial moist stage, are shown diagrammatically in Fig. 3.1.

The *water absorption* is determined by measuring the decrease in mass of a saturated and surface-dry sample after oven drying for 24 hours. The ratio of the decrease in mass to the mass of the dry sample, expressed as a percentage, is termed absorption. Standard procedures are described in BS 812: Part 107: (Draft).

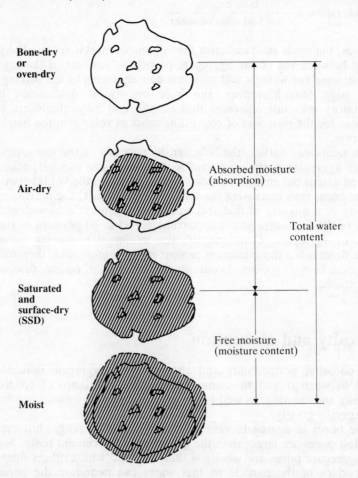

Fig. 3.1: Schematic representation of moisture in aggregate

The assumption that oven-dry aggregate in an actual mix would absorb sufficient water to bring it to the saturated and surface-dry state may not be valid. The amount of water absorbed depends on the order of feeding the ingredients into the mixer and on the coating of coarse aggregate with cement paste. Therefore, a more realistic time for the determination of water absorption is 10 to 30 minutes rather than 24 hours. Moreover, if the aggregate is in an air-dry state, the actual water absorption will be correspondingly less. The actual water absorption of the aggregate has to be deducted from the *total* water requirement of the mix to obtain the *effective water/cement* ratio, which controls both the workability and the strength of concrete.

Moisture content

Since absorption represents the water contained in the aggregate in a saturated, surface-dry condition, we can define the *moisture content* as the water in excess of the saturated and surface-dry condition. Thus, the total water content of a moist aggregate is equal to the sum of absorption and moisture content (see Fig. 3.1).

Aggregate exposed to rain collects a considerable amount of moisture on the surface of the particles, and, except at the surface of the stockpile, keeps this moisture over long periods. This is particularly true of fine aggregate, and the moisture content must be allowed for in the calculation of batch quantities and of the total water requirement of the mix. In effect, the mass of water added to the mix has to be decreased and the mass of aggregate must be increased by an amount equal to the mass of the moisture content. Since the moisture content changes with weather and varies also from one stockpile to another, the moisture content must be determined frequently.

There are several methods available, but the accuracy depends on sampling so that it is important to have a representative sample. In the laboratory, the total moisture content can be determined by means of the oven-drying method, prescribed by BS 812: Part 109: 1990. If A is the mass of an air-tight container, B the mass of the container and sample, and C the mass of the container and sample after drying to a constant mass, the total moisture content (per cent) of the dry mass of aggregate is

$$\frac{B-C}{C-A} \times 100.$$

The ASTM C 70–79 method is based on the measurement of moisture content of aggregate of known specific gravity from the apparent loss in mass on immersion in water (*buoyancy meter test*). The balance can read the moisture content directly if the size of the sample is adjusted according to the specific gravity of the aggregate to such a value that a saturated and surface-dry sample has a standard mass when immersed.

The test is rapid and gives the moisture content to the nearest 0.5 per cent.

Electrical devices have been developed to give instantaneous or continuous reading of the moisture content of aggregate in a storage bin; these devices operate on the basis of the variation in electrical resistance or capacitance with a varying moisture content. In some batching plants, moisture content meters are used in connection with automatic devices which regulate the quantity of water to be added to the mixer, but an accuracy of greater than 1 per cent cannot be achieved.

Bulking of sand

In the case of sand, there is another effect of the presence of moisture, viz. bulking, which is an increase in the volume of a given mass of sand caused by the films of water pushing the sand particles apart. While bulking *per se* does not affect the proportioning of materials by mass, in the case of volume batching, bulking results in a smaller mass of sand occupying the fixed volume of the measuring box. Volume batching represents bad practice, and no more than the preceding warning is needed.

Unsoundness due to volume changes

The physical causes of large or permanent volume changes of aggregate are freezing and thawing, thermal changes at temperatures above freezing, and alternating wetting and drying. If the aggregate is *unsound,* such changes in physical conditions result in a deterioration of the concrete in the form of local scaling, so-called pop-outs, and even extensive surface cracking. Unsoundness is exhibited by porous flints and cherts, especially lightweight ones with a fine-textured pore structure, by some shales, and by other particles containing clay minerals.

Methods of testing and classifying drying shrinkage of aggregate in concrete are given in BS 812: Part 120: 1989, and BS 812: Part 121: 1989 and ASTM C 88–83 prescribe a test for unsoundness in which the aggregate is subjected alternately to immersion in a sulphate solution and to drying, the process of which causes disruption of the particles due to the pressure generated by the formation of salt crystals. The degree of unsoundness is expressed by the reduction in particle size after a specified number of cycles. Other tests consist of subjecting the aggregate to cycles of freezing and thawing. However, the conditions of all these tests do not really represent those when the aggregate is part of the concrete, that is when the behaviour of the aggregate is influenced by the presence of the surrounding cement paste. Hence, only a service record can satisfactorily prove the durability of any aggregate.

For frost damage to occur, there must be critical conditions of water content and lack of drainage. These are governed by the size distribution, shape and continuity of the pores in the aggregate, because these characteristics of the pores control the rate and amount of absorption and the rate at which water can escape from the aggregate particles. Indeed, these features are more important than merely the total volume of pores as reflected by the magnitude of absorption. BS 812: Part 124: 1989 prescribes a method of assessing frost-heave of aggregate compacted into cylindrical specimens at a predetermined moisture content and density.

Thermal properties

There are three thermal properties that may be significant in the performance of concrete: coefficient of thermal expansion, *specific heat,* and conductivity. The last two are of interest in mass concrete to which insulation is applied (see page 170), but usually not in ordinary structural work. The *coefficient of thermal expansion* of aggregate determines the corresponding value for concrete, but its influence depends on the aggregate content of the mix and on the mix proportions in general (see page 250). If the coefficient of thermal expansion of aggregate differs by more than 5.5×10^{-6} per °C (3×10^{-6} per °F) from that of cement paste, then durability of concrete subjected to freezing and thawing may be detrimentally affected. Smaller differences between the thermal expansion of cement paste and of aggregate are probably not detrimental within the temperature range of, say, 4 to 60 °C (40 to 140 °F), because of the modifying effects of shrinkage and creep.

Table 3.4 shows that the coefficient of thermal expansion of the more common aggregate-producing rocks lies between 5 and 13×10^{-6} per °C (3 and 7×10^{-6} per °F). For hydrated Portland cement paste, the coefficient normally lies between 11 and 16×10^{-6} per °C (6 and 9×10^{-6} per °F), the value depending on the degree of saturation.

Table 3.4: **Linear coefficient of thermal expansion of different rock types**

Rock type	Thermal coefficient of linear expansion	
	10^{-6} per °C	10^{-6} per °F
Granite	1.8 to 11.9	1.0 to 6.6
Diorite, andesite	4.1 to 10.3	2.3 to 5.7
Gabbro, basalt, diabase	3.6 to 9.7	2.0 to 5.4
Sandstone	4.3 to 13.9	2.4 to 7.7
Dolomite	6.7 to 8.6	3.7 to 4.8
Limestone	0.9 to 12.2	0.5 to 6.8
Chert	7.4 to 13.1	4.1 to 7.3
Marble	1.1 to 16.0	0.6 to 8.9

Deleterious substances

There are three broad categories of deleterious substances that may be found in aggregate: *impurities* which interfere with the processes of hydration of cement, *coatings* preventing the development of good bond between aggregate and cement paste, and certain individual particles which are *weak* or *unsound* in themselves. These harmful effects are distinct from those due to the development of chemical reactions between the aggregate and the cement paste (see Chapter 14). The aggregate may also contain sulphate or chloride salts; the methods of determining their contents are prescribed by BS 812: Part 118: (Draft) and BS 812: Part 117: 1988, respectively.

Organic impurities

Natural aggregates may be sufficiently strong and resistant to wear and yet may not be satisfactory for concrete-making if they contain organic impurities which interfere with the hydration process. The organic matter consists of products of decay of vegetable matter in the form of humus or organic loam, which is usually present in sand rather than in coarse aggregate, and is easily removed by washing.

The effects of organic matter can be checked by the colorimetric test of ASTM C 40–79. The acids in the sample are neutralized by a 3 per cent solution of NaOH, prescribed quantities of the aggregate and of the solution being placed in a bottle. The mixture is vigorously shaken to allow the intimate contact necessary for chemical reaction, and then left to stand for 24 h, when the organic content can be judged by the colour of the solution: the greater the organic content the darker the colour. If the colour of the liquid above the test sample is not darker than the standard yellow colour specified, the sample can be assumed to contain only a harmless amount of organic impurities. On the other hand, if the colour is darker than the standard, the aggregate has a rather high organic content which may or may not be harmful. Hence, further tests are necessary: concrete test specimens are made using the suspect aggregate and their strength is compared with the strength of concrete of the same mix proportions made with an aggregate of known quality.

Clay and other fine material

Clay may be present in aggregate in the form of surface coatings which interfere with the bond between the aggregate and the cement paste. In addition, silt and crusher dust may be present either as surface coatings or as loose material. Even in the latter form, silt and fine dust should not be present in large quantities because, owing to their fineness and therefore large surface area, they increase the amount of water necessary to wet all the particles in the mix.

In view of the above, BS 882: 1983 limits the content of all three materials together to not more than:

15 per cent by mass of crushed rock fines (8 per cent for use in heavy-duty floor finishes),
10 per cent by mass of crushed rock all-in aggregate,
3 per cent by mass of crushed rock, uncrushed or partially crushed sand or crushed gravel fines, and
1 per cent by mass of uncrushed, partially crushed or crushed gravel.

ASTM C 33–84 lays down similar requirements, but distinguishes between concrete subject to abrasion and other concretes. In the former case, the amount of material passing a 75 μm (No. 200 ASTM) test sieve is limited to 3 per cent of the mass of sand, instead of the 5 per cent value permitted for other concretes; the corresponding value for coarse aggregate is laid down as 1 per cent for all types of concrete. In the same standard, the contents of clay lumps and friable particles are specified separately, the limits being 3 per cent in fine aggregate, and, for coarse aggregate, 3 and 5 per cent for concretes subjected to abrasion and other concretes, respectively.

It should be noted that different test methods are prescribed in different specifications so that the results are not directly comparable.

The clay, silt and fine dust contents of fine aggregate can be determined by the *decantation* method described in BS 812: Part 103: 1985 whilst a *wet-sieve* method can be used for coarse aggregate, as prescribed by the same standard and by ASTM C 117–84.

Salt contamination

Sand won from the seashore or from a river estuary contains salt, which can be removed by washing in fresh water. Special care is required with sand deposits just above the high-water mark because they contain large quantities of salt (sometimes over 6 per cent by mass of sand). This can be exceedingly dangerous in reinforced concrete where corrosion of steel may result. However, in general, sand from the sea bed which has been washed, even in sea water, does not contain harmful quantities of salts.

There is a further consequence of salt in the aggregate. It will absorb moisture from the air and cause *efflorescence* – unsightly white deposits on the surface of the concrete (see page 267).

Unsoundness due to impurities

There are two types of unsound aggregate particles: those that fail to maintain their integrity due to non-durable impurities, and those that lead to disruptive action on freezing or even on exposure to water, i.e.

due to changes in volume as a result of changes in physical conditions. The latter has already been discussed (see page 56).

Shale and other particles of low density are regarded as unsound, and so are soft inclusions, such as clay lumps, wood and coal, as they lead to pitting and scaling. If present in large quantities (over 2 to 5 per cent of the mass of the aggregate) these particles may adversely affect the strength of concrete and should certainly not be permitted in concrete which is exposed to abrasion. The presence of coal and other materials of low density can be determined by the method prescribed by ASTM C 123–83.

Mica, and gypsum and other sulphates should be avoided, as well as sulphides (iron pyrites and marcasite). The permissible quantities of unsound particles laid down by ASTM C 33–84 are summarized in Table 3.5.

Table 3.5: **Permissible quantities of unsound particles prescribed by ASTM C 33–84**

Type of particles	Maximum content, per cent of mass	
	In fine aggregate	In coarse aggregate
Friable particles	3.0 ⎫	
Soft particles	– ⎬	3.0 to 10.0[a]
Coal	0.5 to 1.0[b]	0.5 to 1.0[b]
Chert that will readily disintegrate	–	3.0 to 8.0[c]

[a] Including chert.
[b] Depending on importance of appearance.
[c] Depending on exposure.

Sieve analysis

The process of dividing a sample of aggregate into fractions of same particle size is known as a sieve analysis, and its purpose is to determine the grading or size distribution of the aggregate. A sample of air-dried aggregate is graded by shaking or vibrating a nest of stacked sieves, with the largest sieve at the top, for a specified time so that the material retained on each sieve represents the fraction coarser than the sieve in question but finer than the sieve above.

Table 3.6 lists the sieve sizes normally used for grading purposes according to BS 812: Part 1: 1975 and ASTM C 136–84. Also shown are the previous designations of the nearest size. It should be remembered that 5 mm (or $\frac{3}{16}$ in., No. 4 ASTM) is the dividing line between the fine and coarse aggregate.

Table 3.6: **BS and ASTM sieve sizes normally used for grading of aggregate**

Coarse aggregate

BS			ASTM		
Aperture		**Previous designation**	**Aperture**		**Previous designation**
mm	in.		mm	in.	
75	3	3 in.	75	3	3 in.
–	–	–	63	2.5	2.5 in.
50	2	2 in.	50	2	2 in.
37.5	1.5	1½ in.	37.5	1.5	1½ in.
–	–	–	25.0	1	1 in.
20.0	0.786	¾ in.	19.0	0.75	¾ in.
–	–	–	12.5	0.50	½ in.
14.0	0.551	½ in.	–	–	–
10.0	0.393	⅜ in.	9.5	0.374	⅜ in.

Fine aggregate

BS			ASTM		
Aperture		**Previous designation**	**Aperture**		**Previous designation**
mm or μm	in.		mm or μm	in.	
5.0 mm	0.196	3/16 in.	4.75 mm	0.187	No. 4
2.36 mm	0.0937	No. 7	2.36 mm	0.0937	No. 8
1.18 mm	0.0469	No. 14	1.18 mm	0.0469	No. 16
600 μm	0.0234	No. 25	600 μm	0.0234	No. 30
300 μm	0.0117	No. 52	300 μm	0.0117	No. 50
150 μm	0.0059	No. 100	150 μm	0.0059	No. 100

Grading curves

The results of a sieve analysis can be reported in tabular form, as shown in Table 3.7. Column (2) shows the mass retained on each sieve, whilst column (3) is the same quantity expressed as a percentage of the total mass of the sample. Hence, working from the finest size upwards, the

Table 3.7: **Example of sieve analysis**

Sieve size		Mass retained g	Percentage retained	Cumulative percentage passing	Cumulative percentage retained
BS (1)	ASTM (1)	(2)	(3)	(4)	(5)
10.0 mm	$\frac{3}{8}$ in.	0	0.0	100	0
5.00 mm	4	6	2.0	98	2
2.36 mm	8	31	10.1	88	12
1.18 mm	16	30	9.8	78	22
600 μm	30	59	19.2	59	41
300 μm	50	107	34.9	24	76
150 μm	100	53	17.3	7	93
<150 μm	<100	21	6.8	–	–

Total = 307

Total = 246

Fineness modulus =   2.46

cumulative percentage (to the nearest one per cent) passing each sieve can be calculated (column (4)), and it is this percentage that is used in the plotting of the grading curve. Such a curve is plotted on a grading chart, where the ordinates represent the cumulative percentage passing and the abscissae are the sieve apertures plotted to a logarithmic scale, which gives a constant spacing for the standard series of sieves. This is illustrated in Fig. 3.2 which represents the data of Table 3.7.

Fineness modulus

A single factor computed from the sieve analysis is sometimes used, particularly in the US. This is the fineness modulus (FM), defined as the sum of the cumulative percentages *retained* on the sieves of the standard series, divided by 100. The standard series consists of sieves, each twice the size of the preceding one, viz.: 150, 300, 600 μm, 1.18, 2.36, 5.00 mm (ASTM No. 100, 50, 30, 16, 8, 4) and up to the largest sieve size present. It should be remembered that, when all particles in a sample are coarser than, say, 600 μm (No. 30 ASTM), the cumulative percentage retained on 300 μm (No. 50 ASTM) should be entered as 100; the same value, of course, would be entered for 150 μm (No. 100 ASTM). For the example of Table 3.7, the fineness modulus is 2.47 (column (5)). The grading curve is plotted in Fig. 3.2.

Usually, the fineness modulus is calculated for the fine aggregate rather than for coarse aggregate. Typical values range from 2.3 and 3.0, a higher

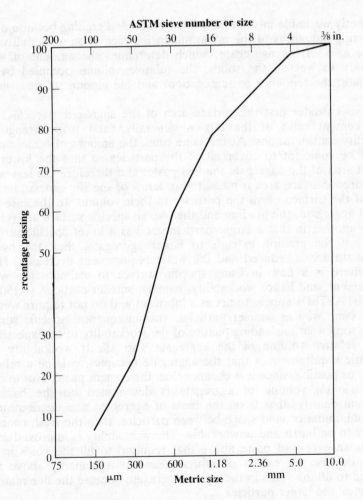

ASTM sieve number or size

Fig. 3.2: Example of a grading curve (see Table 3.7)

value indicating a coarser grading. The usefulness of the fineness modulus lies in detecting slight variations in the aggregate from the *same* source, which could affect the workability of the fresh concrete.

Grading requirements

We have seen how to find the grading of a sample of aggregate, but it still remains to determine whether or not a particular grading is suitable to produce a 'good' concrete. In the first instance, grading is of importance only in so far as it affects workability, because strength is independent of the grading. However, high strength requires a maximum compaction with a reasonable amount of work, which can only be achieved with a sufficiently workable mix. In fact, there is no *ideal* grading because of the interacting influences of the main influencing factors on workability: the surface area of the aggregate, which determines the amount of water necessary to wet all the solids; the relative volume occupied by the aggregate; the tendency to segregation; and the amount of fines in the mix.

Let us consider first the surface area of the aggregate particles. The water/cement ratio of the mix is generally fixed from strength or durability considerations. At the same time, the amount of cement paste has to be sufficient to cover all of the particles so that the lower the surface area of the aggregate the less paste, and therefore the less water is required. Surface area is measured in terms of specific surface, i.e. the ratio of the surface of all the particles to their volume. In the case of a graded aggregate, the grading and the overall specific surface are related to one another in that a larger particle size has a lower specific surface. Hence, if the grading extends to larger aggregate, then the overall specific surface is reduced and the water requirement decreases. However, there is a flaw in using specific surface to estimate the water requirement, and hence workability, namely, smaller particles ($<150\,\mu$m (No. 100 ASTM)) appear to act as a lubricant and do not require wetting in the same way as coarser particles. In consequence, specific surface gives a somewhat misleading picture of the workability to be expected.

The relative volume of the aggregate also affects workability. An economic requirement is that the aggregate occupies as large a relative volume as possible since it is cheaper than the cement paste. However, if the maximum volume of aggregate is determined on the basis of maximum density, that is on the basis of aggregate size distribution to give a minimum of void space between particles, then the fresh concrete is likely to be harsh and unworkable. The workability is improved when there is an excess of paste above that required to fill the voids in the sand, and also an excess of mortar (sand plus cement) above that required to fill the voids in the coarse aggregate because the fine material 'lubricates' the larger particles.

The third factor is the tendency of concrete to segregate, discussed on page 81. Unfortunately, the requirements of workability and segregation are partly incompatible because the easier it is for particles of different sizes to pack, with smaller particles fitting into the voids between larger ones, the easier it is also for the small particles to be displaced out of the voids, i.e. to segregate in the dry state. In actual fact, it is the mortar that has to be prevented from passing out of the voids in the coarse aggregate in order that the concrete be satisfactory.

The fourth factor influencing workability is the presence of the amount of material smaller than $300\,\mu$m (No. 50 ASTM) sieve. To be satisfactorily workable without harshness, the mix should contain the volume of fines ($<125\,\mu$m (No. 120 ASTM)) given in the table on the next page. In that table, the absolute volume of fines includes those of the aggregate, cement and any filler; also, one-half of the volume of entrained air can be taken as equivalent to fines and should be included in the volume of fines.

Maximum aggregate size		Absolute volume of fines as fraction of volume of concrete
mm	in.	
8	0.315	0.165
16	0.630	0.140
32	1.260	0.125
63	2.480	0.110

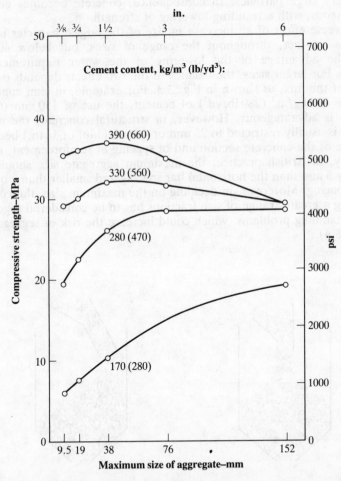

Fig. 3.3: Influence of maximum size of aggregate on the 28-day compressive strength of concretes of different richness
(From: E. C. HIGGINSON, G. B. WALLACE and E. L. ORE, Effect of maximum size of aggregate on compressive strength of mass concrete, *Symp. on Mass Concrete, Amer. Concr. Inst. Sp. Publicn. No.* 6, pp. 219–56 (1963).)

Maximum aggregate size

It has been mentioned before that the larger the aggregate particle the smaller the surface area to be wetted per unit mass (i.e. specific surface). Thus, extending the grading of aggregate to a larger maximum size lowers the water requirement of the mix so that, for specified workability and richness of mix, the water/cement ratio can be reduced with a consequent increase in strength. However, there is a limit of maximum aggregate size above which the decrease in water demand is offset by the detrimental effects of a lower bond area and of discontinuities introduced by the very large particles. In consequence, concrete becomes grossly heterogeneous, with a resulting lowering of strength.

The adverse effect of an increase in size of the largest particles in the mix exists, in fact, throughout the range of sizes, but below 40 mm ($1\frac{1}{2}$ in.) the advantage of the lowering of the water requirement is dominant. For larger sizes, the balance of the two effects depends on the richness of the mix, as shown in Fig. 3.3. For example, in lean concrete containing 170 kg/m³ (280 lb/yd³) of cement, the use of 150 mm (6 in.) aggregate is advantageous. However, in structural concrete, the maximum size is usually restricted to 25 mm or 40 mm (1 in. or $1\frac{1}{2}$ in.) because of the size of the concrete section and of spacing of reinforcement; more specifically, in British practice, the maximum aggregate size should be smaller by 5 mm than the horizontal bar spacing and smaller than $\frac{2}{3}$ of the vertical spacing. Moreover, in deciding on the maximum size, the cost of stockpiling a greater range of size fractions has to be considered together with the handling problems, which could increase the risk of segregation (see Fig. 3.4).

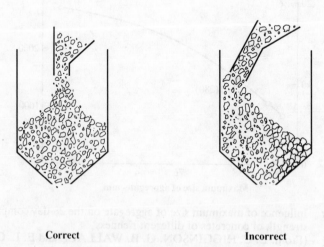

Correct   Incorrect

Fig. 3.4: Placing aggregate in a hopper
(Based on: *ACI Manual of Concrete Practice.*)

Practical gradings

From the brief review in the previous sections, it can be seen how important it is to use aggregate with a grading such that a reasonable workability and minimum segregation are obtained in order to produce a strong and economical concrete. The process of calculation of the proportions of aggregates of different size to achieve the desired grading comes within the scope of mix design (see Chapter 19), but in this section, there are given the recommended grading limits known to meet the requirements discussed previously.

BS 882: 1983 and ASTM C 33–84 specify the grading limits for fine aggregate as shown in Table 3.8. The former standard lays down overall limits and, in addition, specifies that not more than one in ten consecutive samples shall have a grading outside the limits for any *one* of the coarse, medium and fine gradings labelled C, M and F, respectively. However, fine aggregate not complying with the BS 882: 1983 requirements may be used, provided that concrete of the required quality can be produced. The ASTM C 33–84 limits are much narrower than the overall limits of BS 882: 1983, and the former standard allows reduced percentages passing the sieves 300 μm and 150 μm (No. 50 and No. 100 ASTM) when the cement content is above 297 kg/m^3 (500 lb/yd^3) or if air entrainment is used with at least 237 kg/m^3 (400 lb/yd^3) of cement.

Table 3.8: **BS and ASTM grading requirements for fine aggregate**

Sieve size		Percentage by mass passing sieve				
		BS 882:1983				ASTM C 33–84
BS	ASTM No.	Overall limits	Additional limits*			
			C	M	F	
10 mm	$\frac{3}{8}$ in.	100	–	–	–	100
5 mm	$\frac{3}{16}$ in.	89–100	–	–	–	95–100
2.36 mm	8	60–100	60–100	65–100	80–100	80–100
1.18 mm	16	30–100	30–90	45–100	70–100	50–85
600 μm	30	15–100	15–54	25–80	55–100	25–60
300 μm	50	5–70	5–40	5–48	5–70	10–30
150 μm	100	0–15†	–	–	–	2–10

* C = coarse; M = medium; F = fine.
† For crushed rock sands the permissible limit is increased to 20 per cent, except when used for heavy duty floors.

The requirements of BS 882: 1983 for the grading of coarse aggregate are reproduced in Table 3.9: values are given both for graded aggregate and for nominal one-size fractions. For comparison, some of the limits of

Table 3.9: Grading requirements for coarse aggregate according to BS 882: 1983 (amended in 1984)

Sieve size		Percentage by mass passing BS sieve							
		Nominal size of graded aggregate			Nominal size of single-sized aggregate				
mm	in.	40 to 5 mm ($\frac{1}{2}$ in. to $\frac{3}{16}$ in.)	20 to 5 mm ($\frac{3}{4}$ in. to $\frac{3}{16}$ in.)	14 to 5 mm ($\frac{1}{2}$ in. to $\frac{3}{16}$ in.)	40 mm ($1\frac{1}{2}$ in.)	20 mm ($\frac{3}{4}$ in.)	14 mm ($\frac{1}{2}$ in.)	10 mm ($\frac{3}{8}$ in.)	5 mm ($\frac{3}{16}$ in.)
50.0	2	100	–		100	–	–	–	–
37.5	$1\frac{1}{2}$	90–100	100		85–100	100	–	–	–
20.0	$\frac{3}{4}$	35–70	90–100	100	0–25	85–100	100	–	–
14.0	$\frac{1}{2}$	–	–	90–100	–	–	85–100	100	–
10.0	$\frac{3}{8}$	10–40	30–60	50–85	0–5	0–25	0–50	85–100	100
5.00	$\frac{3}{16}$	0–5	0–10	0–10	–	0–5	0–10	0–25	50–100
2.36	No. 7	–	–	–		–	–	0–5	0–30

Table 3.10: **Some of the grading requirements for coarse aggregate according to ASTM C 33–84**

Sieve size		Percentage by mass passing sieve				
		Nominal size of graded aggregate			Nominal size of single-sized aggregate	
mm	in.	37.5 to 4.75 mm ($1\frac{1}{2}$ to $\frac{3}{16}$ in.)	19.0 to 4.75 mm ($\frac{3}{4}$ to $\frac{3}{16}$ in.)	12.5 to 4.75 mm ($\frac{1}{2}$ to $\frac{3}{16}$ in.)	63 mm ($2\frac{1}{2}$ in.)	37.5 mm ($1\frac{1}{2}$ in.)
75	3	–	–	–	100	–
63.0	$2\frac{1}{2}$	–	–	–	90–100	–
50.0	2	100	–	–	35–70	100
38.1	$1\frac{1}{2}$	95–100	–	–	0–15	90–100
25.0	1	–	100	–	–	20–55
19.0	$\frac{3}{4}$	35–70	90–100	100	0–5	0–15
12.5	$\frac{1}{2}$	–	–	90–100	–	–
9.5	$\frac{3}{8}$	10–30	20–55	40–70	–	0–5
4.75	$\frac{3}{16}$	0–5	0–10	0–15	–	–
2.36	No. 8	–	0–5	0–5	–	–

Table 3.11: **Grading requirements for all-in aggregate according to BS 882:1983 (amended in 1984)**

Sieve size		Percentage by mass passing sieve of nominal size			
mm	in.	40 mm ($1\frac{1}{2}$ in.)	20 mm ($\frac{3}{4}$ in.)	10 mm ($\frac{3}{8}$ in.)	5 mm ($\frac{3}{16}$ in.)
50	2	100	–	–	–
37.5	$1\frac{1}{2}$	95–100	100	–	–
20.0	$\frac{3}{4}$	45–80	95–100	–	–
14.0	$\frac{1}{2}$	–	–	100	–
10.0	$\frac{3}{8}$	–	–	95–100	–
5.0	$\frac{3}{16}$	25–50	35–55	30–65	70–100
2.36	No. 7	–	–	20–50	25–70
1.18	No. 14	–	–	15–40	15–45
600 μm	No. 25	8–30	10–35	10–30	5–25
300 μm	No. 52	–	–	5–15	3–20
150 μm	No. 100	0–8*	0–8*	0–8*	0–15

* Increased to 10 per cent for crushed rock fines.

69

ASTM C 33–84 are given in Table 3.10. The actual grading requirements depend to some extent on the shape and surface characteristics of the particles. For instance, sharp, angular particles with rough surfaces should have a slightly finer grading in order to reduce the possibility of interlocking and to compensate for the high friction between the particles.

BS 882: 1983 includes the grading requirements for all-in aggregate (see page 41); Table 3.11 gives the details.

Gap-graded aggregate

As mentioned earlier, aggregate particles of a given size pack so as to form voids that can be penetrated only if the next smaller size of particles is sufficiently small. This means that there must be a minimum difference between the sizes of any two adjacent particle fractions. In other words, sizes differing only slightly cannot be used side by side, and this has led to advocacy of gap-graded aggregate, as distinct from *continuously graded* conventional aggregate. On the grading curve, gap-grading is represented by a horizontal line over the range of sizes omitted (see Fig. 3.5).

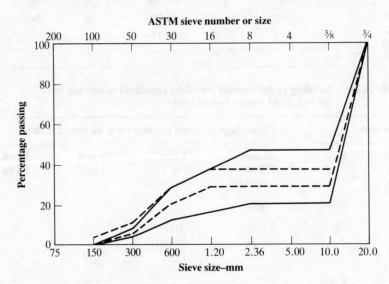

Fig. 3.5: Typical gap gradings

Gap-graded aggregate can be used in any concrete, but there are particular uses: preplaced aggregate concrete (see page 143) and exposed aggregate concrete where a pleasing finish is obtained, since a large quantity of only one size of coarse aggregate becomes exposed after treatment. However, to avoid segregation, gap-grading is recommended mainly for mixes of relatively low workability that are to be compacted by vibration; good control and care in handling are essential.

Bibliography

3.1 ACI COMMITTEE 221, Guide for use of normal weight aggregates in concrete, *ACI Journal*, pp. 115–39 (March/April 1984).

3.2 E. E. BERRY and V. M. MALHOTRA, *Fly ash in Concrete*, pp. 247 (Canada Centre for Mineral and Energy Technology, Nov. 1984)

Problems

3.1 What is meant by surface texture of aggregate?
3.2 What is meant by sphericity of aggregate?
3.3 Can an aggregate particle be both flaky and elongated?
3.4 Why do we determine the elongation index?
3.5 Why do we determine the flakiness index?
3.6 What may be the consequences of impurities in aggregate?
3.7 What is meant by soundness of aggregate?
3.8 What is the property of sea-dredged aggregates which requires special attention?
3.9 How does the shape of aggregate particles influence the properties of fresh concrete?
3.10 What is bulking of sand?
3.11 How would you determine whether aggregate contains organic material?
3.12 What are the consequences of organic material in concrete?
3.13 Define the fineness modulus of aggregate.
3.14 What is angularity number?
3.15 What is a gap-graded mix?
3.16 What are the advantages of a gap-graded mix?
3.17 How is gap-grading noticed on a grading curve?
3.18 How is the quality of bond assessed?
3.19 Discuss the influence of aggregate grading on density (unit weight in air) of concrete.
3.20 What is the maximum size of fine aggregate?
3.21 What is oversize?
3.22 What is undersize?
3.23 What are the common deleterious materials which may be found in aggregate?
3.24 Why is grading of aggregate important with regard to the properties of hardened concrete?
3.25 Why is the grading of aggregate important with regard to the properties of fresh concrete?
3.26 How would you assess the shape of aggregate particles?
3.27 How does the shape of aggregate particles affect the properties of fresh concrete?
3.28 How can the shape of aggregate particles be relevant to the properties of hardened concrete?

71

3.29 What is the influence of the fineness modulus on the properties of concrete mixes?

3.30 What is meant by the saturated and surface-dry and bone-dry conditions of aggregate? Define absorption and moisture content.

3.31 What do you understand by the term aggregate grading?

3.32 How does the grading of aggregate affect the water requirement of the mix?

3.33 Explain the difference between apparent specific gravity and bulk specific gravity of aggregate.

3.34 What are some of the common deleterious materials in natural aggregates?

3.35 How would you assess the strength of aggregate?

3.36 Explain the ten per cent fines value.

3.37 Define toughness of aggregate.

3.38 How would you assess resistance of aggregate to wear?

3.39 How would you measure the apparent specific gravity of coarse aggregate? State a typical value for natural aggregate.

3.40 What are bulk density and voids ratio?

3.41 What is buoyancy meter test?

3.42 State a typical value of coefficient of thermal expansion of common aggregate.

3.43 What are the effects of clay and very fine material on the properties of concrete?

3.44 How can aggregate cause efflorescence in concrete?

3.45 How does the maximum size of aggregate affect the workability of concrete with a given water content?

3.46 How does the variation in moisture content of the aggregate affect the workability of fresh concrete and the strength of hardened concrete?

3.47 Is there an ideal grading for aggregate? Discuss this with reference to workability of fresh concrete.

3.48 Calculate: (i) the apparent specific gravity, (ii) the bulk specific gravity, (iii) the apparent particle density, and (iv) the bulk particle density of sand, given the following data:

mass of sand (oven-dry)	= 480 g
mass of sand (SSD)	= 490 g
mass of pycnometer full of water	= 1400 g
mass of pycnometer plus sand and topped up with water	= 1695 g

Answer: (i) 2.59
(ii) 2.51
(iii) 2594 kg/m^3
(iv) 2513 kg/m^3

3.49 If the mass of a vessel full of water is 15 kg (33 lb), the mass of the empty vessel is 5 kg (11 lb) and the mass of the vessel with compacted coarse aggregate is 21 kg (46 lb), calculate the bulk density and voids ratio of the coarse aggregate.

Answer: 1600 kg/m^3 (99.8 lb/ft^3); 0.38

3.50 Calculate the absorption of the sand used in Question 3.48. If the sand in the stockpile has a total water content of 3.5 per cent, what is the moisture content?

Answer: 2.1 per cent; 1.4 per cent

73

4

Quality of water

When we consider the strength of concrete in Chapter 6, the vital influence of the *quantity* of water in the mix on the strength of the resulting concrete will become clear. At this stage, we are concerned only with the individual ingredients of the concrete mix: cement, aggregate, and water, and it is the *quality* of the latter that is the subject matter of this chapter.

The quality of the water is important because impurities in it may interfere with the *setting* of the cement, may adversely affect the *strength* of the concrete or cause *staining* of its surface, and may also lead to *corrosion* of the reinforcement. For these reasons, the suitability of water for mixing and curing purposes should be considered. Clear distinction must be made between the effects of *mixing* water and the *attack* on hardened concrete by aggressive waters because some of the latter type may be harmless or even beneficial when used in mixing.

Mixing water

In many specifications, the quality of water is covered by a clause saying that water should be fit for drinking.

Such water very rarely contains dissolved solids in excess of 2000 parts per million (ppm), and as a rule less than 1000 ppm. For a water/cement ratio of 0.5 by mass, the latter content corresponds to a quantity of solids equal to 0.05 per cent of the mass of cement, and thus any effect of the common solids (considered as aggregate) would be small. If the silt content is higher than 2000 ppm, it is possible to reduce it by allowing the water to stand in a settling basin before use. However, water used to wash out truck mixers is satisfactory as mixing water (because the solids in it are proper concrete ingredients), provided of course that it was satisfactory to begin with. ASTM C 94–83 allows the use of wash water, but, obviously, different cements and different admixtures should not be involved.

The criterion of potability of water is not absolute: drinking water may be unsuitable as mixing water when the water has a high concentration of sodium or potassium and there is a danger of alkali–aggregate reaction (see page 273).

While the use of potable water is generally safe, water not fit for drinking may often also be satisfactorily used in making concrete. As a rule, any water with a pH (degree of acidity) of 6.0 to 8.0 which does not taste saline or brackish is suitable for use, but a dark colour or a smell do not necessarily mean that deleterious substances are present. Natural waters that are slightly acidic are harmless, but water containing humic or other organic acids may adversely affect the hardening of concrete; such water, as well as highly alkaline water, should be tested.

Two, somewhat peripheral, comments may be made. The presence of algae in mixing water results in air entrainment with a consequent loss of strength. Hardness of water does not affect the efficiency of air-entraining admixtures.

In some countries, it may be difficult to obtain sufficient quantities of fresh water and only *brackish water* is available. Such water contains chlorides and sulphates. When the chloride ion content does not exceed 500 ppm, or SO_3 ion content does not exceed 1000 ppm, the water is harmless, but water with even higher salt contents has been used satisfactorily. BS 3148: 1980 recommends limits on chloride and on SO_3 as above, and also recommends that alkali carbonates and bicarbonates should not exceed 1000 ppm.

Occasionally, the use of *sea water* as mixing water has to be considered. Sea water has, typically, a total salinity of about 3.5 per cent (78 per cent of the dissolved solids being NaCl and 15 per cent $MgCl_2$ and $MgSO_4$). Such water leads to a slightly higher early strength but a lower long-term strength; the loss of strength is usually not more than 15 per cent and can therefore be tolerated. The effects on setting time have not been clearly established but these are unimportant if water is acceptable from strength considerations; BS 3148: 1980 suggests a tolerance of 30 minutes in the initial setting time.

Sea water (or any water containing large quantities of chlorides) tends to cause persistent dampness and efflorescence (see page 267). Such water should not be used where appearance of the concrete is of importance or where a plaster finish is to be applied.

In the case of *reinforced* concrete, sea water increases the risk of corrosion of the reinforcement, especially in tropical countries (see page 275). Corrosion has been observed in structures exposed to humid air when the cover to reinforcement is inadequate or the concrete is not sufficiently dense so that the corrosive action of residual salts in the presence of moisture can take place. On the other hand, when reinforced concrete is permanently in water, either sea or fresh, the use of sea water in mixing seems to have no ill-effects. However, in practice, it is generally considered inadvisable to use sea water for mixing.

Curing water

Generally, water satisfactory for mixing is also suitable for curing purposes (see Chapter 10). However, iron or organic matter may cause staining, particularly if water flows slowly over concrete and evaporates rapidly. In some cases, discoloration is of no significance, and any water suitable for mixing, or even slightly inferior in quality, is acceptable for curing. However, it is essential that curing water be free from substances that attack hardened concrete. For example, concrete is attacked by water containing free CO_2. Flowing pure water, formed by melting ice or by condensation, and containing little CO_2, dissolves $Ca(OH)_2$ and causes surface erosion. This topic is discussed further in Chapter 14. Curing with sea water may lead to attack of reinforcement.

Tests on water

A simple way of determining the suitability of water for mixing is to compare the setting time of cement and the strength of mortar cubes using the water in question with the corresponding results obtained using known 'good' water or distilled water; there is no appreciable difference between the behaviour of distilled and ordinary drinking water. BS 3148: 1980 suggests a tolerance of 10 per cent to allow for chance variations in strength. Such tests are also recommended when water, for which no service record is available, contains dissolved solids in excess of 2000 ppm, or, in the case of alkali carbonate or bicarbonate, in excess of 1000 ppm. When unusual solids are present these tests are also advisable.

Whether or not staining will occur due to impurities in the curing water cannot be determined on the basis of chemical analysis and should be checked by a performance test involving simulated wetting and evaporation.

Bibliography

4.1 F. M. LEA, *The Chemistry of Cement and Concrete* (London, Arnold, 1970).

4.2 W. J. McCOY, Mixing and curing water for concrete, *ASTM Sp. Tech. Publicn. No.* 169B, pp. 765–73 (1978).

Problems

4.1 What is meant by sulphate ion concentration in water?
4.2 How is the solids content in water expressed?
4.3 Specify water for use as concrete mix water.
4.4 Can wash water from a concrete mixer be used as mix water?
4.5 Comment on the use of brackish water for various types of construction.

4.6 What are the requirements for water to be used for curing concrete?
4.7 Is drinking water always suitable as mix water?
4.8 Why are we concerned about the solids content in mix water?
4.9 What are the dangers of using sea water as mixing water?
4.10 Describe a test for the suitability of water for mixing.

5

Fresh concrete

Having considered the ingredients of concrete, we should now address ourselves to the properties of freshly mixed concrete.

Since the long-term properties of hardened concrete: strength, volume stability, and durability are seriously affected by its degree of compaction, it is vital that the consistence or workability of the fresh concrete be such that the concrete can be properly compacted and also that it can be transported, placed, and finished sufficiently easily without segregation, which would be detrimental to such compaction.

Workability

The strict definition of workability is the amount of useful internal work necessary to produce full compaction. The useful internal work is a physical property of concrete *alone* and is the work or energy required to overcome the internal friction between the individual particles in the concrete. In practice, however, additional energy is required to overcome the surface friction between concrete and the formwork or the reinforcement. Also, wasted energy is consumed by vibrating the form and in vibrating the concrete which has already been compacted. Thus, in practice, it is difficult to measure the workability as defined, and what we measure is workability which is applicable to the particular method adopted.

Another term used to describe the state of fresh concrete is *consistence,* which is the firmness of form of a substance or the ease with which it will flow. In the case of concrete, consistence is sometimes taken to mean the degree of wetness; within limits, wet concretes are more workable than dry concretes, but concretes of the same consistence may vary in workability.

Because the strength of concrete is adversely and significantly affected by the presence of voids in the compacted mass, it is vital to achieve a maximum possible *density*. This requires a sufficient workability for virtually full compaction to be possible using a reasonable amount of work under the given conditions. The need for compaction is apparent from Fig. 5.1, which demonstrates the increase in compressive strength with an increase in the density. It is obvious that the presence of voids in

concrete reduces the density and greatly reduces the strength: 5 per cent
of voids can lower the strength by as much as 30 per cent.

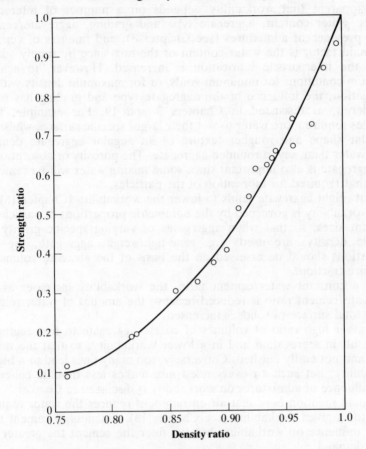

Fig. 5.1: Relation between strength ratio and density ratio
(From: W. H. GLANVILLE, A. R. COLLINS and D. D.
MATTHEWS, The grading of aggregates and workability of
concrete, *Road Research Tech. Paper No. 5,* London, H.M.S.O.
(1950). Crown copyright)

Voids in hardened concrete are, in fact, either bubbles of *entrapped air*
or spaces left after *excess water* has been removed. The volume of the
latter depends solely on the water/cement ratio of the mix whereas the
presence of air bubbles is governed by the grading of the fine particles in
the mix and by the fact that the bubbles are more easily expelled from a
wetter mix than from a dry one. It follows, therefore, that for any given
method of compaction there may be an optimum water content of the
mix at which the *sum of volumes* of air bubbles and of water space will be
a minimum, and the density will be a maximum. However, the optimum
water content may vary for different methods of compaction.

Factors affecting workability

It is apparent that workability depends on a number of interacting factors: water content, aggregate type and grading, aggregate/cement ratio, presence of admixtures (see Chapter 8), and fineness of cement. The main factor is the water content of the mix since by simply adding water the interparticle lubrication is increased. However, to achieve optimum conditions for minimum voids, or for maximum density with no segregation, the influence of the aggregate type and grading has to be considered, as discussed in Chapters 3 and 19. For example, finer particles require more water to wet their larger specific surface, whilst the irregular shape and rougher texture of an angular aggregate demand more water than, say, a rounded aggregate. The porosity or absorption of the aggregate is also important since some mixing water will be removed from that required for lubrication of the particles.

Lightweight aggregate tends to lower the workability (Chapter 18). In fact, workability is governed by the *volumetric* proportions of particles of different sizes, so that when aggregates of varying specific gravity (or particle density) are used, e.g. semi-lightweight aggregate, the mix proportions should be assessed on the basis of the absolute volume of each size fraction.

For a constant water/cement ratio, the workability increases as the aggregate/cement ratio is reduced because the amount of water relative to the total surface of solids is increased.

A rather high ratio of volumes of coarse aggregate to fine aggregate can result in segregation and in a lower workability, so that the mix is *harsh* and not easily finished. Conversely, too many fines lead to a higher workability, but such an oversanded mix makes less durable concrete. The influence of admixtures on workability is discussed in Chapter 8, but we should mention here that air entrainment reduces the water requirement for a given workability (see Chapter 15). Fineness of cement is of minor influence on workability but the finer the cement the greater the water demand.

There are two other factors which affect workability: time and temperature. Freshly mixed concrete stiffens with time but this should not be confused with the setting of cement. It is simply that some of the mixing water is absorbed by the aggregate, some is lost by evaporation (particularly if the concrete is exposed to the sun or wind), and some is removed by initial chemical reactions. The stiffening of concrete is effectively measured by a loss of workability with time, known as *slump loss,* which varies with richness of the mix, type of cement, temperature of concrete, and initial workability. Because of this change in apparent workability or consistence and because we are really interested in the workability at the time of placing, i.e. some time after mixing, it is preferable to delay the appropriate test until, say, 15 minutes after mixing.

A higher temperature reduces the workability and increases the slump loss. In practice, when the ambient conditions are unusual, it is best to make actual site tests in order to determine the workability of the mix.

Cohesion and segregation

In considering the workability of concrete, it was pointed out that concrete should not segregate, i.e. it ought to be cohesive; the absence of segregation is essential if full compaction is to be achieved. Segregation can be defined as separation of the constitutents of a heterogeneous mixture so that their distribution is no longer uniform. In the case of concrete, it is the differences in the size of particles (and sometimes in the specific gravity of the mix constituents) that are the primary cause of segregation, but its extent can be controlled by the choice of suitable grading and by care in handling.

There are two forms of segregation. In the first, the coarser particles tend to separate out since they travel further along a slope or settle more than finer particles. The second form of segregation, occurring particularly in wet mixes, is manifested by the separation of grout (cement plus water) from the mix. With some gradings, when a lean mix is used, the first type of segregation occurs if the mix is too dry; the addition of water would improve the cohesion of the mix, but when the mix becomes too wet the second type of segregation would take place.

The influence of grading on segregation was discussed in detail in Chapter 3, but the *actual* extent of segregation depends on the method of handling and placing of concrete. If the concrete does not have far to travel and is transferred directly from the skip or the wheelbarrow to the final position in the formwork, the danger of segregation is small. On the other hand, dropping concrete from a considerable height, passing along a chute, particularly with changes of direction, and discharging against an obstacle, all encourage segregation so that under such circumstances a particularly cohesive mix should be used. With a correct method of handling, transporting and placing, the likelihood of segregation can be greatly reduced: there are many practical rules and these should be learnt by experience.

It must be stressed, nevertheless, that concrete should always be placed direct in the position in which it is to remain and must not be allowed to flow or be worked *along* the form. This prohibition includes the use of a vibrator to spread a heap of concrete over a larger area. Vibration provides a most valuable means of compacting concrete, but, because a large amount of work is being done on the concrete, the danger of segregation (in placing, as distinct from handling) is increased with improper use of a vibrator. This is particularly so when vibration is allowed to continue too long: with many mixes, separation of coarse aggregate toward the bottom of the form and of the cement paste toward the top may result. Such concrete would obviously be weak, and the *laitance* (scum) on its surface would be too rich and too wet so that a crazed surface with a tendency to *dusting* (see page 82) might result.

The danger of segregation can be reduced by the use of air entrainment (see Chapter 15). Conversely, the use of coarse aggregate whose specific gravity is appreciably greater than that of the fine aggregate can lead to increased segregation.

Segregation is difficult to measure quantitatively but is easily detected when concrete is handled on a site in any of the ways listed earlier as undesirable. A good picture of cohesion of the mix is obtained by the flow table test (see page 89). As far as proneness to segregation due to over-vibration is concerned, a practical test is to vibrate a concrete cube or cylinder for about 10 minutes and then strip it to observe the distribution of coarse aggregate: any segregation will be easily seen.

Bleeding

Bleeding, known also as *water gain,* is a form of segregation in which some of the water in the mix tends to rise to the surface of freshly placed concrete. This is caused by the inability of the solid constituents of the mix to hold all of the mixing water when they settle downwards. Bleeding can be expressed quantitatively as the total settlement (reduction in height) per unit height of concrete, and the bleeding capacity as well as the rate of bleeding can be determined experimentally using the test of ASTM C 232–71 (reapproved 1977). When the cement paste has stiffened sufficiently, bleeding of concrete ceases.

As a result of bleeding, the top of every lift (layer of concrete placed) may become too wet, and, if the water is trapped by superimposed concrete, a porous and weak layer of non-durable concrete will result. If the bleeding water is remixed during the finishing of the top surface, a weak wearing surface will be formed. This can be avoided by delaying the finishing operations until the bleeding water has evaporated, and also by the use of wood floats and by avoidance of over-working the surface. On the other hand, if evaporation of water from the surface of the concrete is faster than the bleeding rate, plastic shrinkage cracking may result (see page 255).

In addition to accumulating at the upper surface of the concrete, some of the rising water becomes trapped on the underside of large aggregate particles or of reinforcement, thus creating zones of poor bond. This water leaves behind voids and, since all these voids are oriented in the same direction, the permeability of the concrete in a horizontal plane may be increased. A small number of voids is nearly always present, but appreciable bleeding must be avoided as the danger of frost damage may be increased (see Chapter 15). Bleeding is often pronounced in thin slabs, such as roads, in which frost generally represents a considerable danger.

Bleeding need not necessarily be harmful. If it is undisturbed (and the water evaporates) the effective water/cement ratio may be lowered with a resulting increase in strength. On the other hand, if the rising water carries with it a considerable amount of the finer cement particles, a layer of laitance will be formed. If this is at the top of a slab, a porous surface will result with a permanently 'dusty' surface. At the top of a lift, a plane of weakness would form and the bond with the next lift would be inadequate. For this reason, laitance should always be removed by brushing and washing.

Although dependent on the water content of the mix, the tendency to bleeding depends largely on the properties of cement. Bleeding is lower with finer cements and is also affected by certain chemical factors: there is less bleeding when the cement has a high alkali content, a high C_3A content, or when calcium chloride is added, although the two latter factors may have other undesirable effects. A higher temperature, within the normal range, increases the rate of bleeding, but the total *bleeding capacity* is probably unaffected. Rich mixes are less prone to bleeding than lean ones, and a reduction in bleeding is obtained by the addition of pozzolans or of aluminium powder. Air entrainment effectively reduces bleeding so that finishing can follow casting without delay.

Workability tests

Unfortunately, there is no acceptable test which will measure directly the workability as defined earlier. The following methods give a measure of workability which is applicable only with reference to the particular method. However, these methods have found universal acceptance and their merit is chiefly that of simplicity of operation with an ability to detect variations in the uniformity of a mix of given nominal proportions.

Slump test

There are some slight differences in the details of procedure used in different countries, but these are not significant. The prescriptions of ASTM C 143–78 are summarized below.

The mould for the slump test is a frustum of a cone, 305 mm (12 in.) high. The base of 203 mm (8 in.) diameter is placed on a smooth surface with the smaller opening of 102 mm (4 in.) diameter at the top, and the container is filled with concrete in three layers. Each layer is tamped 25 times with a standard 16 mm ($\frac{5}{8}$ in.) diameter steel rod, rounded at the end, and the top surface is struck off by means of a screeding and rolling motion of the tamping rod. The mould must be firmly held against its base during the entire operation; this is facilitated by handles or foot-rests brazed to the mould.

Immediately after filling, the cone is slowly lifted, and the unsupported concrete will now slump – hence the name of the test. The decrease in the height of the *centre*[1] of the slumped concrete is called slump, and is measured to the nearest 5 mm ($\frac{1}{4}$ in.). In order to reduce the influence on

---

[1] BS 1881: Part 102: 1983 requires the slump to be measured to the highest part of the concrete.

slump of the variation in the surface friction, the inside of the mould and its base should be moistened at the beginning of *every* test, and prior to lifting of the mould the area immediately around the base of the cone should be cleaned from concrete which may have dropped accidentally.

If instead of slumping evenly all round, as in a true slump (Fig. 5.2), one-half of the cone slides down an inclined plane, a *shear slump* is said to have taken place, and the test should be repeated. If shear slump persists, as may be the case with harsh mixes, this is an indication of lack of cohesion of the mix.

Mixes of stiff consistence have a *zero slump,* so that in the rather dry range no variation can be detected between mixes of different work-ability. There is no problem with rich mixes, their slump being sensitive

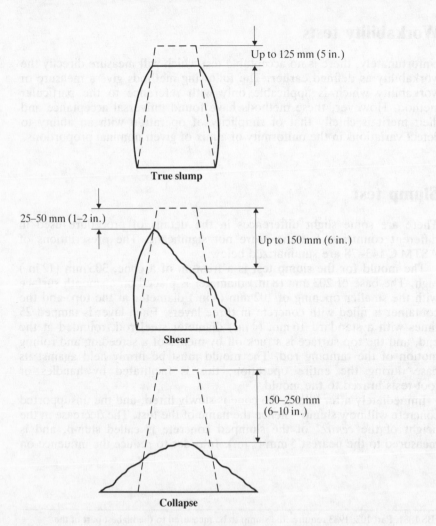

Fig. 5.2: Slump: true, shear, and collapse

to variations in workability. However, in a lean mix with a tendency to harshness, a true slump can easily change to the shear type, or even to collapse (Fig. 5.2), and widely different values of slump can be obtained in different samples from the same mix; thus, the slump test is unreliable for lean mixes.

The order of magnitude of slump for different workabilities is given in Table 5.1 (see also Table 19.4). It should be remembered, however, that with different aggregates the same slump can be recorded for different workabilities, as indeed the slump bears no unique relation to the workability as defined earlier.

Despite these limitations, the slump test is very useful on site as a

Table 5.1: **Workability, slump, and compacting factor of concretes with 19 or 38 mm ($\frac{3}{4}$ or $1\frac{1}{2}$ in.) maximum size of aggregate**

Degree of workability	Slump		Compacting factor	Use for which concrete is suitable
	mm	in.		
Very low	0–25	0–1	0.78	Roads vibrated by power-operated machines. At the more workable end of this group, concrete may be compacted in certain cases with hand-operated machines.
Low	25–50	1–2	0.85	Roads vibrated by hand-operated machines. At the more workable end of this group, concrete may be manually compacted in roads using aggregate of rounded or irregular shape. Mass concrete foundations without vibration or lightly reinforced sections with vibration.
Medium	25–100	2–4	0.92	At the less workable end of this group, manually compacted flat slabs using crushed aggregates. Normal reinforced concrete manually compacted and heavily reinforced sections with vibration.
High	100–175	4–7	0.95	For sections with congested reinforcement. Not normally suitable for vibration.

(Building Research Establishment, Crown copyright)

check on the day-to-day or hour-to-hour variation in the materials being fed into the mixer. An increase in slump may mean, for instance, that the moisture content of aggregate has unexpectedly increased; another cause would be a change in the grading of the aggregate, such as a deficiency of sand. Too high or too low a slump gives immediate warning and enables the mixer operator to remedy the situation. This application of the slump test, as well as its simplicity, is responsible for its widespread use.

Compacting factor test

Although there is no generally accepted method of directly measuring workability, i.e. the amount of work necessary to achieve full compaction, probably the best test yet available uses the inverse approach: the degree of compaction achieved by a standard amount of work is determined. The work applied includes perforce the work done to overcome the surface friction but this is reduced to a minimum, although probably the actual friction varies with the workability of the mix.

The degree of compaction, called the compacting factor, is measured by the density ratio, i.e. the ratio of the density actually achieved in the test to the density of the same concrete fully compacted.

The test, known as the compacting factor test, was developed in the UK and is described in BS 1881: Part 103: 1983 and in ACI Standard 211.3–75 (revised 1980). The apparatus consists essentially of two hoppers, each in the shape of a frustum of a cone, and one cylinder, the three being above one another. The hoppers have hinged doors at the bottom, as shown in Fig. 5.3. All inside surfaces are polished to reduce friction.

The upper hopper is filled with concrete, this being placed gently so that, at this stage, no work is done on the concrete to produce compaction. The bottom door of the hopper is then released and the concrete falls into the lower hopper. This hopper is smaller than the upper one and is, therefore, filled to overflowing and thus always contains approximately the same amount of concrete in a standard state; this reduces the influence of the personal factor in filling the top hopper. The bottom door of the lower hopper is released and the concrete falls into the cylinder. Excess concrete is cut by two floats slid across the top of the mould, and the net mass of concrete in the known volume of the cylinder is determined.

The density of the concrete in the cylinder is now calculated, and the ratio of this density to the density of the fully compacted concrete is defined as the compacting factor. The latter density can be obtained by actually filling the cylinder with concrete in four layers, each tamped or vibrated, or alternatively can be calculated from the absolute volumes of the mix ingredients (see page 380).

Table 5.1 lists values of the compacting factor for different workabilities. Unlike the slump test, the variation in the workability of dry concrete are reflected in a *large* change in the compacting factor, i.e. the

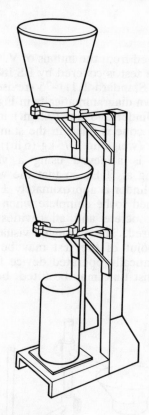

Fig. 5.3: Compacting factor apparatus

test is more sensitive at the low workability end of the scale than at high workability. However, very dry mixes tend to stick in one, or both, hoppers, and the material has to be eased gently by poking with a steel rod. Moreover, it seems that for concrete of very low workability the actual amount of work required for full compaction depends on the richness of the mix while the compacting factor does not: leaner mixes need more work than richer ones. This means that the implied assumption that all mixes with the same compacting factor require the same amount of useful work is not always justified. Nevertheless, the compacting factor test undoubtedly provides a good measure of workability.

The compacting factor apparatus shown in Fig. 5.3 is about 1.2 m (4 ft) high, and is not very convenient to use on site. Thus, although yielding reliable results, the compacting factor apparatus is not often used outside precast concrete works and large sites.

An automatic compacting factor test apparatus has also been developed. Here, the cylinder is supported by a spring balance which can be calibrated for a given mix so as to read workability directly, or even to indicate the excess or deficiency of water in kilogrammes per batch.

Vebe test

The name Vebe is derived from the initials of V. Bährner of Sweden who developed the test. The test is covered by BS 1881: Part 104: 1983 and is referred to also in ACI Standard 211.3–75 (revised 1980).

The apparatus is shown diagrammatically in Fig. 5.4. A standard slump cone is placed in a cylinder 240 mm (9.5 in.) in diameter and 200 mm (8 in.) high. The slump cone is filled in the standard manner, removed, and a disc-shaped rider (weighing 2.75 kg (6 lb)) is placed on top of the concrete. Compaction is achieved using a vibrating table with an eccentric weight rotating at 50 Hz so that the vertical amplitude of the table with the empty cylinder is approximately ±0.35 mm (±0.014 in.).

Compaction is assumed to be complete when the transparent rider is totally covered with concrete and all cavities in the surface of the concrete have disappeared. This is judged visually, and the difficulty of establishing the end point of the test may be a source of error. To overcome it an automatically operated device for recording the movement of the plate against time may be fitted, but this is not a standard procedure.

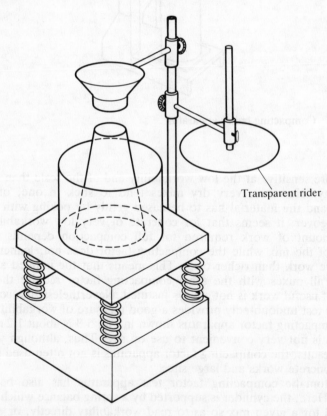

Transparent rider

Fig. 5.4: Vebe apparatus

It is assumed that the input of energy required for full compaction is a measure of workability of the mix, and this is expressed in *Vebe seconds,* i.e. the time required for the operation. Sometimes, a correction for the change in the volume of concrete from V_2 before vibration to V_1 after vibration is applied, the time being multiplied by V_2/V_1.

Vebe is a good laboratory test, particularly for very dry mixes. This is in contrast to the compacting factor test where error may be introduced by the tendency of some dry mixes to stick in the hoppers. The Vebe test also has the additional advantage that the treatment of concrete during the test is comparatively closely related to the method of placing in practice.

Flow table test

This test has recently become more widespread in its use, particularly for flowing concrete made with superplasticizing admixtures (see page 156). The apparatus, shown in Fig. 5.5, consists essentially of a wooden board covered by a steel plate with a total mass of 16 kg (about 35 lb). This board is hinged along one side to a base board, each board being a 700 mm (27.6 in.) square. The upper board can be lifted up to a stop so that the free edge rises 40 mm (1.6 in.). Appropriate markings indicate the location of the concrete to be deposited on the table.

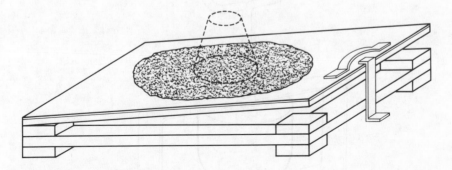

Fig. 5.5: Flow table test

The table top is moistened and a frustum of a cone of concrete, lightly tamped by a wooden tamper in a prescribed manner, is placed using a mould 200 mm (8 in.) high with a bottom diameter of 200 mm (8 in.) and a top diameter of 130 mm (about 5 in.). Before lifting the mould, excess concrete is removed, the surrounding table top is cleaned, and after an interval of 30 sec. the mould is slowly removed. The table top is lifted and allowed to drop, avoiding a significant force against the stop, 15 times, each cycle taking approximately 4 sec. In consequence, the concrete spreads and the maximum spread parallel to the two edges of the table is measured. The average of these two values, given to the nearest millimetre, represents the flow. A value of 400 indicates a medium workability, and 500 a high workability. Concrete should at this stage appear uniform and cohesive or else the test is considered inappropriate for the given mix. Thus the test offers an indication of the cohesiveness of the mix.

Full details of the test are given in BS 1881: Part 105: 1984. The German Standard DIN 1048: Part 1 also describes this test, together with a compaction test, in which a mould 200 mm (8 in.) square and 400 mm (16 in.) high is loosely filled with concrete, which is then fully compacted. The ratio of the initial height to the final height is a measure of consistence; this is related to the reciprocal of the compacting factor.

Ball penetration test

This is a simple field test consisting of the determination of the depth to which a 152 mm (6 in.) diameter metal hemisphere, weighing 14 kg (30 lb), will sink under its own weight into fresh concrete. A sketch of the apparatus, devised by J. W. Kelly and known as the *Kelly ball,* is shown in Fig. 5.6.

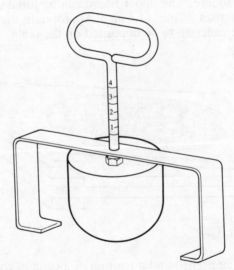

Fig. 5.6: Kelly ball

The use of this test is similar to that of the slump test, that is for routine checking of consistence for control purposes. The test is covered by ASTM Standard C 360–82 and is rarely used in the UK. It is, however, worth considering the Kelly ball test as an alternative to the slump test, over which it has some advantages. In particular, the ball test is simpler and quicker to perform and, what is more important, it can be applied to concrete in a wheelbarrow or actually in the form. In order to avoid boundary effects, the depth of the concrete being tested should be not less than 200 mm (8 in.), and the least lateral dimension 460 mm (18 in.).

As would be expected, there is no simple correlation between penetration and slump, since neither test measures any basic property of concrete but only the response to specific conditions. However, when a particular mix is used, a linear relation can be found. In practice, the ball test is essentially used to measure variations in the mix, such as those due to a variation in the moisture content of the aggregate.

Comparison of tests

It should be said at the outset that no unique relation between the results of the various tests should be expected as each test measures the

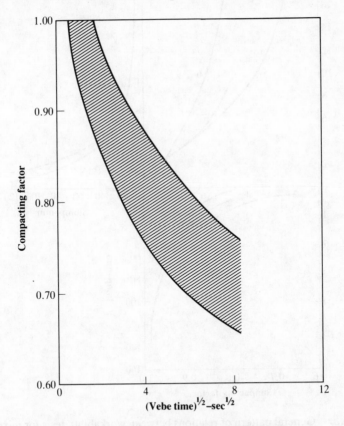

Fig. 5.7: Relation between compacting factor and Vebe time
(From: A. R. CUSENS, The measurement of the workability of dry concrete mixes, *Mag. Concr. Res.*, 8, No. 22, pp. 23–30 (March 1956).)

behaviour of concrete under different conditions. The particular uses of each test have been mentioned.

The compacting factor is closely related to the reciprocal of workability, and the Vebe time is a direct function of workability. The Vebe test measures the properties of concrete under vibration as compared with the free-fall conditions of the compacting factor test.

An indication of the relation between the compacting factor and the Vebe time is given by Fig. 5.7, but this applies only to the mixes used, and the relation must not be assumed to be generally applicable since it depends on factors such as the shape and texture of the aggregate or

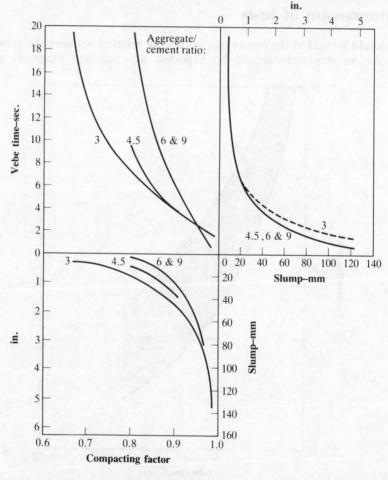

Fig. 5.8: General pattern of relations between workability tests for mixes of varying aggregate/cement ratios
(From: J. D. DEWAR, Relations between various workability control tests for ready-mixed concrete, *Cement Concr. Assoc. Tech. Report TRA*/375 (London, Feb. 1964).)

presence of entrained air, as well as on mix proportions. For specific mixes, the relation between compacting factor and slump has been obtained, but such a relation is also a function of the properties of the mix. A general indication of the pattern of the relation between the compacting factor, Vebe time, and slump is shown in Fig. 5.8. The influence of the richness of the mix (or aggregate/cement ratio) in two of these relations is clear. The absence of influence in the case of the relation between slump and Vebe time is illusory because slump is insensitive at one end of the scale (low workability) and Vebe time at the other (high workability); thus two asymptotic lines with a small connecting part are present.

As already stated, the ideal test for workability has yet to be devised. For this reason, it is worth stressing the value of visual inspection of workability and of assessing it by patting with a trowel in order to see the ease of finishing. Experience is clearly necessary but, once it has been acquired, the 'by eye' test, particularly for the purpose of checking uniformity, is both rapid and reliable.

Density (unit mass or unit weight in air) of fresh concrete

It is common to determine the density of compacted fresh concrete when measuring workability or the air content (see Chapter 15). Density is easily obtained by weighing the compacted fresh concrete in a standard container of known volume and mass; ASTM C 138–81 and BS 1881: Part 107: 1983 describe the procedures. If the density, ρ, is known, the volume of concrete can be found from the mass of the ingredients. When these are expressed as quantities in one batch put into the mixer, we can calculate the *yield* of concrete per batch.

Let the masses per batch of water, cement, fine aggregate, and coarse aggregate be, respectively, W, C, A_f, and A_c. Then, the volume of compacted concrete obtained from one batch (or yield) is

$$V = \frac{C + A_f + A_c + W}{\rho}. \tag{5.1}$$

Also, the cement content (i.e. mass of cement per unit volume of concrete) is

$$\frac{C}{V} = \rho - \frac{A_f + A_c + W}{V}. \tag{5.2}$$

Bibliography

5.1   ACI COMMITTEE 211.3–75, (revised 1987), Standard practice for selecting proportions for no-slump concrete, Part 1, *ACI Manual of Concrete Practice,* pp. 19, 1990.

5.2   T. C. POWERS, *The Properties of Fresh Concrete* (Wiley, 1968).

Problems

5.1   What is mass concrete?
5.2   Discuss the use of a flow table.
5.3   For what mixes is slump not a good test?
5.4   For what mixes is Vebe not a good test?
5.5   What is meant by consistence of a mix?
5.6   What is the relation between cohesiveness and segregation?
5.7   What is meant by segregation of a concrete mix?
5.8   What is meant by bleeding of concrete?
5.9   What is meant by honeycombing?
5.10 Give examples of mixes with the same slump but different workabilities.
5.11 What is the significance of bleeding in construction which proceeds in several lifts?
5.12 What are the factors affecting the workability of concrete?
5.13 Why is it important to control the workability of concrete on site?
5.14 Discuss the advantages and disadvantages of the Vebe test.
5.15 Discuss the factors affecting consistence of concrete.
5.16 Discuss the factors affecting cohesion of concrete.
5.17 Discuss the factors affecting bleeding of concrete.
5.18 Explain what is meant by bleeding of concrete.
5.19 What are the workability requirements for concrete with congested reinforcement?
5.20 What is the relation between bleeding and plastic settlement?
5.21 Why is slump not a direct measure of workability?
5.22 What is meant by lean concrete?
5.23 What is meant by a lean mix?
5.24 Why is absence of segregation important?
5.25 Discuss the applicability of the various workability tests to concretes of different levels of workability.
5.26 What type of slump in a slump test is unsatisfactory?
5.27 Why does workability decrease with time?
5.28 Define workability of concrete.
5.29 How is the compacting factor measured?
5.30 What is meant by yield?
5.31 A 1:1.8:4.5 mix by mass has a water/cement ratio of 0.6. Calculate the cement content of the concrete if its compacted density is 2400 kg/m^3 (150 lb/ft^3).

Answer: 304 kg/m^3 (512 lb/yd^3)

6

Strength of concrete

Strength of concrete is commonly considered to be its most valuable property, although in many practical cases other characteristics, such as durability, impermeability and volume stability, may in fact be more important. Nevertheless, strength usually gives an overall picture of the quality of concrete because it is directly related to the structure of cement paste.

Strength, as well as durability and volume changes of hardened cement paste, appears to depend not so much on the chemical composition as on the physical structure of the products of hydration of cement and on their relative volumetric proportions. In particular, it is the presence of flaws, discontinuities and pores which is of significance, and to understand their influence on strength it is pertinent to consider the mechanics of fracture of concrete under stress. However, since our knowledge of this fundamental approach is inadequate, it is necessary to relate strength to measurable parameters of the structure of hydrated cement paste. It will be shown that a primary factor is *porosity,* i.e. the relative volume of pores or voids in the cement paste. These can be viewed as sources of weakness. Other sources of weakness arise from the presence of the aggregate, which itself may contain flaws in addition to being the cause of microcracking at the interface with the cement paste. Unfortunately, the porosity of the hydrated cement paste and microcracking are difficult to quantify in a useful manner so that for engineering purposes it is necessary to resort to an empirical study of the effects of various factors on strength of concrete. In fact, it will be seen that the overriding factor is the water/cement ratio, with the other mix proportions being only of secondary importance.

Fracture mechanics approach

Fracture mechanics is the study of stress and strain behaviour of homogeneous, brittle materials. It is possible to consider concrete as a brittle material, even though it exhibits a small amount of apparent plasticity (see page 114), because fracture under short-term loading takes place at a moderately low total strain: a strain of 0.001 to 0.005 at failure has been suggested as the limit of brittle behaviour. On the other hand, concrete can hardly be considered to be homogeneous because the properties of its constituents are different and it is, to some extent, anisotropic. Nevertheless, the fracture mechanics approach helps to understand the mechanism of failure of concrete. Clearly, only the basic principles are considered in this book.

Tensile strength considerations

Even when we eliminate one source of heterogeneity, viz. the aggregate, we find that the actual tensile strength of the hydrated cement paste is very much lower than the *theoretical strength* estimated on the basis of molecular cohesion of the atomic structure, and calculated from the energy required to create new surfaces by fracture of a perfectly homogeneous and flawless material. This theoretical strength has been estimated to be 1000 times higher than the actual measured strength.

The discrepancy between the theoretical and actual strengths can be explained by the presence of flaws or cracks, as postulated by Griffith, which lead to very high stress concentrations at their tips under load (see Fig. 6.1) so that localized microscopic fracture can occur when the average (nominal) stress in the whole material is comparatively low. The concentration of stress at the crack tip is, in fact, three-dimensional but the greatest weakness is when the orientation of a crack is normal to the

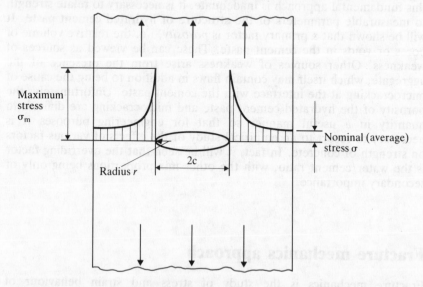

Fig. 6.1: Stress concentration at the tip of a crack in a brittle material under tension

direction of the applied load, as shown in Fig. 6.1. Also, the maximum stress is greater the longer and sharper the crack, i.e. the greater the value of c and the smaller the value of r, as shown by the relation

$$\frac{\sigma_m}{\sigma} = 2\left(\frac{c}{r}\right)^{\frac{1}{2}}. \tag{6.1}$$

The notation is explained in Fig. 6.1.

When the external load increases, the maximum stress σ_m increases until it reaches the failure stress of the material containing the crack, known as the *brittle fracture strength* of the material, σ_f. This is given by

$$\sigma_f = \left(\frac{WE}{\pi c}\right)^{\frac{1}{2}}. \tag{6.2}$$

where W is the work required to cause fracture, and E is the modulus of elasticity. At this stage, new surfaces are formed, the crack extends and there is a release of elastic energy stored in the material. If this energy is sufficient to continue the propagation of the crack, then there exists the condition for an imminent failure of the whole material. On the other hand, if the energy released is too low, the crack is arrested until the external load is increased.

According to the brittle fracture theory, failure is initiated by the largest crack which is oriented in the direction normal to the applied load, and thus the problem is one of statistical probability of the occurrence of such a crack. This means that size and, possibly, shape of the specimen are factors in strength because, for example, there is a higher probability that a larger specimen contains a greater number of critical cracks which can initiate failure.

In a truly brittle material, the energy released by the onset of crack propagation is sufficient to continue this propagation, because, as the crack extends (c increases), the maximum stress increases (Eq. (6.1)) and the brittle fracture strength decreases (Eq. (6.2)). In consequence, the process accelerates. However, in the case of cement paste, the energy released at the onset of cracking may not be sufficient to continue the propagation of a crack because it may be blocked by the presence of an 'obstacle': a large pore, the unhydrated remnant of a cement particle, or the presence of a more ductile material which requires more energy to cause fracture.

The above argument presupposes a uniform (nominal) distribution of stress. In the case of non-uniform stress, the propagation of a crack is blocked, *additionally*, by the surrounding material at a lower stress; this occurs, for example, in flexure. Consequently, whatever the type of stress distribution, the external load has to be increased before a different crack or flaw is subjected to the same process, so that eventually there is a linking of independent cracks before total failure occurs.

The structure of the cement paste is complex and there exist several sources of flaws and discontinuities even before the application of an external load: up to 50 per cent of the volume of the cement paste may consist of pores (see page 106). The presence of aggregate aggravates the situation, as already mentioned. The cracks of the various sources are randomly distributed in concrete and vary in size and orientation. Consequently, concrete is weaker than the cement paste, which it contains. The actual failure paths usually follow the interfaces of the largest aggregate particles, cut through the cement paste, and occasionally also through the aggregate particles themselves.

Behaviour under compressive stress

In the preceding section, we considered failure under the action of a uniaxial tensile force, and indeed Griffith's work applies to such a case. Concrete is used mainly so as to exploit its good compressive strength, and we should now consider the fracture mechanics approach for a material under bi- and triaxial stress and under uniaxial compression.

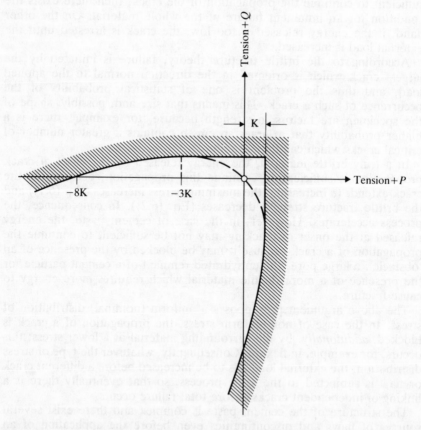

Fig. 6.2: Orowan's criteria of fracture under biaxial stress

Even when two unequal principal stresses are compressive, the stress along the edge of an internal flaw is tensile at some points so that fracture can take place. The fracture criteria are represented graphically in Fig. 6.2 for a combination of two principal stresses P and Q, where K is the tensile strength in direct tension. Fracture occurs under a combination of P and Q such that the point representing the state of stress crosses the curve outwards onto the shaded side. We can see that, when uniaxial compression is applied, the compressive strength is $8K$, i.e. eight times

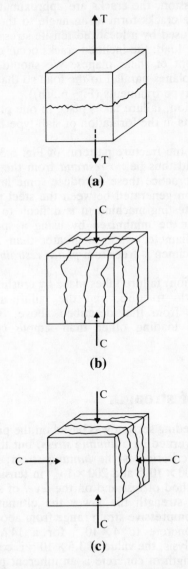

(a)

(b)

(c)

Fig. 6.3: Fracture patterns of concrete under: (a) uniaxial tension, (b) uniaxial compression, and (c) biaxial compression

the direct tensile strength; this value is of the correct order for the observed ratio of compressive to tensile strengths of concrete (see Chapter 10). There are, however, some difficulties in reconciling certain aspects of Griffith's hypothesis with the observed direction of cracks in concrete compression specimens.

Figure 6.3 shows the observed fracture patterns of concrete under different states of stress. Under uniaxial tension, fracture occurs more or less in a plane normal to the direction of the load.

Under uniaxial compression, the cracks are approximately parallel to the applied load but some cracks form at an angle to the applied load. The parallel cracks are caused by a localized tensile stress in a direction normal to the compressive load; the inclined cracks occur due to collapse caused by the development of shear planes. We should note that the cracks are formed in *two* planes parallel to the load so that the specimen disintegrates into column-type fragments (Fig. 6.3(b)).

Under biaxial compression, failure takes place in one plane parallel to the applied load and results in the formation of slab-type fragments (Fig. 6.3(c)).

It should be noted that the fracture patterns of Fig. 6.3 are for direct stresses only. There should thus be *no restraint* from the platens of the testing machine while in practice these introduce some lateral compression because of the friction generated between the steel platen and the concrete. In an ordinary testing machine, it is difficult to eliminate this friction, but its effect can be minimized by using a specimen whose length/width (or length/diameter) ratio is greater than 2, so that the central position of the specimen is free from *platen restraint* (see Chapter 16).

Under triaxial compression, failure takes place by crushing. We are no longer dealing with brittle fracture, and the failure mechanism is, therefore, quite different from that described above. The failure of concrete under types of loading other than simple compression is considered in Chapter 11.

Practical criteria of strength

The discussion in the preceding section was based on the premise that the strength of concrete is governed by a *limiting stress,* but there are strong indications that the real criterion is the *limiting strain*; this is usually assumed to be between 100×10^{-6} and 200×10^{-6} in tension. The actual value depends on the method of test and on the *level* of strength of the concrete: the higher the strength the lower the ultimate strain. The corresponding values of compressive strain range from about 2×10^{-3} for a 70 MPa (10 000 psi) concrete to 4×10^{-3} for a 14 MPa (2000 psi) concrete. In structural analysis, the value of 3.5×10^{-3} is commonly used.

While the notional strength of concrete is an inherent property of the material, in practice, strength is a function of the stress system which is acting. Ideally, it should be possible to express all the failure criteria under all possible stress combinations by a single stress parameter, such as strength in uniaxial tension. However, such a solution has not yet been found, although there have been many attempts to develop empirical relations for failure criteria which would be useful in structural design.

As mentioned earlier, we are not able to express the various factors in the strength of concrete, such as mix proportions, in the form of an equation of strength. All we have is an accumulation of observations at the engineering and empirical levels. We shall use this approach in the discussion of the main factors influencing the strength of concrete.

The most important *practical* factor is the water/cement ratio, but the *underlying* parameter is the number and size of pores in the hardened cement paste. This was referred to on page 97. In fact, the water/cement ratio of the mix mainly determines the porosity of the hardened cement paste, as is demonstrated in the next section.

Porosity

Fresh cement paste is a plastic network of particles of cement in water but, once the paste has set, its apparent or gross volume remains *approximately* constant. As shown in Chapter 2, the paste consists of hydrates of the various cement compounds and of $Ca(OH)_2$, and the gross volume available for all these products of hydration consists of the sum of the absolute volume of the *dry* cement and of the volume of the mix water (assuming that there is no loss of water due to bleeding or evaporation). In consequence of hydration, the mix water takes one of three forms: combined water, gel water and capillary water.

Figure 6.4 illustrates the proportions by volume of the constituents of cement paste before and during hydration of cement. The hydrated cement, or *cement gel,* consists of the *solid products of hydration* plus the water which is held physically or is adsorbed on the large surface area of the hydrates; this water is called *gel water,* and is located between the solid products of hydration in so-called *gel pores.* These are very small (about 2 nm (80×10^{-9} in.) in diameter). It has been established that the volume of gel water is 28 per cent of the volume of cement gel.

In addition to the gel water, there exists water which is combined chemically or physically with the products of hydration, and is thus held very firmly. The quantity of *combined water* can be determined as the non-evaporable water content,[1] and in fully hydrated cement represents about 23 per cent of the mass of dry cement.

Now, the solid products of hydration occupy a volume which is less

---

[1] The demarcation between non-evaporable and evaporable water is usually based on the loss of water upon drying at 105 °C (220 °F).

101

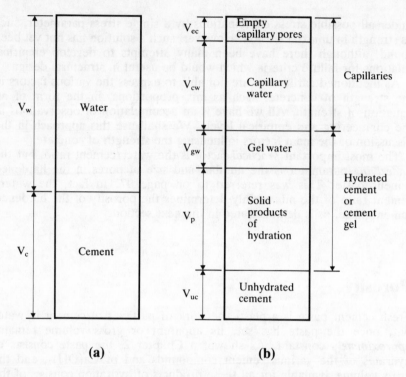

Fig. 6.4: Diagrammatic representation of the volumetric porportions:
(a) before hydration (degree of hydration, $h = 0$), and
(b) during hydration (degree of hydration, h)

than the sum of the absolute volumes of the original dry cement (which has hydrated) and of the combined water; hence, there is a residual space within the gross volume of the paste. For fully hydrated cement, with no excess water above that required for hydration, this residual space represents about 18.5 per cent of the original volume of dry cement. The residual space takes the form of voids or *capillary pores,* which can be empty or full of water, depending on the quantity of the original mix water and depending also on whether additional water could ingress during hydration. Capillary pores are much larger than gel pores (diameter of about 1 μm (40×10^{-6} in.)).

If the mix contained more water than necessary for full hydration, there will be present capillary pores in excess of the 18.5 per cent volume mentioned above; these are full of water.

Let us consider the volume changes due to the hydration of cement, the fresh cement paste being assumed to be fully compacted (see Fig. 6.4). We recall that the mass of combined water is 23 per cent of the mass of dry cement which has hydrated fully. Therefore, if the proportion of hydrated cement, i.e. the degree of hydration, is h, then, for the mass of original cement C, the mass of combined water is $0.23 Ch$.

We stated earlier that, when a volume of cement V_c has hydrated fully, a volume of empty capillary pores, V_{ec}, equal to $0.185 V_c$ is formed. The specific gravity (on an absolute basis) of dry cement is about 3.15. Therefore, the mass[2] occupied by the solid V_c is $3.15 V_c$. Hence, for a degree of hydration, h, the volume of empty capillary pores is

$$V_{ec} = 0.185 \, V_c h = 0.185 \, \frac{C}{3.15} h = 0.059 \, Ch. \tag{6.3}$$

Hence, the volume of combined water is

$$(0.23 - 0.059) \, Ch = 0.171 \, Ch.$$

The volume of the solid products of hydration is given by the sum of the volumes of the combined water and of the hydrated cement, i.e.

$$V_p = \frac{Ch}{3.15} + 0.171 \, Ch = 0.488 \, Ch. \tag{6.4}$$

To obtain the volume of gel water, V_{gw}, we use the fact that it always occupies 28 per cent of the volume of cement gel, or, in other words, that the *gel porosity* is

$$p_g = 0.28 = \frac{V_{gw}}{V_p + V_{gw}}. \tag{6.5}$$

Substituting from Eqs (6.4) and (6.5),

$$V_{gw} = 0.190 \, Ch. \tag{6.6}$$

Now, we can derive the volume occupied by the capillary water, V_{cw}, by reference to Fig. 6.4 as

$$V_{cw} = V_c + V_w - [V_{uc} + V_p + V_{gw} + V_{ec}] \tag{6.7}$$

where V_c = volume of original dry cement = $C/3.15$

V_{uc} = volume of unhydrated cement, i.e.

$$V_{uc} = V_c(1 - h) \tag{6.8}$$

and V_p, V_{gw}, and V_{ec} are given by Eqs (6.4), (6.6) and (6.3), respectively.

---

[2] Strictly speaking, to calculate the mass in kg of lb, we should express the density (or unit mass) in kg/m^3 or lb/ft^3. However, if we use the units of gm/cm^3, the density is numerically equal to specific gravity, and this is more convenient. We therefore express mass in grams and volume in cubic centimetres.

After substitution, Eq. (6.7) becomes

$$V_{cw} = V_w - 0.419 \, Ch. \tag{6.9}$$

Using the above equations, the volumetric composition of the cement paste can be estimated at different stages of hydration. Figure 6.5 illustrates the influence of the water/cement ratio on the resulting values. An interesting feature of this figure is that there is a minimum water/cement ratio necessary to achieve full hydration (approximately 0.36 by mass) because, below this value, there is insufficient space to accommodate all the products of hydration (see page 108). This situation

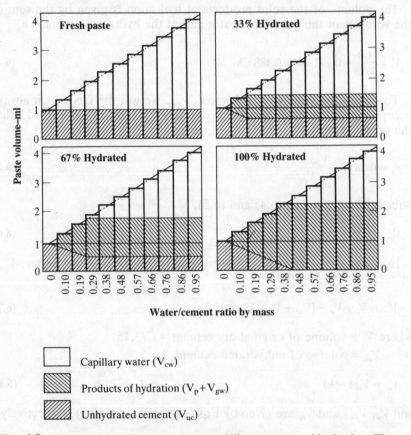

Capillary water (V_{cw})

Products of hydration ($V_p + V_{gw}$)

Unhydrated cement (V_{uc})

Fig. 6.5: Composition of cement paste at different stages of hydration. The percentage indicated applies only to pastes with enough water-filled space to accommodate the products of hydration at the degree of hydration indicated
(From: T. C. POWERS, The non-evaporable water content of hardened Portland cement paste: its significance for concrete research and its method of determination, *ASTM Bul. No.* 158, pp. 68–76 (May 1949).)

applies to a cement paste cured under water, i.e. when there is an external source of water which can be imbibed into the empty capillaries where hydration takes place.

Conversely, when the original mix is sealed, i.e. it has no access to external water, a higher minimum water/cement ratio is required for full hydration. This is so because hydration can proceed only if the capillary pores contain enough water to ensure a sufficiently high internal relative humidity, and not only the amount necessary for the chemical reactions.

The total volume of the capillary pores or voids is a fundamental factor in determining the properties of hardened concrete. This volume is given by Eqs (6.9) and (6.3):

$$V_{cw} + V_{ec} = V_w - 0.36\,Ch = \left[\frac{W}{C} - 0.36\,h\right]C. \tag{6.10}$$

The expression is now in terms of the original water/cement ratio by mass, W/C. It is usual to express the volume of the capillary pores as a fraction of the total volume of the hydrated cement paste; this is called the *capillary porosity*, p_c, and is

$$p_c = \frac{\left[\dfrac{W}{C} - 0.36\,h\right]C}{V_c + V_w}$$

or

$$p_c = \frac{\dfrac{W}{C} - 0.36\,h}{0.317 + \dfrac{W}{C}}. \tag{6.11}$$

We can now calculate the *total porosity* of the cement paste, p_t, as the ratio of the sum of the volumes of gel pores and of capillary pores to the total volume of cement paste:

$$p_t = \frac{0.190\,Ch + \left[\dfrac{W}{C} - 0.36\,h\right]C}{0.317 + \dfrac{W}{C}}$$

whence

$$p_t = \frac{\dfrac{W}{C} - 0.17\,h}{0.317 + \dfrac{W}{C}}. \tag{6.12}$$

Equations (6.11) and (6.12) demonstrate that porosity depends upon

105

the water/cement ratio *and* on the degree of hydration. In fact, the term W/C in the numerator of these equations is the main influencing factor on porosity, as can be seen from Fig. 6.6. This figure illustrates also the decrease in porosity with an increase in the degree of hydration. The magnitude of porosity is such that, for the usual range of water/cement ratios, the cement paste is only about 'half solid'. For instance, at a water/cement ratio of 0.6, the total volume of pores is between 47 and 60 per cent of the total volume of the cement paste, depending on the degree of hydration.

The expression for porosity derived earlier assumes that the fresh cement paste is fully compacted, i.e. it contains no accidental or entrapped air. If such air is present or if air entrainment is used, then Eqs

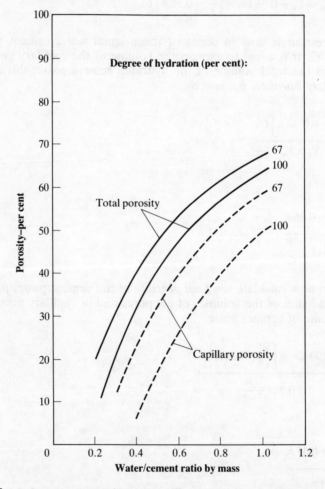

Fig. 6.6: Influence of water/cement ratio and degree on hydration on capillary and total porosities of cement paste, as given by Eqs (6.11) and (6.12)

(6.11) and (6.12) become, respectively,

$$p_c = \frac{\dfrac{W}{C} + \dfrac{a}{C} - 0.36}{0.317 + \dfrac{W}{C} + \dfrac{a}{C}} \qquad \textbf{(6.13)}$$

and

$$p_t = \frac{\dfrac{W}{C} + \dfrac{a}{C} - 0.17\,h}{0.317 + \dfrac{W}{C} + \dfrac{a}{C}} \qquad \textbf{(6.14)}$$

where a = volume of air in the fresh cement paste.

The relation between the water/cement ratio and porosity of hardened cement paste is now clear. There is a corresponding relation between porosity and strength, and this is independent of whether the capillary pores are full of water or empty. Figure 6.7 shows the relation between porosity and strength for cement pastes in which extremely high strengths

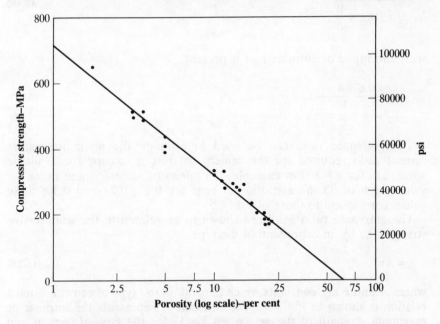

Fig. 6.7: Relation between compressive strength and logarithm of porosity of cement paste compacts for various treatments of pressure and high temperature
(From: D. M. ROY and G. R. GOUDA, Porosity – strength relation in cementitious materials with very high strengths, *J. Amer. Ceramic Soc.*, 53, No. 10, pp. 549–50 (1973).)

were obtained by a high pressure so as to achieve good compaction at very low water/cement ratios.

It is of interest to note that the relation between strength and porosity is not unique to concrete but is also applicable to metals and some other materials.

Gel/space ratio

An alternative parameter to porosity is the gel/space ratio, x, which is defined as the ratio of the volume of the cement gel to the sum of the volumes of cement gel and of capillary pores, i.e.

$$x = \frac{V_p + V_{gw}}{(V_p + V_{gw}) + (V_{cw} + V_{ec})}. \tag{6.15}$$

Using the previously derived expression, Eq. (6.15) becomes

$$x = \frac{0.678\,h}{0.318\,h + \dfrac{W}{C}} \tag{6.16}$$

or, if entrapped or entrained air is present,

$$x = \frac{0.678\,h}{0.318\,h + \dfrac{W}{C} + \dfrac{a}{C}}. \tag{6.17}$$

The gel/space ratio can be used to estimate the minimum water/cement ratio required for the cement gel just to occupy the available space, i.e. for $x = 1$. For example, the minimum water/cement ratios for values of h of 33, 67 and 100 per cent are 0.12, 0.24 and 0.36; these values correspond to those of Fig. 6.5.

The gel/space ratio has been shown to be related to the compressive strength, f_c, by an expression of the type

$$f_c = Ax^b \tag{6.18}$$

where A and b are constants which depend on the type of cement. Such a relation is shown in Fig. 6.8. The constant A represents the intrinsic or maximum strength of the gel (when $x = 1$) for the type of cement and type of specimen used. In other words, the maximum possible strength is achieved in a fully-hydrated cement paste having a water/cement ratio of 0.36 and compacted in the usual manner. However, the gel/space ratio concept is limited in its application because, as already stated, higher strengths can be achieved with partially hydrated cement pastes with

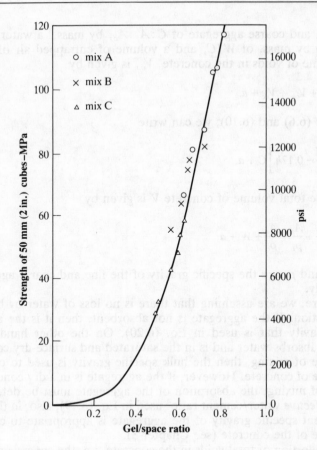

Fig. 6.8: Relation between the compressive strength of mortar and gel/space
ratio
(From: T. C. POWERS, Structural and physical properties of
hardened Portland cement, *J. Amer. Ceramic Soc.*, 41, pp. 1–6 (Jan.
1958).)

lower water/cement ratios but subjected to high pressure in order to
reduce porosity.

Total voids in concrete

In the preceding section, we expressed the total volume of voids, i.e. the
volume of pores and of accidental or entrapped air, as a proportion of the
volume of cement gel including voids (Eq. 6.14). However, the volume of
voids as a proportion of the volume of concrete is also of interest.

Let us consider concrete having mix proportions of cement, fine aggregate, and coarse aggregate of $C:A_f:A_c$, by mass,[3] a water/cement ratio, also by mass, of W/C, and a volume of entrapped air of a. The total volume of voids in the concrete, V_v, is given by

$$V_v = V_{gw} + V_{cw} + V_{ec} + a.$$

Using Eqs (6.6) and (6.10), we can write

$$V_v = \left[\frac{W}{C} - 0.17h\right]C + a. \tag{6.19}$$

Now, the total volume of concrete V is given by

$$V = \frac{C}{3.15} + \frac{A_f}{\rho_f} + \frac{A_c}{\rho_c} + W + a \tag{6.20}$$

where ρ_f and ρ_c are the specific gravity of the fine and coarse aggregate, respectively.

As before, we are assuming that there is no loss of water by bleeding or segregation. If the aggregate is not absorbent, then it is the absolute specific gravity that is used in Eq. (6.20). On the other hand, if the aggregate absorbs water and is in the saturated and surface-dry condition at the time of mixing, then the bulk specific gravity is used to calculate the volume of concrete. However, if the aggregate is in a dry condition at the time of mixing, the absorption of the aggregate must be determined and the *effective* water/cement ratio[4] used in Eq. (6.19); also, in this case, the apparent specific gravity of the aggregate is appropriate to calculate the volume of the concrete (see Chapter 3).

The proportion of total voids in the concrete, i.e. the *concrete porosity*, P, can be derived from Eqs (6.19) and (6.20):

$$P = \frac{V_v}{V} = \frac{\dfrac{W}{C} - 0.17\,h + \dfrac{a}{C}}{0.317 + \dfrac{1}{\rho_f}\dfrac{A_f}{C} + \dfrac{1}{\rho_c}\dfrac{A_c}{C} + \dfrac{W}{C} + \dfrac{a}{C}}. \tag{6.21}$$

As a specific example, let us consider concrete with mix proportions of $1:2:4$ by mass and a water/cement ratio of 0.55. The air content has been measured as 2.3 per cent of the volume of the concrete and the specific gravity of fine and coarse aggregate is 2.60 and 2.65, respectively.

---

[3] This is the standard way of describing mix proportions; for instance, a $1:2:4$ mix consists of 1 part of cement, 2 parts of fine aggregate, and 4 parts of coarse aggregate, all by mass. Similarly a $1:6$ mix consists of 1 part cement and 6 parts of total aggregate.
[4] See page 55.

Now, the air content is given by Eq. (6.20) as

$$\frac{a}{V} = \frac{a}{\dfrac{1}{3.15} + \dfrac{2}{2.6} + \dfrac{4}{2.65} + 0.55 + a} = \frac{2.3}{100}.$$

Hence, the volume of entrapped air is $a = 0.074$.

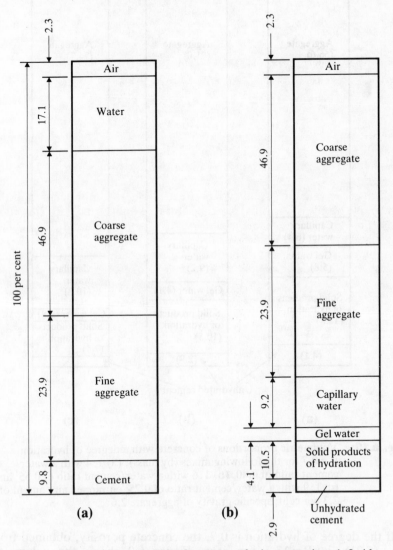

Fig. 6.9: Volumetric proportions of concrete of mix proportions $1:2:4$ by mass with a water/cement ratio of 0.55 and entrapped air content of 2.3 per cent: (a) before hydration, and (b) when the degree of hydration is $h = 0.7$

111

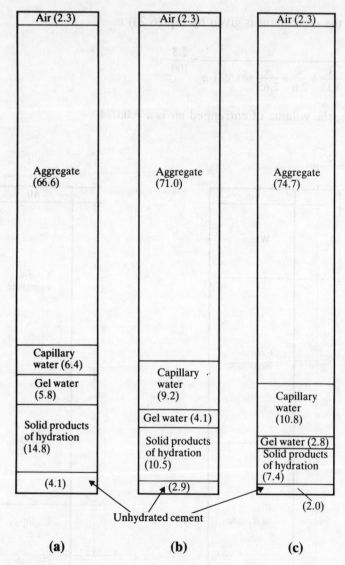

Fig. 6.10: Volumetric proportions of concrete with a degree of hydration $h = 0.7$ for the following mixes (by mass): (a) $1:4$ with a water/cement ratio of 0.40, (b) $1:6$ with a water/cement ratio of 0.55, and (c) $1:9$ with a water/cement ratio of 0.75; entrapped air content of 2.3 per cent; specific gravity of aggregate 2.6

If the degree of hydration is 0.7, the concrete porosity, obtained from Eq. (6.21), is $P = 15.7$ per cent. Figure 6.9 shows the volumetric proportions of the concrete when mixed and at 70 per cent hydration ($h = 0.7$); the latter were obtained using Eqs (6.3), (6.4) and (6.6) to (6.8). The corresponding situation for a $1:4$ mix with a water/cement ratio of 0.40 is shown in Fig. 6.10(a); for a 1:6 mix with a water/cement ratio of 0.55 in Fig. 6.10(b); and for a 1:9 mix with a water/cement ratio of 0.75 in Fig. 6.10(c). In all cases, the degree of hydration is $h = 0.7$, and the specific gravity of aggregate 2.6.

The discussion in this and the preceding sections has made it clear that porosity is a primary factor influencing the strength of concrete. However, it is not only the total volume of pores but also some other features of the pores that are of significance, albeit they are difficult to quantify; these are discussed below.

Pore size distribution

As we have stated, capillary pores are much larger than gel pores, although, in fact, there is a whole range of pore sizes throughout the hardened cement paste. When only partly hydrated, the paste contains an interconnected system of capillary pores. The effect of this is a lower strength and, through increased permeability, a higher vulnerability to freezing and thawing and to chemical attack. This vulnerability depends also on the water/cement ratio.

These problems are avoided if the degree of hydration is sufficiently high for the capillary pore system to become segmented through partial blocking by newly developed cement gel. When this is the case, the capillary pores are interconnected only by the much smaller gel pores, which are impermeable. An indication of the minimum period of curing required for the capillary pores to become segmented is given in Table 6.1. However, we should note that the finer the cement the shorter the period of curing necessary to produce a given degree of hydration at a

Table 6.1: **Approximate curing period required to produce the degree of hydration at which capillaries become segmented**

Water/cement ratio by mass	Degree of hydration, per cent	Curing period required
0.40	50	3 days
0.45	60	7 days
0.50	70	14 days
0.60	92	6 months
0.70	100	1 year
over 0.70	100	impossible

From: T. C. POWERS, L. E. COPELAND and H. M. MANN, Capillary continuity or discontinuity in cement pastes, *J. Portl. Cem. Assoc. Research and Development Laboratories*, 1, No. 2, pp. 38–48 (May 1959).

113

given water/cement ratio. Table 6.1 shows that to achieve durable concrete, shorter periods of curing are required for mixes with lower water/cement ratios but, of course, such mixes have higher strengths because of their lower porosity. Permeability and durability are discussed in Chapter 14.

Microcracking and stress–strain relation

So far we have concentrated on the properties of hardened cement paste, but we should now take note of the presence of aggregate, i.e. consider concrete. It has been shown that very fine *bond cracks* exist at the interface between coarse aggregate and hydrated cement paste even prior to the application of load. Such microcracking occurs as a result of differential volume changes between the cement paste and the aggregate,

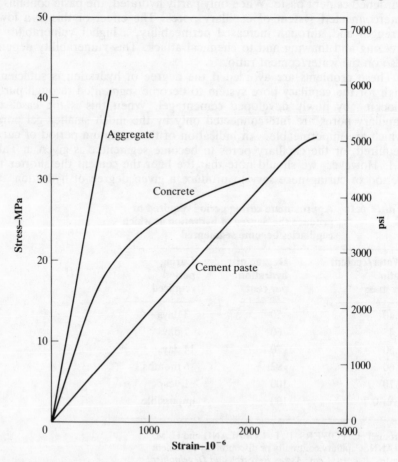

Fig. 6.11: Stress–strain relations for cement paste, aggregate, and concrete

i.e. due to differences in stress–strain behaviour, and in thermal and moisture movements. These cracks remain stable and do not grow under stress up to about 30 per cent of the ultimate strength of the concrete.

Figure 6.11 shows that the stress–strain relations for the aggregate alone and for the cement paste are linear, but the stress–strain relation for concrete becomes curvilinear at higher stresses. This apparent paradox is explained by the development of bond cracks at the interfaces between the two phases, i.e. due to microcracking at stresses above 30 per cent of the ultimate strength. At this stage, the microcracks begin to increase in length, width and number. In consequence, the strain increases at a faster rate than the stress. This is the stage of *slow propagation* of microcracks, which are probably stable under a sustained load (although for prolonged periods the strain increases because of creep (see Chapter 12)). This development of microcracks, together with creep, contributes to the ability of concrete to redistribute local high stresses to regions of lower stress, and thus to avoid early localized failure.

At 70 to 90 per cent of the ultimate strength, cracks open through the mortar matrix (cement paste and fine aggregate) and thus bridge the bond cracks so that a continuous crack pattern is formed. This is the stage of *fast propagation* of cracks, and, if the load is sustained, failure will probably occur with the passage of time; we call this static fatigue (see Chapter 11). Of course, if the load is increased, rapid failure at the nominal ultimate strength will take place.

The foregoing is a description of the stress–strain behaviour in compression as indicated by measuring the *axial* (compressive) strain when the load is increased at a constant rate of stress until failure occurs at the maximum stress. If the *lateral* strain is observed, a corresponding extension is observed (Fig. 6.12). The ratio of lateral strain to axial strain

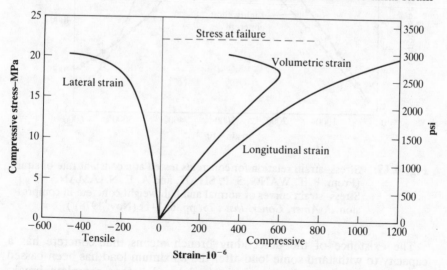

Fig. 6.12: Strains in a prism tested to failure in compression

(i.e. Poisson's ratio) is constant for stresses below approximately 30 per cent of the ultimate strength. Beyond this point, Poisson's ratio increases slowly, and at 70 to 90 per cent of the ultimate strength, it increases rapidly due to the formation of mainly vertical unstable cracks. At this stage, the specimen is no longer a continuous body, as shown by the *volumetric strain* curve of Fig. 6.12: there is a change from a slow contraction in volume to a rapid increase in volume.

The progress of cracking can also be detected by ultrasonic tests (see Chapter 16) and by acoustic emission tests. As cracking develops, the transverse ultrasonic pulse-velocity decreases and the level of sound emitted increases, both of these exhibiting large changes prior to failure.

As previously stated, the stress–strain curve of the type shown in Fig. 6.11 is found when concrete is loaded in uniaxial compression, with stress increasing at a constant rate. This is, for instance, the case in standard compression tests on cubes or cylinders. However, if the specimen is loaded at a constant *rate of strain,* a descending part of the stress–strain curve is obtained before failure. (This type of test requires the use of a testing machine with a stiff frame, and it is the displacement rather than the load that has to be controlled.) Figure 6.13 shows complete stress–strain curves for this type of loading.

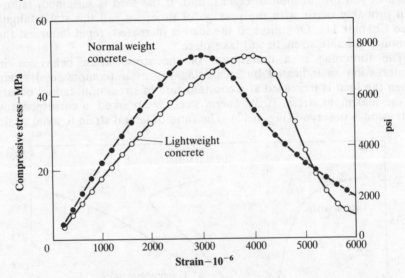

Fig. 6.13: Stress–strain relation for concretes tested at a constant rate of strain (From: P. T. WANG, S. P. SHAH, and A. E. NAAMAN, Stress–strain curves of normal and lightweight concrete in compression. *J. Amer. Concr. Inst.,* 75, pp. 603–11 (Nov. 1978).)

The existence of the descending branch means that concrete has a capacity to withstand some load after the maximum load has been passed because the linking of microcracks is delayed before complete breakdown. A steeper descending curve for lightweight aggregate concrete (see

Fig. 6.13) implies that it has a more brittle nature than concrete made with normal weight aggregate. In high-strength concrete, both the ascending and descending parts of the curve are steeper: again this implies a more brittle type of behaviour.

The area enclosed by the complete stress–strain curve represents the work necessary to cause failure or *fracture toughness*.

Factors in strength of concrete

Although porosity is a primary factor influencing strength, it is a property difficult to measure in engineering practice, or even to calculate since the degree of hydration is not easily determined (assuming, of course, that the water/cement ratio is known). Similarly, the influence of aggregate on microcracking is not easily quantified. For these reasons, the main influencing factors on strength are taken in practice as: water/cement ratio, degree of compaction, age, and temperature. However, there are also other factors which affect strength: aggregate/cement ratio, quality of the aggregate (grading, surface texture, shape, strength, and stiffness), and the maximum size of the aggregate. These factors are regarded as of secondary importance when usual aggregates up to a maximum size of 40 mm ($1\frac{1}{2}$ in.) are used.

Water/cement ratio, degree of compaction, and age

In ordinary construction, it is not possible to expel all the air from the concrete so that, even in fully-compacted concrete, there are some entrapped air voids. Table 6.2 gives typical values. Assuming full compaction, and at a given age and normal temperature, strength of concrete can be taken to be inversely proportional to the water/cement ratio.[5] This is the so-called *Abrams' 'law'*. Figure 6.14 illustrates this statement and shows also the effects of partial compaction on strength.

Abrams' 'law' is a special case of a general rule formulated empirically by Feret:

$$f_c = K\left[\frac{V_c}{V_c + V_w + a}\right]^2 \tag{6.22}$$

where f_c is strength of concrete,

V_c, V_w and a are absolute volumes of cement, water, and entrapped air, respectively, and

K is a constant.

---

[5] For example, see Tables 19.1 and 19.2.

Table 6.2: **Appropriate entrapped air content for different sizes of aggregate, according to ACI 211.1–84**

Nominal maximum size of aggregate		Entrapped air content, per cent
mm	**in.**	
10	$\frac{3}{8}$	3
12.5	$\frac{1}{2}$	2.5
20	$\frac{3}{4}$	2
25	1	1.5
40	$1\frac{1}{2}$	1
50	2	0.5
70	3	0.3
150	6	0.2

It will be recalled that, at a given degree of hydration, the water/cement ratio determines the porosity of the cement paste. Thus, the relation of Eq. (6.22) accounts for the influence of the *total* volume of voids on strength, i.e. gel pores, capillary pores and entrapped air.

With an increase in age, the degree of hydration generally increases so

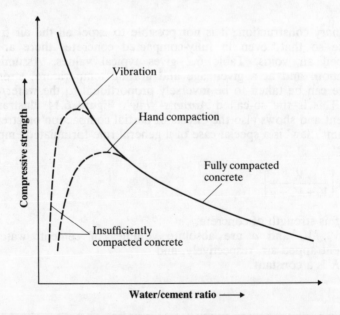

Fig. 6.14: Relation between strength and water/cement ratio of concrete

118

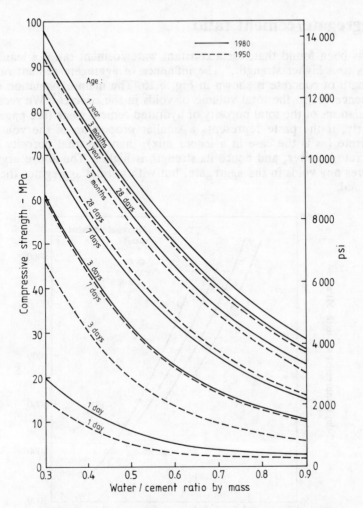

Fig. 6.15: Influence of age on compressive strength of ordinary Portland (Type I) cement concrete at different water/cement ratios. The data are typical for cements manufactured in 1950 and 1980
(Based on Concrete Society Technical Report No. 29, Changes in the properties of ordinary Portland cement and their effects on concrete, 1986 and D.C. Teychenné, R.E. Franklin and H. Erntroy, Design of Normal Concrete Mixes, Building Research Establishment, Department of the Environment, London, HMSO, 1986.)

that strength increases; this effect is shown in Fig. 6.15 for concretes made with ordinary Portland (Type I) cement. It should be emphasized that strength depends on the *effective* water/cement ratio, which is calculated on the basis of the mix water less the water absorbed by the aggregate; in other words, the aggregate is assumed to 'use up' some water so as to reach a saturated and surface-dry condition at the time of mixing.

119

Aggregate/cement ratio

It has been found that, for a constant water/cement ratio, a leaner mix leads to a higher strength.[6] The influence of aggregate/cement ratio on strength of concrete is shown in Fig. 6.16. The main explanation of this influence lies in the total volume of voids in the *concrete*. We recall our calculations of the total porosity of hydrated cement paste (see page 101). Clearly, if the paste represents a smaller proportion of the volume of concrete (as is the case in a leaner mix), then the total porosity of the concrete is lower, and hence its strength is higher. The above argument ignores any voids in the aggregate, but with normal aggregates these are minimal.

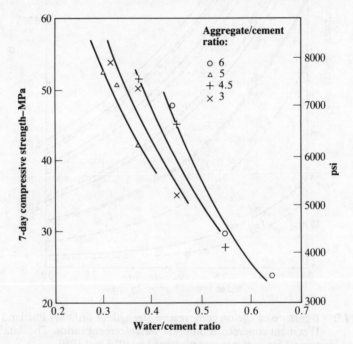

Fig. 6.16: Influence of the aggregate/cement ratio on strength of concrete (From: B. G. SINGH, Specific surface of aggregates related to compressive and flexural strength of concrete, *J. Amer. Concr. Inst.*, 54, pp. 897–907 (April 1958).)

---

[6] *Lean concrete* is considered to mean mixes with high aggregate/cement ratios (generally not above 10) and should be distinguished from a *lean-mix concrete base*; the latter, used in road construction, may have an aggregate/cement ratio as high as 20 and is suitable for compaction by rolling (see Chapter 19).

Aggregate properties

As stated earlier, the influence of the aggregate properties on strength is of secondary importance. Some of these are discussed in Chapter 11, and, here, only the shape of the aggregate is considered. The stress at which significant cracking commences is affected by the shape of the aggregate: smooth gravel leads to cracking at lower stresses than rough and angular crushed aggregate, other things being equal. The effect, similar in tension and compression, is due to a better bond and less microcracking with an angular crushed aggregate.

In fact, the influence of the aggregate shape is more apparent in the modulus of rupture test than in the uniaxial compressive or tensile tests, probably because of the presence of a stress gradient which delays the progress of cracking leading to ultimate failure. Hence, concrete with an angular-shaped aggregate will have a higher flexural strength than when rounded-shaped aggregate is used, especially in mixes with low water/cement ratios. However, in practical mixes of the same workability, a rounded-shaped aggregate requires less water than an angular-shaped aggregate, and therefore the flexural strengths of the two concretes are similar.

Bibliography

6.1 L. E. COPELAND and J. C. HAYES, The determination of non-evaporable water in hardened cement paste, *ASTM Bull.* No. 194, pp. 70–4 (Dec. 1953).

6.2 A. M. NEVILLE, *Properties of Concrete* (London, Longman, 1981).

6.3 T. C. POWERS, Structure and physical properties of hardened Portland cement paste, *J. Amer. Ceramic Soc.*, **41**, pp. 1–6 (Jan. 1958).

6.4 T. C. POWERS and T. L. BROWNYARD, Studies of the physical properties of hardened Portland cement paste (Nine parts), *J. Amer. Concr. Inst.*, **43** (Oct. 1946 to April 1947).

6.5 S. P. SHAH and G. WINTER, Inelastic behaviour and fracture of concrete, Symp. on Causes, Mechanism, and Control of Cracking in Concrete, *Amer. Concr. Inst. Sp. Publicn.* No. 20, pp. 5–28 (1968).

6.6 G. J. VERBECK, Hardened concrete – pore structure, *ASTM Sp. Tech. Publicn.* No. 169, pp. 136–42 (1955).

Problems

6.1 What are the types of composite materials? Describe two simple relevant models and comment on their validity.

6.2 Discuss crack propagation in concrete.

6.3 Describe Griffith's model for cracking of concrete.

6.4 What is meant by a flaw in cement paste?

6.5 What is meant by a crack arrester in concrete?

6.6 Comment on the statement that concrete is not a brittle material.

6.7 What are the volume concentration models for the prediction of the modulus of elasticity of concrete? Explain how they have been derived. Comment on their validity.

6.8 Explain the influence of water/cement ratio on strength of concrete.

6.9 What is meant by strain capacity?

6.10 What is meant by ultimate strain?

6.11 Sketch the failure patterns for concrete specimens subjected to uniaxial tension, uniaxial compression and biaxial compression, assuming no end restraint.

6.12 What are the various types of water in hydrated cement paste?

6.13 What is fracture toughness?

6.14 What is meant by non-evaporable water?

6.15 What is the difference between gel pores and capillary pores?

6.16 What is the minimum water/cement ratio for full hydration of cement?

6.17 Why is porosity important with regard to: (a) strength, (b) durability?

6.18 State the difference between capillary porosity, gel porosity, total porosity and concrete porosity.

6.19 Define the gel/space ratio.

6.20 Discuss the effect of curing on the capillary pore system and how this affects durability.

6.21 Describe the stress–strain characteristics of concrete up to failure. Is there any difference between the stress–strain characteristics of aggregate and of cement paste?

6.22 Define Poisson's ratio.

6.23 What is volumetric strain?

6.24 How does the application of constant rate of strain affect the stress–strain curve for concrete?

6.25 Discuss the effects of the degree of compaction and age on strength of concrete.

6.26 What is the effective water/cement ratio?

6.27 Calculate the modulus of elasticity of concrete using:
  (i) the composite soft model, and
  (ii) the composite hard model,
  assuming the aggregate occupies 70 per cent of the volume of concrete; the moduli of elasticity of cement paste and aggregate are 25 and 50 GPa (3.62×10^6 and 7.25×10^6 psi), respectively.

Answer: (i) 38.5 GPa (5.6×10^6 psi)
    (ii) 42.5 GPa (6.4×10^6 psi)

6.28 A mix has an aggregate/cement ratio of 6 and a concrete porosity of 17 per cent. Assuming there is no entrapped air, calculate the water/cement ratio of the mix, given that the specific gravity of aggregate is 2.6 and the degree of hydration is 90 per cent.

Answer: 0.72

6.29 Calculate the concrete porosity if 2.0 per cent of air is accidentally entrapped in the mix of question 6.28, having a water/cement ratio of 0.72.

Answer: 18.6 per cent

7

Mixing, handling, placing, and compacting concrete

We have considered, so far, what could be called a recipe for concrete. We know the properties of the ingredients, although not much about their proportions; we also know the properties of the mixture: the fresh concrete; and now we should look at the practical means of producing fresh concrete and placing it in the forms so that it can harden into the structural or building material: the hardened concrete, usually referred to simply as concrete. The sequence of operations is as follows. The correct quantities of cement, aggregate, and water, possibly also of admixture, are batched and mixed in a concrete mixer. This produces fresh concrete, which is transported from the mixer to its final location. The fresh concrete is then placed in the forms, and compacted so as to achieve a dense mass which is allowed, and helped, to harden. Let us consider the various operations in turn.

Mixers

The mixing operation consists essentially of rotation or stirring, the objective being to coat the surface of all the aggregate particles with cement paste, and to blend all the ingredients of concrete into a uniform mass; this uniformity must not be disturbed by the process of discharging from the mixer.

The usual type of mixer is a batch mixer, which means that one batch of concrete is mixed and discharged before any more materials are put into the mixer. There are four types of batch mixers.

A *tilting drum mixer* is one whose drum in which mixing takes place is tilted for discharging. The drum is conical or bowl-shaped with internal vanes, and the discharge is rapid and unsegregated so that these mixers are suitable for mixes of low workability and for those containing large-size aggregate.

A *non-tilting drum mixer* is one in which the axis of the mixer is always horizontal, and discharge takes place by inserting a chute into the drum or by reversing the direction or rotation of the drum (a *reversing drum mixer*). Because of a slow rate of discharge, some segregation may occur,

124

a part of the coarse aggregate being discharged last. This type of mixer is charged by means of a loading skip, which is also used with the larger tilting drum mixers, and it is important that the *whole* charge be transferred from the skip into the mixer every time.

A *pan-type mixer* is a forced-action mixer, as distinct from drum mixers which rely on the free fall of concrete inside the drum. The pan mixer consists essentially of a circular pan rotating about its axis with one or two stars of paddles rotating about a vertical axis *not* coincident with the axis of the pan; sometimes the pan is static and the axis of the star travels along a circular path about the axis of the pan. In either case, the concrete in every part of the pan is thoroughly mixed, and scraper blades ensure that mortar does not stick to the sides of the pan. The height of the paddles can be adjusted to prevent the formation of a coating of mortar on the bottom of the pan. Pan mixers are particularly efficient with stiff and cohesive mixes and are, therefore, often used for precast concrete, as well as for mixing small quantities of concrete or mortar in the laboratory.

A *dual drum mixer* is sometimes used in highway construction. Here, there are two drums in series, concrete being mixed part of the time in one and then transferred to the other for the remainder of the mixing time before discharging. In the meantime, the first drum is recharged so that initial mixing takes place without inter-mixing of the batches. In this manner the yield of concrete can be doubled, which is a considerable advantage in the case of highway construction where space or access is often limited. Triple drum mixers are also used.

It may be relevant to mention that in drum-type mixers no scraping of the sides takes place during mixing so that a certain amount of mortar adheres to the sides of the drum and remains until the mixer is cleaned. It follows that, at the beginning of concreting, the first mix would leave behind a proportion of its mortar, the discharge consisting largely of coated coarse aggregate particles. This initial batch should be discarded. As an alternative, a certain amount of mortar (concrete less coarse aggregate) may be introduced into the drum prior to mixing the concrete, a procedure known as *buttering*. The mortar in excess of that stuck in the mixer can be used in construction, e.g. by placing at a cold joint. The necessity of buttering should not be forgotten when using a laboratory mixer.

The size of a mixer is described by the volume of concrete *after* compaction (BS 1305: 1974), as distinct from the volume of the unmixed ingredients in a loose state, which is up to 50 per cent greater than the compacted volume. Mixers are made in a variety of sizes from 0.04 m^3 (1.5 ft^3) for laboratory use up to 13 m^3 (17 yd^3). If the quantity mixed represents only a small fraction of the mixer capacity, the operation will be uneconomic, and the resulting mix may be not uniform; this is bad practice. Overloading the mixer by up to 10 per cent is generally harmless, but, if greater, a uniform mix will not be obtained; this is very bad practice.

All the mixers described so far are batch mixers, but there exist also *continuous mixers*, which are fed automatically by a continuous weigh-batching system. The mixer itself may be of drum-type or may be in the form of a screw moving in a stationary housing. Specialized mixers are used in shotcreting (see page 141) and for mortar for preplaced aggregate concrete (see page 143).

Charging the mixer

There are no general rules on the order of feeding the ingredients into the mixer as this depends on the properties of the mixer and of the mix. Usually, a small amount of water is fed first, followed by all the solid materials, preferably fed uniformly and simultaneously into the mixer. If possible, the greater part of the water should also be fed during the same time, the remainder being added after the solids. However, when using very dry mixes in drum mixers it is necessary to feed the coarse aggregate just after the small initial water feed in order to ensure that the aggregate surface is sufficiently wetted. Moreover, if coarse aggregate is absent to begin with, the finer ingredients can become lodged in the head of the mixer – an occurrence known as *head pack*. If water or cement is fed too fast or is too hot there is a danger of forming cement balls, sometimes as large as 75 mm (3 in.) in diameter.

With small laboratory pan mixers and very stiff mixes, the sand should be fed first, then a part of the coarse aggregate, cement and water, and finally the remainder of the coarse aggregate so as to break up any nodules of mortar.

Uniformity of mixing

In any mixer, it is essential that a sufficient interchange of materials occurs between different parts of the chamber, so that uniform concrete is produced. The efficiency of the mixer can be measured by the variability of samples from the mix. ASTM C 94–83 prescribes samples to be taken from about points $\frac{1}{6}$ and $\frac{5}{6}$ of the discharge of a batch, and the differences in the properties of the two samples should not exceed any of the following:

(a) density of concrete: 16 kg/m³ (1 lb/ft³)
(b) air content: 1 per cent
(c) slump: 25 mm (1 in.) when average is less than 100 mm (4 in.), and 40 mm (1.5 in.) when average is 100 to 150 mm (4 to 6 in.)
(d) percentage of aggregate retained on 4.75 mm ($\frac{3}{16}$ in.) sieve: 6 per cent
(e) density of air-free mortar: 1.6 per cent
(f) compressive strength (average 7-day value of 3 cylinders): 7.5 per cent.

BS 3963: 1974 lays down a performance test of mixers using a specified mix. Tests are made on *two* samples from each quarter of a batch, and each sample is subjected to wet analysis (see page 302), in accordance with BS 1881: Part 125: 1988, to determine:

(a) water content as percentage of solids (to 0.1 per cent),
(b) fine aggregate content as percentage of total aggregate (to 0.5 per cent),
(c) cement as percentage of total aggregate (to 0.01 per cent),
(d) water/cement ratio (to 0.01).

The *sampling accuracy* is assured by a limit on the average *range* of pairs, and if two samples in a pair differ unduly then their results are discarded. The *mixer performance* is judged by the difference between the highest and lowest *average* of pairs for each batch using three separate test batches; thus one bad mixing operation does not condemn a mixer. The maximum acceptable variabilities of the percentages listed earlier are prescribed by BS 1305: 1974 for batch-type mixers.

Mixing time

On site, there is often a tendency to mix concrete as rapidly as possible, and, hence, it is important to know the minimum mixing time necessary to produce a concrete of uniform composition and, consequently, of reliable strength. The optimum mixing time depends on the type and size of mixer, on the speed of rotation, and on the quality of blending of ingredients during charging of the mixer. Generally, a mixing time of less than 1 to $1\frac{1}{4}$ min produces appreciable non-uniformity in composition and

Table 7.1: **Recommended minimum mixing times**

Capacity of mixer		Mixing time, min
m³	yd³	
0.8	up to 1	1
1.5	2	$1\frac{1}{4}$
2.3	3	$1\frac{1}{2}$
3.1	4	$1\frac{3}{4}$
3.8	5	2
4.6	6	$2\frac{1}{4}$
7.6	10	$3\frac{1}{4}$

ACI 304–73 (reaffirmed 1983) and ASTM Standard C 94–83.

127

a significantly lower strength; mixing beyond 2 min causes no significant improvement in these properties.

Table 7.1 gives typical values of mixing times for various capacities of mixers, the mixing time being reckoned from the time when *all* the solid materials have been charged into the mixer; water should be added not later than one-quarter of the mixing time. The values in Table 7.1 refer to usual mixers but many modern larger mixers perform satisfactorily with a mixing time of 1 to $1\frac{1}{2}$ min, whilst in high-speed pan mixers the time can be as short as 35 sec. On the other hand, when lightweight aggregate is used, the mixing time should not be less than 5 min, sometimes divided in 2 min of mixing of aggregate and water, followed by 3 min with cement added. In the case of air-entrained concrete, a mixing time of less than 2 or 3 minutes may cause inadequate entrainment of air.

Prolonged mixing

If mixing takes place over a long period, evaporation of water from the mix can occur, with a consequent decrease in workability and an increase in strength. A secondary effect is that of *grinding* of the aggregate, particularly if soft: the grading thus becomes finer and the workability lower. The friction effect also produces an increase in the temperature of the mix. In the case of air-entrained concrete, prolonged mixing reduces the air content by about $\frac{1}{6}$ of its value per hour (depending on the type of air-entraining agent), while a delay in placing without continuous agitating (see page 129) causes a drop in air content by about $\frac{1}{10}$ of its value per hour.

Intermittent remixing up to between 3 and 6 hours is harmless as far as strength and durability are concerned, but workability decreases unless the loss of moisture from the mixer is prevented. Adding water to restore workability, known as *re-tempering*, will possibly lower the strength and increase shrinkage, but the effect depends on how much the added water contributes to the *effective* water/cement ratio of the concrete (see page 55).

Ready-mixed concrete

If instead of being batched and mixed on site, concrete is delivered for placing from a central plant, it is referred to as ready-mixed or pre-mixed concrete. This type of concrete is used extensively as it offers numerous advantages in comparison with orthodox methods of manufacture:

(a) close quality control of batching which reduces the variability of the desired properties of hardened concrete;

(b) use on congested sites or in highway construction where there is little space for a mixing plant and aggregate stockpiles;

(c) use of agitator trucks to ensure care in transportation, thus preventing segregation and maintaining workability;

(d) convenience when small quantities of concrete or intermittent placing is required.

The cost of ready-mixed concrete, since it is a bought commodity, may be somewhat higher than that of site-mixed concrete, but this is often offset by savings in site organization, in supervisory staff, and in cement content. The latter arises from better control so that a smaller allowance need be made for chance variations.

There are two principal categories of ready-mixed concrete: *central-mixed* and *transit-mixed* or *truck-mixed*. In the first category, mixing is done at a central plant and then the concrete is transported in an agitator truck. In the second category, the materials are batched at a central plant but are mixed in the truck either in transit or immediately prior to discharging the concrete at the site. Transit-mixing permits a longer haul and is less vulnerable in case of delay, but the truck capacity is smaller than that of the same truck which contains pre-mixed concrete. To overcome the disadvantage of a reduced capacity, sometimes concrete is partially mixed at the central plant and the mixing is completed *en route*; this is known as *shrink-mixed* concrete.

It should be explained that *agitating* differs from mixing solely by the speed of rotation of the mixer: the agitating speed is between 2 and 6 rpm, compared with the mixing speed of 4 to 16 rpm. It may be noted that the speed of mixing affects the rate of stiffening of the concrete whilst the number of revolutions controls the uniformity of mixing. A limit of 300 revolutions for both mixing and agitating is laid down by ASTM C 94–83 or, alternatively, the concrete must be placed within $1\frac{1}{2}$ hours of mixing. In the case of transit-mixing, water need not be added till nearer the commencement of mixing, but, according to BS 5328: 1990, the time during which cement and moist aggregate are allowed to remain in contact should be limited to 2 hours. These limits tend to be rather on the safe side, and exceeding them need not adversely affect the strength of concrete, provided the mix remains sufficiently workable for full compaction. The effects of prolonged mixing and re-tempering of ready-mixed concrete are the same as for site-mixed concrete (see page 128).

British Standard BS 5328: 1981 prescribes methods of specifying concrete, including ready-mixed concrete.

Handling

There are many methods of transporting concrete from the mixer to the site and, in fact, one such method was discussed in the previous section. The choice of method obviously depends on economic considerations and on the quantity of concrete to be transported. There are many pos-

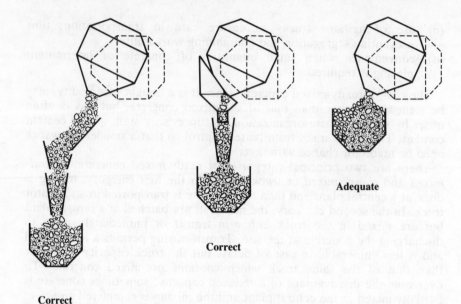

Fig. 7.1: Control of segregation on discharge of concrete from a mixer
(Based on *ACI Manual of Concrete Practice.*)

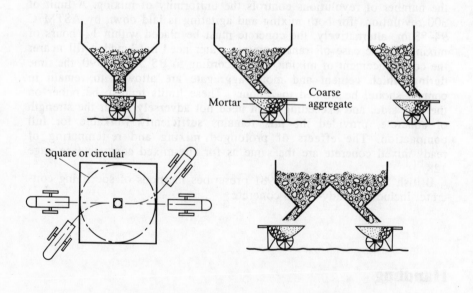

Fig. 7.2: Control of segregation in discharge of concrete from a hopper
(Based on *ACI Manual of Concrete Practice.*)

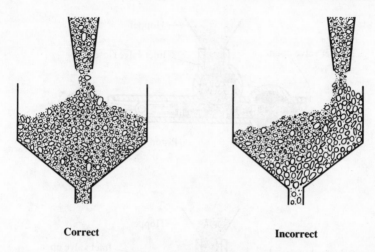

Correct   Incorrect

Fig. 7.3: Control of segregation on filling concrete buckets
(Based on *ACI Manual of Concrete Practice.*)

sibilities, ranging from wheelbarrows, buckets, skips, and belt conveyors to special trucks and to pumping but, in all cases, the important requirement is that the mix should be suitable for the particular method chosen, i.e. it should remain cohesive and should not segregate. Bad handling methods which promote segregation must obviously be avoided (see Figs 7.1 to 7.3). In this chapter only pumping will be discussed since it is rather specialized.

Pumped concrete

Nowadays, large quantities of concrete can be transported by means of pumping through pipelines over quite large distances to locations which are not easily accessible by other means. The system consists essentially of a hopper into which concrete is discharged from the mixer, a concrete pump, and the pipes through which the concrete is pumped.

Many pumps are of the direct-acting, horizontal *piston* type with semi-rotary valves set so as to ensure the passage of the largest particles of aggregate (see Fig. 7.4). Concrete is fed into the pump by gravity and, partly, by suction due to the movement of the piston, whilst the valves open and close intermittently so that the concrete moves in a series of impulses but the pipe always remains full; the use of two pistons produces a steadier flow. Outputs of up to $60 \, \text{m}^3$ ($78 \, \text{yd}^3$) per hour can be achieved through 220 mm (9 in.) diameter pipes.

There exist also small portable pumps of the peristaltic type, sometimes called *squeeze pumps* (see Fig. 7.5). Concrete placed in a collecting hopper is fed by rotating blades into a flexible pipe connected to the pumping chamber, which is under a vacuum of about 600 mm (26 in.) of mercury. The vacuum ensures that, except when being squeezed by a

131

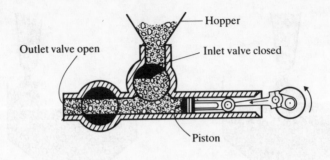

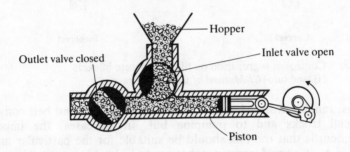

Fig. 7.4: Direct-acting concrete pump
(Based on *ACI Manual of Concrete Practice*.)

roller, the pipe shape remains cylindrical and thus permits a continuous flow of concrete. Two rotating rollers progressively squeeze the flexible pipe and thus move the concrete into the delivery pipe. Squeeze pumps are often truck-mounted and may deliver concrete through a folding boom. Outputs of up to $20 \, m^3$ $(25 \, yd^3)$ per hour can be obtained with 75 mm (3 in.) diameter pipes.

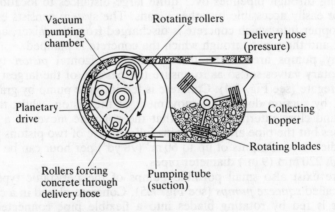

Fig. 7.5: Squeeze-type concrete pump
(Based on *ACI Manual of Concrete Practice*.)

132

Squeeze pumps transport concrete for distances up to 90 m (300 ft) horizontally or 30 m (100 ft) vertically. However, using piston pumps, concrete can be moved up to about 450 m (1500 ft) horizontally or 40 m (140 ft) vertically. The ratio of horizontal distance to the lift depends on the consistence of the mix and on the velocity of the concrete in the pipe: the greater the velocity the smaller that ratio. Relay pumping is possible for greater distances. Sharp bends and sudden changes of pipe section should be avoided.

The pipe diameter must be at least three times the maximum aggregate size. Rigid or flexible pipe can be used but the latter causes additional frictional losses and cleaning problems. Aluminium pipes should not be used because this metal reacts with the alkalis in cement to form hydrogen, which then creates voids in the hardened concrete with consequent loss of strength.

The mix required to be pumped must not be harsh or sticky, nor too dry or too wet, i.e. its consistence is critical. A slump of between 40 and 100 mm ($1\frac{1}{2}$ and 4 in.) or a compacting factor of 0.90 to 0.95, or Vebe time of 3 to 5 sec is generally recommended for the mix in the hopper. Pumping causes partial compaction so that at delivery the slump may be decreased by 10 to 25 mm ($\frac{1}{2}$ to 1 in.). The requirements of consistence are necessary to avoid excessive frictional resistance in the pipe with too-dry mixes, or segregation with too-wet mixes. In particular, the percentage of fines is important since too little causes segregation and too much causes undue frictional resistance and possible blockage of the pipeline. The optimum situation is when there is a minimum frictional resistance against the pipe walls and a minimum content of voids within the mix. This is achieved when there is a continuity of aggregate grading. For concretes with maximum aggregate size of 20 mm ($\frac{3}{4}$ in.), the optimum fine aggregate content lies between 35 and 40 per cent, and the material finer than 300 μm (No. 50 ASTM) should represent 15 to 20 per cent of the mass of fine aggregate. Also, the proportion of fine aggregate which passes the 150 μm (No. 100 ASTM) sieve should be about 3 per cent, this material being sand or a suitable additive (tuff or trass) so as to provide continuity in grading down to the cement fraction.

Pumping of lightweight aggregate concrete can be achieved using special admixtures (pumping aids) to overcome problems of loss of workability due to the high water absorption of the porous particles. Air-entrained concrete is usually pumped only over short distances of about 45 m (150 ft) because the entrained air becomes compressed and loss of workability results.

Placing and compacting

The operations of placing and of compacting are interdependent and are carried out almost simultaneously. They are most important for the purpose of ensuring the requirements of strength, impermeability, and durability of the hardened concrete in the *actual* structure. As far as

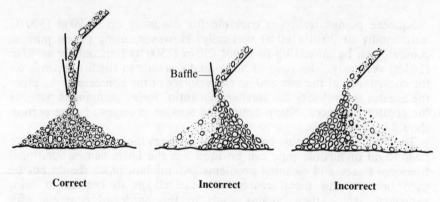

Correct   Incorrect   Incorrect

Fig. 7.6: Control of segregation at the end of concrete chutes
(Based on *ACI Manual of Concrete Practice*.)

placing is concerned, the main objective is to deposit the concrete as close as possible to its final position so that segregation is avoided and the concrete can be fully compacted (see Figs 7.6 to 7.9). To achieve this objective, the following rules should be borne in mind:

(a) hand shovelling and moving concrete by immersion or poker vibrators should be avoided;

(b) the concrete should be placed in uniform layers, not in large heaps or sloping layers;

(c) the thickness of a layer should be compatible with the method of vibration so that entrapped air can be removed from the bottom of each layer;

(d) the rates of placing and of compaction should be equal;

(e) where a good finish and uniform colour are required on columns

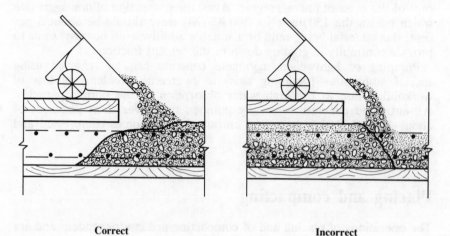

Correct   Incorrect

Fig. 7.7: Placing concrete from buggies
(Based on *ACI Manual of Concrete Practice*.)

134

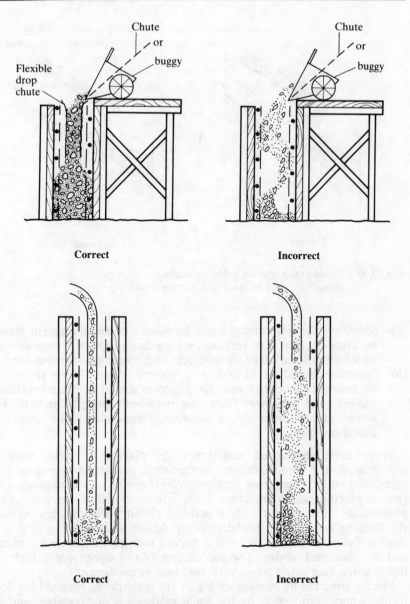

Fig. 7.8: Placing concrete in a deep wall
(Based on *ACI Manual of Concrete Practice*.)

and walls, the forms should be filled at a rate of at least 2 m (6 ft) per hour, avoiding delays (Long delays can result in the formation of cold joints.);

(f) each layer should be fully compacted before placing the next one, and each subsequent layer should be placed whilst the underlying layer is still plastic so that monolithic construction is achieved;

135

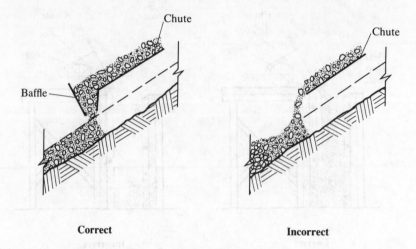

Correct   Incorrect

Fig. 7.9:   Placing concrete on a sloping surface
               (Based on *ACI Manual of Concrete Practice.*)

(g)   collision between concrete and formwork or reinforcement should be avoided. For deep sections, a long down pipe or tremie ensures accuracy of location of the concrete and minimum segregation;

(h)   concrete should be placed in a vertical plane. When placing in horizontal or sloping forms, the concrete should be placed vertically against, and not away from, the previously placed concrete. For slopes greater than 10°, a slip-form screed should be used (see Bibliography).

There exist specialized techniques for placing concrete, such as slip-forming, the tremie method, shotcreting, preplaced aggregate concrete, and roller compacted concrete. *Slip-forming* is a continuous process of placing and compaction, using low workability concrete whose proportions must be carefully controlled. Both horizontal and vertical slip-forming is possible, the latter being slower and requiring formwork until sufficient strength has been achieved to support the new concrete and the formwork above. The capital cost of the equipment is high but this is more than offset by its very high rate of production.

Placing concrete by *tremie* (see Fig. 7.10) is particularly suited for deep forms, where compaction by the usual methods is not possible, and for underwater concreting. In the tremie method, high workability concrete is fed by gravity through a vertical pipe which is gradually raised. The mix should be cohesive, without segregation or bleeding, and usually has a high cement content, a high proportion of fines, and contains a workability aid (such as pozzolan or an admixture).

As stated in Chapter 6, the purpose of compaction is to remove as much of the entrapped air as possible so that the hardened concrete has a minimum of voids, and, consequently, is strong, durable and of low permeability. Low slump concrete contains more entrapped air than high

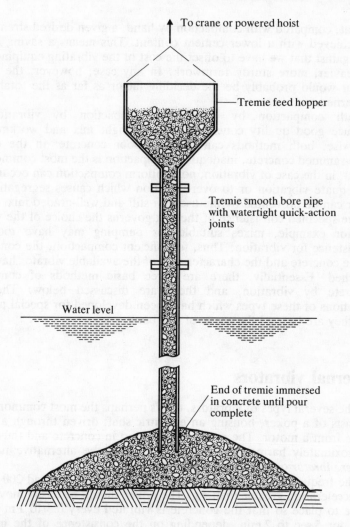

To crane or powered hoist

Tremie feed hopper

Tremie smooth bore pipe
with watertight quick-action
joints

Water level

End of tremie immersed
in concrete until pour
complete

Fig. 7.10:   Underwater concreting
(Based on CONCRETE SOCIETY, Underwater concreting,
Technical Report, No. 3, pp. 13 (London, 1971).)

slump concrete, and, hence, the former requires more effort to compact
it satisfactorily. This effort is provided mainly by the use of vibrators.

Vibration of concrete

The process of compacting concrete by vibration consists essentially of
the elimination of entrapped air and forcing the particles into a closer
configuration. Extremely dry and stiff mixes can be vibrated satisfactorily

so that, compared with compaction by hand, a given desired strength can be achieved with a lower cement content. This means a saving in cost, but against that we have to offset the cost of the vibrating equipment and of heavier, more sturdy formwork. In any case, however, the cost of labour would probably be the deciding factor as far as the total cost is concerned.

Both compaction by hand and compaction by vibration can produce good quality concrete, with the right mix and workmanship. Likewise, both methods can produce poor concrete: in the case of hand-rammed concrete, inadequate compaction is the most common fault whilst, in the case of vibration, non-uniform compaction can occur due to inadequate vibration or to over-vibration which causes segregation; the latter can be prevented by the use of a stiff and well-graded mix.

The specified consistence of the mix governs the choice of the vibrator as, for example, mixes suitable for pumping may have too-wet a consistence for vibration. Thus, for efficient compaction, the consistence of the concrete and the characteristics of the available vibrator have to be matched. Essentially, there are three basic methods of compacting concrete by vibration, and these are discussed below. There are variations of these types which have been developed for special purposes but they are beyond the scope of this book.

Internal vibrators

Of the several types of vibrators, this is perhaps the most common one. It consists of a poker, housing an eccentric shaft driven through a flexible drive from a motor. The poker is immersed in concrete and thus applies approximately harmonic forces to it; hence, the alternative names of *poker vibrator* or *immersion vibrator*.

The frequency of vibration usually varies between 70 and 200 Hz with an acceleration greater than 4 g. The poker should be easily moved from place to place so that the concrete is vibrated every 0.5 to 1 m (or 2 to 3 ft) for 5 sec to 2 min, depending on the consistence of the mix. The actual completion of compaction can be judged by the appearance of the surface of the concrete, which should be neither honeycombed nor contain an excess of mortar. Gradual withdrawal of the poker at the rate of about 80 mm per sec (3 in./sec) is recommended so that the hole left by the vibrator closes fully by itself without any air being trapped. The vibrator should be immersed, quickly, through the entire depth of the freshly deposited concrete and into the layer below if this is still plastic or can be made plastic (see Fig. 7.11). In this manner, monolithic concrete is obtained, thus avoiding a plane of weakness at the junction of the two layers, possible settlement cracks, and the internal effects of bleeding. It should be noted that, with a lift greater than about 0.5 m (2 ft), the vibrator may not be fully effective in expelling air from the lower part of the layer.

Unlike other types, internal vibrators are comparatively efficient since

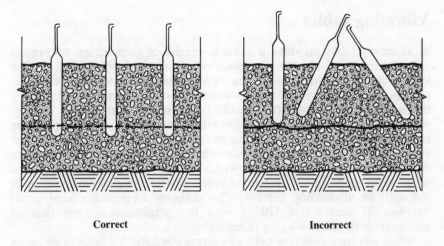

Correct   **Incorrect**

Fig. 7.11: Placing of poker vibrators
(Based on *ACI Manual of Concrete Practice.*)

all the work is done directly *on* the concrete. They are made in sizes
down to 20 mm ($\frac{3}{4}$ in.) in diameter so that they are useful for heavily
reinforced and relatively inaccessible sections. However, an immersion
vibrator will not expel air from the form boundary so that 'slicing' along
the form by means of a flat plate on edge is necessary. The use of
absorptive linings to the form is helpful in this respect but expensive.

External vibrators

This type of vibrator is rigidly clamped to the formwork which rests on an
elastic support, so that both the form and the concrete are vibrated. As a
result, a considerable proportion of the work done is used in vibrating the
formwork, which has to be strong and tight so as to prevent distortion
and leakage of grout.

The principle of the external vibrator is the same as that of an internal
one, but the frequency is between 50 and 150 Hz; sometimes, manufacturers quote the number of impulses, i.e. half-cycles. External vibrators
are used for precast or thin *in situ* sections having a shape or thickness
which is unsuitable for internal vibrators.

The concrete has to be placed in layers of suitable depth as air cannot
be expelled through too great a depth of concrete, and the position of the
vibrator may have to be changed as concreting progresses. Portable,
non-clamped external vibrators may be used at sections not otherwise
accessible, but their range of compaction is very limited. One such
vibrator is an electric hammer, sometimes used for compacting concrete
test specimens.

Vibrating tables

A vibrating table provides a reliable means of compaction of precast concrete units and has the advantage of ensuring uniform vibration. The system can be considered as a case of formwork clamped to the vibrator, as opposed to that of an external vibrator, but the principle of vibrating the concrete and formwork together is the same. Generally, a rapidly-rotating eccentric weight makes the table vibrate with a circular motion but, by having two shafts rotating in opposite directions, the horizontal component of vibration can be neutralized, so that the table transmits a simple harmonic motion in the vertical direction only. There exist also some small, good-quality vibrating tables operated by an electro-magnet fed with an alternating current. The range of frequencies used varies between 25 and about 120 Hz, and the amplitude is such that an acceleration of about 4 to 7 g is achieved.

When vibrating concrete units of varying sizes and for laboratory use, a table with variable amplitude and, preferably, also variable frequency is desirable, although in practice the frequency is rarely varied. Ideally, an increasing frequency and a decreasing amplitude should be used as the consolidation of concrete progresses because the induced movement should correspond to the spacing of the particles: once partial compaction has occurred, the use of a higher frequency permits a greater number of adjusting movements in a given time. Vibration of too large an amplitude relative to the inter-particle space results in the concrete being in a constant state of flow so that full compaction is never achieved. Unfortunately, however, it is not possible to predict the optimum amplitudes and frequencies required for a given mix.

Revibration

The preceding sections refer to vibration of concrete immediately after placing, so that consolidation is generally completed before the concrete has stiffened. However, it was mentioned on page 138 that, in order to ensure a good bond between lifts, the underlying lift should be revibrated provided it is still plastic or can regain a plastic state. This successful application of revibration raises the question of whether revibrating concrete is generally advantageous or otherwise.

In fact, revibration at 1 to 2 hours after placing increases the compressive strength of concrete by up to 15 per cent, but the actual values depend on the workability of the mix. In general, the improvement in strength is more pronounced at earlier ages, and is greatest in concretes liable to high bleeding since the trapped water is expelled by revibration. For the same reason, the bond between concrete and reinforcement is greatly improved. There is also a possible relief of plastic shrinkage stresses around the large aggregate particles.

Despite these advantages, revibration is not widely used as it involves an additional step in the production of concrete, and hence an increased cost; also, if applied too late, revibration can damage the concrete.

Shotcrete

This is the name given to mortar or concrete conveyed through a hose and pneumatically projected at high velocity onto a backup surface. The force of the jet impacting on the surface compacts the material so that it can support itself without sagging or sloughing even on a vertical face or overhead. Shotcrete is more formally called *pneumatically applied mortar* or *concrete*; it is also known as *gunite,* although in the US this name applies only to shotcrete placed by the dry mix process. In the UK, the term of *sprayed concrete* is used, but generally it is mortar rather than concrete that is employed, i.e. the maximum size of aggregate is 5 mm.

Shotcrete is used for thin, lightly reinforced sections, such as shells or folded plate roofs, tunnel linings and prestressed concrete tanks. Shotcrete is also used in repair of deteriorated concrete, in stabilizing rock slopes, in encasing steel for fireproofing, and as a thin overlay on concrete, masonry or steel.

It is the method of placing that provides significant advantages of shotcreting in the above-mentioned applications. At the same time, considerable skill and experience are required so that the quality of shotcrete depends to a large extent on the operator, especially on his skill in the control and actual placing by the nozzle.

Since shotcrete is sprayed on a backup surface and then gradually built up to a thickness of up to 100 mm (4 in.), only one side of formwork is needed; this represents economy because no form ties or supports are needed. On the other hand, the cement content of shotcrete is high and the necessary equipment and mode of placing are more expensive than in the case of conventional concrete.

There are two basic processes by which shotcrete is applied. The more common is the *dry mix process,* in which cement and damp aggregate are intimately mixed and fed into a mechanical feeder or gun (see Fig. 7.12(a)). The mixture is then transferred at a known rate by a distributor into a stream of compressed air in a hose leading to the delivery nozzle. Inside the nozzle is fitted a perforated manifold, through which pressurized water is introduced for mixing with the other ingredients, before the mixture is projected at high velocity.

In the *wet mix process,* all the ingredients, including the mixing water, are pre-mixed (see Fig. 7.12(b)). The mixture is then introduced into the chamber of the delivery equipment, and from there conveyed pneumatically or by a positive displacement pump of the type shown in Fig. 7.5. Compressed air (or in the case of a pneumatically conveyed mix, additonal air) is injected at the nozzle to provide a high nozzle velocity.

The wet mix process gives a better control of the quantity of mixing water, which is metered at the pre-mix stage, and of any admixture used. Also, the wet mix process leads to less dust being produced so that the working conditions are better than with the dry mix process.

Either process can produce an excellent end product, but the dry mix process is more suitable with porous lightweight aggregate and is also capable of greater delivery lengths. This process can be used with a flash set accelerator (e.g. washing soda), which is necessary when the backup

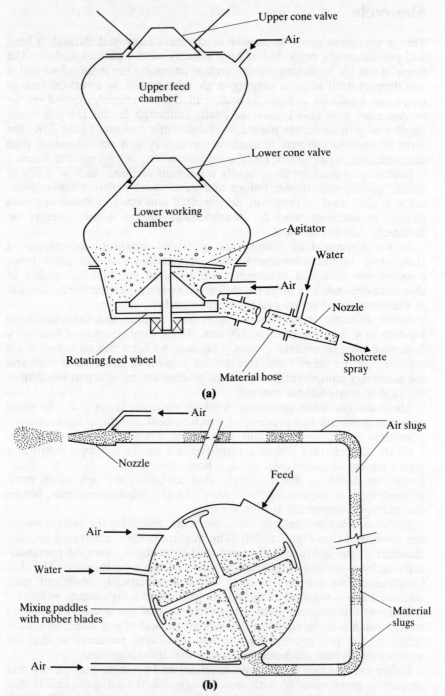

Fig. 7.12: Typical layout of shotcreting: (a) dry mix process, (b) wet mix process

surface is covered with running water; the accelerator adversely affects the strength but makes repair work possible.

The projected shotcrete has to have a relatively dry consistence so that the material can support itself in any position; at the same time, the mix has to be wet enough to achieve compaction without excessive rebound.

It is clear then that not all the shotcrete projected on a surface remains in position, because the coarsest particles are prone to *rebound* from the surface. The proportion of material rebounded is greatest in the initial layers and is greater for soffits (up to 50 per cent) than for floors and slabs (up to 15 per cent). The significance of rebound is not so much in the waste of material as in the danger from accumulation of the material in a position where it will be incorporated in the subsequent layers of shotcrete; also, the loss of aggregate results in a mix which exhibits increased shrinkage. To avoid pockets of rebounded material in inside corners, at the base of walls, behind reinforcement or embedded pipes, or on horizontal surfaces, great care in the placing of shotcrete is necessary and use of small-size reinforcement is desirable.

The usual range of water/cement ratios is 0.35 to 0.50, and there is little bleeding. The usual *mortar* mix is 1:3.5 to 1:4.5, with sand of the same grading as for a conventional mortar. In the case of *concrete,* the maximum aggregate size is 25 mm (1 in.) but the coarse aggregate content is lower than in conventional concrete. There is a greater rebound problem with shotcrete concrete, and its use is small and advantages limited.

Curing of shotcrete is particularly important because of rapid drying in consequence of the large surface/volume ratio, and recommended practices should be followed, as given by ACI 506.3R–82 (see Bibliography).

Preplaced aggregate concrete

This type of concrete, known also as *prepacked, intrusion* or *grouted concrete,* can be placed in locations not easily accessible by, or suitable for, ordinary concreting techniques. It is produced in two stages: coarse aggregate is placed and compacted in the forms, and then the voids, forming about 33 per cent of the overall volume, are filled with mortar. It is clear that the aggregate is gap-graded, typical gradings being given in Table 7.2. To ensure good bond, the coarse aggregate must be free from dirt and dust, since these are not removed in mixing, and it has to be thoroughly wetted or inundated before the mortar is intruded. However, water should not be allowed to stand too long, as algae can grow on the aggregate.

The mortar is pumped under pressure through slotted pipes, usually about 35 mm (1½ in.) in diameter and spaced at 2 m (7 ft) centres. The pipes are gradually withdrawn as the mortar level rises. No internal vibration is used but external vibration at the level of the top of the mortar may improve the exposed surfaces.

Table 7.2: **Typical gradings of aggregate for preplaced agregate concrete**

Coarse aggregate					Fine aggregate		
Sieve size		Cumulative percentage passing			Sieve size		Cumulative percentage passing
mm	in.				metric	ASTM	
150	6	100	–	–	2.36 mm	8	100
75	3	67	100	–	1.18 mm	16	98
38	$1\frac{1}{2}$	40	62	97	600 μm	30	72
19	$\frac{3}{4}$	6	4	9	300 μm	50	34
13	$\frac{1}{2}$	1	1	1	150 μm	100	11

A typical mortar consists (by mass) of two parts of Portland cement, one part of fly ash (PFA) and three to four parts of fine sand, with sufficient water to form a fluid mixture. The purpose of the pozzolan is to reduce bleeding and segregation whilst improving the fluidity of the mortar. A further fluidizing aid is added which also delays the stiffening of the mortar; this aid contains a small amount of aluminium powder, which reacts to produce hydrogen, thus causing a slight expansion before setting occurs. As an alternative, a cement-and-fine-sand mortar can be mixed in a special 'colloid' mixer in which the speed of rotation is very fast, so that the cement remains in suspension until pumping is complete. This type of preplaced aggregate concrete is sometimes called *colloidal* concrete.

Preplaced aggregate concrete is economical in cement (as little as 120 to 150 kg/m³ (200 to 250 lb/yd³) of concrete) but the necessary high water/cement ratio to obtain a sufficient fluidity results in a limited concrete strength (20 MPa (2900 psi)). However, this strength is generally adequate for the usual applications of preplaced aggregate concrete. Moreover, a dense impermeable, durable, and uniform material is obtained.

A particular use of preplaced aggregate concrete is in sections containing a large number of embedded items that have to be precisely located. This arises, for instance, in nuclear shields. There, the danger of segregation of heavy coarse aggregate, especially of steel aggregate, is eliminated because coarse and fine aggregate are placed separately. However, in nuclear shields, pozzolan should not be used because it reduces the density of the concrete.

Because of reduced segregation, preplaced aggregate concrete is also suitable for underwater construction. Other applications are in the construction of water-retaining structures, and in large monolithic blocks, and also in repair work, mainly because preplaced aggregate concrete has a lower shrinkage and a lower permeability (and, hence, a higher resistance to freezing and thawing) than ordinary concrete.

Preplaced aggregate concrete can be used when an exposed aggregate finish is required because the coarse aggregate is uniform. In mass

144

construction, there is the advantage that the temperature rise on hydration (see Chapter 9) can be controlled by circulating refrigerated water round the aggregate prior to the placing of mortar. At the other extreme, in cold weather when frost damage is likely, steam can be circulated to pre-heat the aggregate.

Preplaced aggregate concrete appears thus to have many useful features but, because of numerous practical difficulties, considerable skill and experience in application of the process are necessary for good results to be obtained.

Bibliography

7.1   ACI COMMITTEE 304.R–89, Recommended practice for measuring, mixing, transporting and placing concrete, Part 2, *ACI Manual of Concrete Practice*, 1990.

7.2   ACI COMMITTEE 304.2R–71 (revised 1982), Placing concrete by pumping methods, Part 2, *ACI Manual of Concrete Practice*, 1984.

7.3   ACI COMMITTEE 304.3R–75 (reaffirmed 1980), High density concrete: measuring, mixing, transporting and placing, Part 2, *ACI Manual of Concrete Practice*, 1984.

7.4   ACI COMMITTEE 318–89, Building code requirements for reinforced concrete, Part 3, *ACI Manual of Concrete Practice*, 1990, (see also Commentary on above (ACI 318R–89)).

7.5   ACI COMMITTEE 506.2–77 (revised 1983), Specification for materials proportioning and application of shotcrete, Part 5, *ACI Manual of Concrete Practice*, 1990.

7.6   ACI COMMITTEE 506R–85, Guide to shotcrete, Part 5, *ACI Manual of Concrete Practice*, 1990.

7.7   ACI COMMITTEE 506.3R–82, Guide to certification of shotcrete nozzlemen, Part 5, *ACI Manual of Concrete Practice*, 1990.

7.8   G. BLACKLEDGE, Placing and compacting concrete, *Concrete Society Current Practice sheets*: Part 1, Vol. 15, No. 2, pp. 35–6 (Feb. 1981); Part 2, Vol. 15, No. 3, pp. 35–6 (March 1981).

Problems

7.1   Compare internal and external vibration of concrete.

7.2   What are the particular requirements for pumpability of a concrete mix?

7.3   How is the mixing efficiency of a mixer assessed?

7.4   What is the influence of mixing time on the strength of concrete?

145

7.5 Comment on the relation between the maximum aggregate size and the pipe diameter.

7.6 What are the particular problems in pumping lightweight aggregate concrete?

7.7 What are the particular problems in pumping air-entrained concrete?

7.8 Explain the differences between a tilting drum mixer, a non-tilting drum mixer, a pan-type mixer and a dual drum mixer.

7.9 What is a good sequence of feeding the mixer?

7.10 What are the special requirements for mix proportions of concrete which is to be pumped?

7.11 What are the workability requirements for concrete to be pumped?

7.12 What causes blowholes in concrete?

7.13 What are the effects of re-tempering on the properties of resulting concrete?

7.14 What is: (i) buttering, (ii) head pack?

7.15 What is a colloidal mixer?

7.16 How is the performance of a mixer assessed?

7.17 How does pumping affect the workability of the mix?

7.18 What are the two main categories of ready-mixed concrete?

7.19 What are the advantages of using ready-mixed concrete?

7.20 What are the disadvantages of using ready-mixed concrete?

7.21 What is meant by segregation of aggregate in a stockpile?

7.22 What is the difference between agitating and mixing?

7.23 Why are pipes for pumping concrete not made of aluminium?

7.24 By what method would you place concrete under water?

7.25 Describe dry-process shotcreting.

7.26 Describe wet-process shotcreting.

7.27 What should you do with rebound material?

7.28 What are the advantages of placing concrete by pumping?

7.29 What is shrink-mixed concrete?

7.30 What is the main requirement of good handling of concrete?

7.31 What are the advantages and disadvantages of revibration of concrete?

7.32 State alternative terms for preplaced aggregate concrete and give some typical uses of this concrete.

8

Admixtures

Often, instead of using a special cement, it is possible to change some of the properties of the more commonly used cements by incorporating a suitable additive or an admixture. In other cases, such incorporation is the sole means of achieving the desired effect. A great number of proprietary products are available: their desirable effects are described by the manufacturers but some other effects may not be known, so that a cautious approach, including performance tests, is sensible. It should be noted that the terms 'additive' and 'admixture' are often used synonymously, though, strictly speaking, *additive* refers to a substance which is added at the cement manufacturing stage, while *admixture* implies addition at the mixing stage.

This chapter considers mainly chemical admixtures and miscellaneous admixtures. In addition, there exist air-entraining agents whose main purpose is to protect concrete from the deleterious effects of freezing and thawing. This subject will be considered in Chapter 15. Chemical admixtures are essentially water-reducers (plasticizers), set-retarders and accelerators, respectively classified as Type A, B and C according to ASTM C 494–82. The classification of chemical admixtures by BS 5075: Part 1: 1982 is substantially similar: Tables 8.1 and 8.2 list the requirements for various types. Information on miscellaneous admixtures is scant so that they are not covered by national standards. However, useful information is given in a guide of the ACI Committee 212.1R-81.

Accelerators

These are admixtures which accelerate the *hardening* or the development of early strength of concrete; the admixture need not have any specified effect on the setting (or stiffening) time. However, in practice, the setting time is reduced, as prescribed by Type A of ASTM C 494–82 and BS 5075: Part 1: 1982. It should be noted that there exist also *set-accelerating* (or *quick-setting*) admixtures, which specifically reduce the setting time. An example of a quick-setting admixture is sodium carbonate (washing soda) which is used to promote a flash set in

Table 8.1: Specification for various types of admixtures according to BS 5075: Part 1: 1982

Type of admixture	Water reduction, per cent	Compacting factor	Stiffening time — Time from completion of mixing to reach a resistance to penetration* of: 0.5 MPa (70 psi)	3.5 PMa (500 psi)	Compressive strength — Percentage of control mix (minimum)	Age
Accelerating	–	Not more than 0.02 below control mix	More than 1 h.	At least 1 h less than control mix	125 95	24 h 28 days
	8	–		–	–	–
Retarding	–	Not more than 0.02 below control mix		At least 1 h longer than control mix	90 95	7 days 28 days
	8	–		–	–	–
Normal water-reducing	–	At least 0.03 above control mix			90 90	7 days 28 days
	8	Not more than 0.02 below control mix	Within 1 h of control mix	Within 1 h of control mix	110 110	7 days 28 days

Type		Penetration*	Time of setting		Strength (% of control mix)		Age
Accelerating water-reducing	–	At least 0.03 above control mix	–	–	125	125	24 h
	8	Not more than 0.02 below control mix	More than 1 h	At least 1 h less than control mix	90	110	28 days
Retarding water-reducing	–	At least 0.03 above control mix	–	–	90	110	7 days
	8	Not more than 0.02 below control mix	At least 1 h longer than control mix	–	90	110	28 days

* The penetration is determined by a special brass rod of 6.175 mm in diameter. The air content with no water reduction shall not be more than 2 per cent higher than that of the control mix, and not more than a total of 3 per cent.

Table 8.2: Specification for various types of admixtures according to ASTM 494–82

Property	Type A, water-reducing	Type B, retarding	Type C, accelerating	Type D, water-reducing and retarding	Type E, water-reducing and accelerating	Type F, water-reducing, high range	Type G, water-reducing, high range and retarding
Water content, max per cent of control	95	–	–	95	95	88	88
Time of setting, allowable deviation from control, min							
Initial: at least	–	60 later	60 earlier	60 later	60 earlier	–	60 later
not more than	60 earlier nor 90 later	210 later	210 earlier	210 later	210 earlier	60 earlier nor 90 later	210 later
Final: at least	–	–	60 earlier	–	60 earlier	–	–
not more than	60 earlier nor 90 later	210 later	–	210 later	–	60 earlier nor 90 later	210 later
Compressive strength, min per cent of control:[a]							
1 day	–	–	–	–	–	140	125
3 days	110	90	125	110	125	125	125
7 days	110	90	100	110	110	115	115
28 days	110	90	100	110	110	110	110
6 months	100	90	90	100	100	100	100
1 year	100	90	90	100	100	100	100

Flexural strength, min per cent control:[a]							
3 days	100	90	100	110	100	110	110
7 days	100	90	100	100	100	100	100
28 days	100	90	90	100	100	100	100
Length change, max shrinkage (alternative requirements):[b]							
Per cent of control	135	135	135	135	135	135	135
Increase over control	0.010	0.010	0.010	0.010	0.010	0.010	0.010
Relative durability factor, min[c]	80	80	80	80	80	80	80

[a] The compressive and flexural strength of the concrete containing the admixture under test at any test age shall be not less than 90 per cent of that attained at any previous test age. The objective of this limit is to require that the compressive or flexural strength of the concrete containing the admixture under test shall not decrease with age.

[b] Alternative requirements, per cent of control limit applies when length change of control is 0.030 per cent or greater; increase over control limit applies when length change of control is less than 0.030 per cent.

[c] This requirement is applicable only when the admixture is to be used in air-entrained concrete which may be exposed to freezing and thawing while wet.

shotcreting (see page 141); this adversely affects strength but makes urgent repair work possible. Other examples of set-accelerating admixtures are: aluminium chloride, potassium carbonate, sodium fluoride, sodium aluminate, and ferric salts. None of these should be used without a full study of all consequences.

Let us return to consideration of accelerators. The most common one is calcium chloride ($CaCl_2$), which accelerates primarily the *early* strength development of concrete. This admixture is sometimes used when concrete is to be placed at low temperatures (2 to 4 °C (35 to 40 °F)) or when urgent repair work is required because it increases the rate of heat development during the first few hours after mixing. Calcium chloride probably acts as a catalyst in the hydration of C_3S and C_2S, or alternatively the reduction in the alkalinity of the solution promotes the hydration of the silicates. The hydration of C_3A is delayed somewhat, but the normal process of hydration of cement is not changed.

Calcium chloride may be added to rapid-hardening (Type III) as well as to ordinary Portland (Type I) cement, and the more rapid the natural rate of hardening of the cement the greater is the action of the accelerator. Calcium chloride must not, however, be used with high-alumina cement. Figure 8.1 shows the effect of calcium chloride on the early strength of concretes made with different types of cement; the long-term strength is believed to be unaffected.

The quantity of calcium chloride added to the mix must be carefully controlled. To calculate the quantity required it can be assumed that the addition of 1 per cent of anhydrous calcium chloride, $CaCl_2$ (as a fraction of the mass of cement) affects the rate of hardening as much as a rise in temperature of 6 °C (11 °F). A calcium chloride content of 1 to 2 per cent is generally sufficient. The latter figure should not be exceeded unless a test is undertaken using the actual cement, as the effects of calcium chloride depend to a certain degree on the cement composition. Calcium chloride generally accelerates setting and an overdose can cause flash set.

It is important that calcium chloride be uniformly distributed throughout the mix, and this is best achieved by dissolving the admixture in the mixing water. It is preferable to prepare a concentrated aqueous solution using calcium chloride flakes rather than the granular form which dissolves very slowly. The flakes consist of $CaCl_2.2H_2O$ so that 1.37 g of flakes is equivalent to 1 g of $CaCl_2$.

The use of calcium chloride reduces the resistance of cement to sulphate attack, particularly in lean mixes, and the risk of alkali–aggregate reaction is increased for a reactive aggregate. Other undesirable effects are that the addition of calcium chloride increases shrinkage and creep (see Chapters 12 and 13), and there is a lowering of the resistance of air-entrained concrete to freezing and thawing at later ages. However, there is a beneficial effect in increasing the resistance of concrete to erosion and abrasion.

The possibility of corrosion of reinforcing steel by integral calcium chloride has been the subject of controversy for some time. When used in the correct proportions, calcium chloride has been found to cause corrosion in certain cases while in other instances no corrosion occurred.

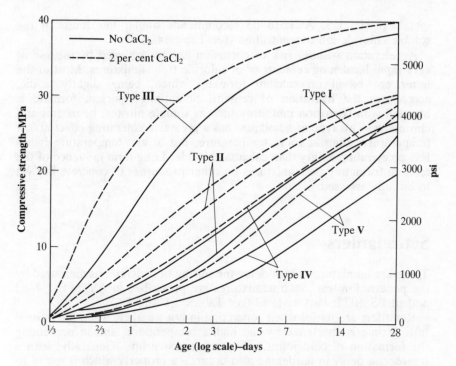

Fig. 8.1: Influence of CaCl$_2$ on the strength of concretes made with different types of cement: ordinary Portland (Type I), modified (Type II), rapid-hardening Portland (Type III), low-heat (Type IV), and sulphate-resisting (Type V)
(Based on: US BUREAU OF RECLAMATION, *Concrete Manual*, 8th Edn. (Denver, Colorado, 1975), and W. H. PRICE, Factors influencing concrete strength, *J. Amer. Concr. Inst.*, 47, pp. 417–32 (Feb. 1951).)

The explanation of the controversy is probably associated with a non-uniform distribution of chloride ions and with the migration of chloride ions in permeable concrete, accompanied by ingress of moisture and oxygen, especially in warm conditions.

Although we are discussing here the addition of calcium chloride, what is relevant with respect to corrosion is the chloride ion, Cl$^-$. All sources of the ion, including for instance its presence on the surface of marine aggregate, should be taken into account. We may note that 1.56 g of CaCl$_2$ corresponds to 1 g of chloride ion.

When concrete is *permanently* dry so that no moisture is present, no corrosion can take place but under other circumstances the possibility of corrosion of reinforcement represents a serious risk to the structure. For this reason, BS 8110: Part 1: 1985 severely restricts the *total* chloride content in reinforced concrete according to the type of cement and concrete application (see Table 14.3). These low limits effectively ban the use of any chloride-based admixtures in concrete containing embedded

153

metal. In the US, ACI 318–83 recommends similar low limits of the soluble chloride ion concentration (see Table 14.4).

Acceleration without risk of corrosion can be achieved by the use of very rapid hardening cements or of chloride-free admixtures. Most of the latter are based on calcium formate, which, being slightly acidic, accelerates the hydration of cement. Sometimes, calcium formate is blended with corrosion inhibitors such as soluble nitrites, benzoates and chromates. This type of admixture has a greater accelerating effect at low temperatures than at room temperature, but at any temperature it has less accelerating ability than calcium chloride. Long-term influence of the calcium formate type admixtures on other properties of concrete has yet to be fully assessed.

Set-retarders

These are admixtures which delay the setting of concrete, as measured by the penetration test. Such admixtures are prescribed in ASTM C 494–82 and in BS 5075: Part 1: 1982 (see Tables 8.1 and 8.2).

Retarders are useful when concreting in hot weather, when the normal setting time is shortened by the higher temperature, and in preventing the formation of cold joints between successive lifts. Generally, with a retarder, a delay in hardening also occurs – a property which is useful to obtain an architectural surface finish of exposed aggregate.

Retardation action is exhibited by the addition of sugar, carbohydrate derivatives, soluble zinc salts, soluble borates and others. In practice, retarders which are also water-reducing are commonly used; these are described in the next section. When used in a carefully controlled manner, about 0.05 per cent of sugar by mass of cement will delay the setting time by about 4 hours. However, the exact effects of sugar depend on the chemical composition of cement, and the performance of sugar, and indeed of any retarder, should be determined by trial mixes with the actual cement to be used in construction. A large quantity of sugar, say 0.2 to 1 per cent of the mass of cement, will virtually prevent the setting of cement, a feature which is useful in case of malfunction of a concrete mixer.

The setting time of concrete is increased by delaying the adding of the retarding admixture to the mix. The increased retardation occurs especially with cements which have a high C_3A content because, once some C_3A has hydrated, it does not absorb the admixture; it is therefore available for action with the calcium silicates.

The mechanism of the retarding action is not known with certainty. The admixtures modify the crystal growth or morphology so that there is a more efficient barrier to further hydration than is the case without a retarder. Eventually the retarder is removed from solution by being incorporated into the hydrated material, but the composition or identity of the hydration products is not changed. This is also the case with set-retarding and water-reducing admixtures.

Compared with an admixture-free concrete, the use of retarding admixtures reduces the early strength but later the rate of strength development is higher, so that the longer-term strength is not much different. Also, retarders tend to increase the plastic shrinkage because the plastic stage is extended, but drying shrinkage is unaffected.

Water-reducers (plasticizers)

These admixtures are used for three purposes:

(a) To achieve a higher strength by decreasing the water/cement ratio at the same workability as an admixture-free mix.
(b) To achieve the same workability by decreasing the cement content so as to reduce the heat of hydration in mass concrete.
(c) To increase the workability so as to ease placing in inaccessible locations.

As Table 8.2 shows, ASTM C 494–82 classifies admixtures which are water-reducing *only* as Type A, but if the water-reducing properties are accompanied by set-retardation, then the admixture is classified as Type D. There exist also water-reducing and accelerating admixtures (Type E). The corresponding BS 5075: Part 1: 1982 requirements are given in Table 8.1.

The principal active components of water-reducing admixtures are surface-active agents which are concentrated at the interface between two immiscible phases and which alter the physico-chemical forces at this interface. The surface-active agents are absorbed on the cement particles, giving them a negative charge, which leads to repulsion between the particles and results in stabilizing their dispersion; air bubbles are also repelled and cannot attach to the cement particles. In addition, the negative charge causes the development of a sheath of oriented water molecules around each particle, thus separating the particles. Hence, there is a greater particle mobility, and water, freed from the restraining influence of the flocculated system, becomes available to lubricate the mix so that workability is increased.

The reduction in the quantity of mixing water which is possible owing to the use of admixtures varies between 5 and 15 per cent. A part of this is, in many cases, due to the entrained air introduced by the admixture. The actual decrease in mixing water depends on the cement content, aggregate type, pozzolans and air-entraining agent if present. Trial mixes are therefore essential to achieve optimum properties, as well as to ascertain any possible undesirable side effects: segregation, bleeding and loss of workability with time (or slump loss).

In contrast to air-entraining agents, water-reducing admixtures do not always improve the cohesiveness of the concrete. Hydroxylated carboxylic acid type admixtures can increase bleeding in high workability concretes but, on the other hand, lignosulphonic acid type admixtures usually improve cohesiveness because they entrain air; however, sometimes it is necessary to use an air-detraining agent to avoid over-air entrainment. It should also be noted that, although setting is retarded by the use of these admixtures, the rate of loss of workability with time is not always reduced; generally, the higher the initial workability, the greater the rate of loss of workability. If this poses a problem, then re-dosing with the admixture can be used, provided that set-retardation is not adversely affected.

The dispersing ability of water-reducing admixtures results in a greater surface area of cement exposed to hydration, and for this reason there is an increase in strength at early ages compared with an admixture-free mix of the same water/cement ratio. Long-term strength may also be improved because of a more uniform distribution of the dispersed cement throughout the concrete. In general terms, these admixtures are effective with all types of cement, although their influence on strength is greater with cements which have a low C_3A or low alkali content. There are no detrimental effects on other long-term properties of concrete, and, when the admixture is used correctly, the durability can be improved.

As with other types of admixtures, the use of reliable dispensing equipment is essential since the dosage levels of the admixture represent only a fraction of one per cent of the mass of cement.

Superplasticizers

These are a more recent and more effective type of water-reducing admixtures known in the US as *high range water reducers* and called Type F by ASTM. There exists also a high range water-reducing and set-retarding admixture, classified as Type G. Table 8.2 lists the requirements of ASTM C 494–82 and Table 8.3 gives the equivalent UK specification of BS 5075: Part 3: 1985.

The dosage levels are usually higher than with conventional water-reducers, and the possible undesirable side effects are considerably reduced. For example, because they do not markedly lower the surface tension of water, superplasticizers do not entrain a significant amount of air.

Superplasticizers are used to produce *flowing* concrete in situations where placing in inaccessible locations, in floor or pavement slabs or where very rapid placing is required. A second use of superplasticizers is in the production of very-high strength concrete, using normal workability but a very low water/cement ratio. Figure 8.2 illustrates these two applications of superplasticizers.

Superplasticizers are sulphonated melamine formaldehyde condensates or sulphonated naphthalene formaldehyde condensates, the latter being probably the more effective in dispersing the cement and generally having also some retarding properties. The dispersing action is mainly promoted by the sulphonic acid being adsorbed on the surface of cement particles, causing them to become negatively charged and thus mutually repulsive. This increases the workability at a given water/cement ratio, typically

156

Table 8.3: Specification for superplasticizing admixtures according to BS 5075: Part 3: 1985

Type of admixture	Water reduction per cent	Workability	Loss of slump	Stiffening time		Compressive strength	
				Time from completion of mixing to reach a resistance to penetration* of:		Percentage of control mix (minimum)	Age
				0.5 MPa (70 psi)	3.5 MPa (500 psi)		
Superplasticizing	–	Flow table: 510 to 620 mm	At 45 min not less than that of control mix at 10 to 15 min. At 4 h not more than that of control mix at 10 to 15 min.	–	–	90 90	7 days 28 days
Superplasticizing	16	Slump: not more than 15 mm below that of control mix	–	Within 1 h of control mix	Within 1 h of control mix	140 125 115	24 h 7 days 28 days
Retarding superplasticizing	–	Flow table: 510 to 620 mm	At 4 h not less than that of control mix at 10 to 15 min.	–		90 90	7 days 28 days
Retarding superplasticizing	16	Slump: not more than 15 mm below that of control mix	–	1 to 4 h longer than control mix	–	125 115	7 days 28 days

* The penetration is determined by a special brass rod 6.175 mm in diameter.
Note: The air content, with or without water reduction, shall not be more than 2 per cent higher than that of the control mix, and not more than a total of 3 per cent.

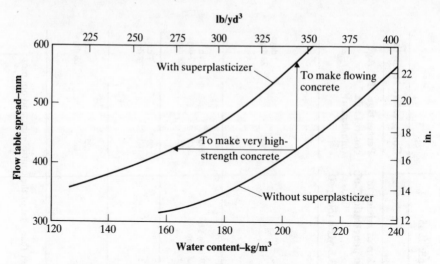

Fig. 8.2: Typical relation between flow table spread and water content of concrete made with and without superplasticizer
(Based on: A. MEYER, Experiences in the use of superplasticizers in Germany, Superplasticizers in concrete, *Amer. Concr. Inst. Sp. Publicn. No.* 62, pp. 21–6 (1979).)

raising the slump from 75 mm (3 in.) to 200 mm (8 in.). In the UK, the high workability is measured by the flow table spread test (see page 89) and values between 500 and 600 mm are typical. The resulting flowing concrete is cohesive and not subject to excessive bleeding or segregation, particularly if very angular, flaky or elongated coarse aggregates are avoided and the fine aggregate content is increased by 4 to 5 per cent. It should be remembered, when designing formwork, that flowing concrete can exert full hydrostatic pressure.

When the aim is to achieve high strength at a given workability, the use of a superplasticizer can result in a water reduction of 25 to 35 per cent (compared with about one-half that value for conventional water-reducing admixtures). In consequence, the use of low water/cement ratios is possible so that very high strength concrete is obtained (see Fig. 8.3). Strengths as high as 100 MPa (15 000 psi) at 28 days, when the water/cement ratio is 0.28, have been achieved. With steam-curing or autoclaving, even higher strengths are possible. For a further improvement in strength at later ages, superplasticizers can be used with partial replacement of cement by fly ash (PFA).

The improved workability produced by superplasticizers is of short duration and thus there is a high rate of slump loss; after some 30 to 90 min the workability returns to normal. For this reason, the superplasticizer should be added to the mix immediately prior to placing; usually, conventional mixing is followed by the addition of the superplasticizer and a short period of *additional* mixing. In the case of ready-mixed concrete, a 2 min re-mixing period is essential. While re-tempering with an additional dose of the superplasticizer is not recommended because of

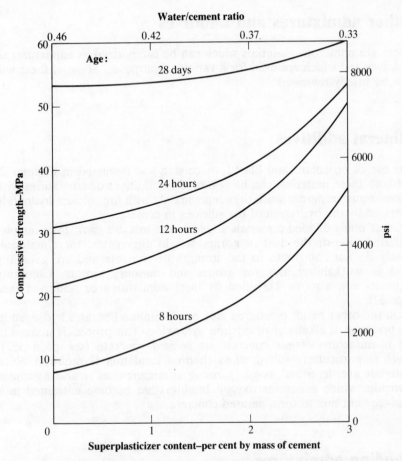

Fig. 8.3: The influence of the addition of superplasticizer on the early strength of concrete made with a cement content of 370 kg/m³ (630 lb/yd³) and cast at room temperature. All concretes of the same workability and made with rapid-hardening Portland (Type III) cement
(From: A. MEYER, Steigerung der Fruhfestigkeit von Beton, *Il Cemento,* 75, No. 3, pp. 271–6 (July–Sept. 1978).)

the risk of segregation, redosing to maintain workability up to 160 min has been successfully used.

Superplasticizers do not significantly affect the setting of concrete except in the case of cements with a very low C_3A content when there may be excessive retardation. Other long-term properties of concrete are not appreciably affected. However, the use of superplasticizers with an air-entraining admixture can sometimes reduce the amount of entrained air and modify the air-void system but specially modified superplasticizers are available which appear to be compatible with conventional air-entraining agents. The only real disadvantage of superplasticizers is their relatively high cost, which is due to the expense of manufacturing a product with a high molecular mass.

Other admixtures and additives

There are numerous materials which can be categorized as admixtures or additives and which are used for a variety of purposes. Some of these will now be briefly reviewed.

Mineral additives

The use of pozzolans and blast-furnace slag was discussed in Chapter 2, but both these materials can be regarded as additives or admixtures with *cementitious properties* since they react mainly with the calcium hydroxide liberated by the hydration of the silicates in cement.

Other finely divided materials added to the mix are *inert,* for example, hydrated lime or the dust of normal-weight aggregates. Inert materials clearly do not contribute to the strength of concrete and are generally used as workability aids for grouts and masonry mortar. Colouring pigments can also be classified as inert admixtures or additives (see page 32).

On the other hand, powdered zinc or aluminium liberates hydrogen in the presence of alkalis or of calcium hydroxide. This process is utilized in the manufacture of *gas* concrete or *aerated* concrete (see page 357), which is particularly suited when thermal insulation is required. Such materials are described as *gas forming* admixtures, as is also hydrogen peroxide, which generates oxygen bubbles that become entrained in a sand–cement mix to form aerated concrete.

Bonding admixtures

These are polymer emulsions (latexes) which improve the adherence of fresh concrete to hardened concrete, and thus are particularly suited for repair work. The emulsion is a colloidal suspension of a polymer in water and, when the emulsion is used to combine concrete with a polymer, a latex-modified concrete (LMC) or a polymer Portland cement concrete is produced. Although costly, polymer latexes improve the tensile and flexural strength, and also the durability, as well as bonding properties (see Chapter 19).

Water-repellent admixtures

Such materials reduce the capillary absorption of fluids by the hardened concrete and therefore lower its permeability. Mineral and vegetable oils or metallic soaps (e.g. calcium stearate) are claimed to exhibit this property but it should be emphasized that these admixtures do *not* prevent water penetration into or through concrete and therefore cannot be used for water-proofing. Also, water-repellents may have an effect on the subsequent surface treatment of concrete.

The various admixtures discussed in this chapter offer many advantages but care is necessary in order to realize the full benefits of admixtures. A reputable supplier will give technical data for the particular application and advise on possible side effects, but trial tests should be carried out using the *actual* constituents of the mix to be used. Also, adequate supervision should be provided at the batching stage so as to ensure correct levels of dosage of the admixture.

Bibliography

8.1   ACI COMMITTEE 212.3R–89, Guide for use of admixtures in concrete, Part 1: Materials and General Properties of Concrete, *ACI Manual of Concrete Practice,* 1990.

8.2   CONCRETE ADMIXTURES HANDBOOK: PROPERTIES SCIENCE AND TECHNOLOGY, (Editor V. S. RAMACHANDRAN) pp. 211–68, (New Jersey, USA, Noyes Publications, 1984).

8.3   DEVELOPMENTS IN THE USE OF SUPERPLASTICIZERS, (Editor V. M. MALHOTRA) *Amer. Concr. Inst. Sp. Publicn.* No. 68, pp. 561 (1981).

8.4   P. C. HEWLETT, Superplasticized Concrete, Parts 1 and 2, *Concrete Society Current Practice Sheets 94 and 95, Concrete,* Vol. 18, Nos 4 and 5, pp. 31–2, April and May 1984.

8.5   CONCRETE ADMIXTURES: USES AND APPLICATIONS, (Editor, M. R. RIXOM) (New York, Construction Press, Longman, 1977).

8.6   SUPERPLASTICIZERS IN CONCRETE, *Amer. Concr. Inst. Sp. Publicn.* No. 62, pp. 427 (1979).

Problems

8.1   What is flowing concrete?
8.2   What are the broad types of admixtures?
8.3   What is the difference between an additive and an admixture?
8.4   Which cement would you not use with calcium chloride?
8.5   Should calcium chloride be used for reinforced concrete in the interior of a building? Give your reasons.

161

8.6 Give an example of: (i) an accelerator, and (ii) a set-accelerating admixture.

8.7 What would you recommend as an accelerating, chloride-free admixture?

8.8 What would you do if a mixing truck got stuck but the mixer continued to operate?

8.9 What are the uses of plasticizers?

8.10 Explain the mechanism of action of retarders.

8.11 What is meant by a retarder?

8.12 What is meant by an accelerator?

8.13 What are the advantages of using calcium chloride in Portland cement concrete?

8.14 What are the disadvantages of using calcium chloride in Portland cement concrete without and with reinforcement?

8.15 Describe the mechanism of action of plasticizers.

8.16 What are the main differences between plasticizers and superplasticizers?

8.17 What are the disadvantages of plasticizers?

8.18 Give examples of plasticizers and superplasticizers.

8.19 State the advantages and disadvantages of superplasticizers.

8.20 What is slump loss?

8.21 Give examples of mineral additives.

8.22 Define an emulsion.

8.23 How would you improve the bond of fresh concrete to hardened concrete?

8.24 Is there such a thing as a waterproofing admixture?

8.25 Outline how you would assess the side effects of any admixture.

9

Temperature problems in concreting

There are some special problems involved in concreting in hot weather, arising both from a higher temperature of the concrete, and, in many cases, from an increased rate of evaporation from the fresh mix. In the case of large volumes or masses of concrete, the problems are associated with possible cracking in consequence of a temperature rise and subsequent fall due to the heat of hydration of cement and of the concomitant restrained volume changes. On the other hand, when concreting in cold weather, precautions are necessary to avoid the ill effects of frost damage in fresh or young concrete. In all these instances, we have to take appropriate steps in mixing, placing and curing of concrete.

Hot-weather problems

A higher temperature of fresh concrete than normal results in a more rapid hydration of cement and leads therefore to accelerated setting and to a lower long-term strength of hardened concrete (see Fig. 9.1) since a less uniform framework of gel is established (see Chapter 10). Furthermore, if high temperature is accompanied by a low relative humidity of the air, rapid evaporation of some of the mix water takes place, causing a higher loss of workability, higher plastic shrinkage, and crazing (see Chapter 13). A high temperature of fresh concrete is also detrimental when placing large concrete volumes because greater temperature differentials can develop between parts of the mass due to the more rapid evolution of the heat of hydration of cement; subsequent cooling induces tensile stresses which may cause thermal cracking (see page 168 and Chapter 13).

Another problem is that air entrainment is more difficult at higher temperatures, although this can be remedied by simply using larger quantities of entraining agent. A related problem is that, if relatively cool concrete is allowed to expand when placed at a higher air temperature, then the air voids expand and the strength of the concrete is reduced. This would occur, for instance, with horizontal panels, but not with vertical ones in steel moulds where expansion is prevented.

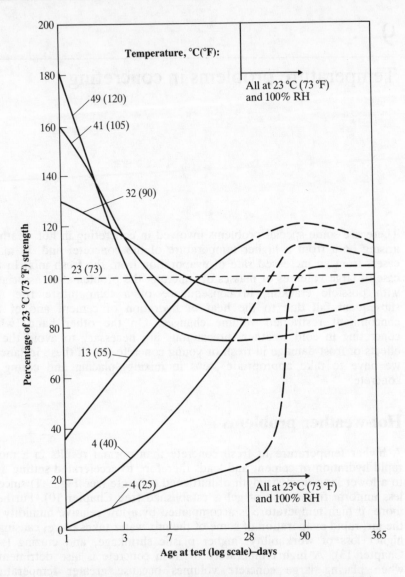

Fig. 9.1: Effect of temperature during the first 28 days on the strength of
concrete (water/cement ratio = 0.41; air content = 4.5 per cent;
ordinary Portland (Type I) cement)
(From: P. KLIEGER, Effect of mixing and curing temperature on
concrete strength, *J. Amer. Concr. Inst.*, 54, pp. 1063–81 (June
1958).)

Curing at high temperatures in dry air presents additional problems as the curing water tends to evaporate rapidly, with a consequent slowing down of hydration. As a result, there is an inadequate development of strength and rapid drying shrinkage takes place, the latter possibly inducing tensile stresses of sufficient magnitude to cause cracking of the hardened concrete (see Chapter 13). It follows that prevention of evaporation from the surface of concrete is essential; methods of achieving this by proper curing are discussed in Chapter 10.

Hot-weather concreting

There are several measures that can be taken to cope with the problems discussed in the previous section. In the first instance, the temperature of the concrete, made on site or delivered, should be kept low, preferably not above 16 °C (60 °F), with an upper limit of 32 °C (90 °F). The temperature of freshly mixed concrete can easily be calculated from that of its ingredients, using the expression

$$T = \frac{0.22(T_a W_a + T_c W_c) + T_w W_w + T_a W_{wa}}{0.22(W_a + W_c) + W_w + W_{wa}} \tag{9.1}$$

where T denotes temperature (°C or °F), W is the mass of the ingredient per unit volume of concrete (kg/m^3 or lb/yd^3), and the suffixes a, c, w, wa refer to dry aggregate, cement, added water, and water absorbed by the aggregate, respectively. The value 0.22 is the approximate ratio of the specific heat of the dry ingredients to that of water, and is applicable in both the SI and the American systems of units.

The *actual* temperature of the concrete will be somewhat higher than indicated by the above expression owing to the mechanical work done in mixing and to the early development of the heat of hydration of cement. Nevertheless, the expression is usually sufficiently accurate.

Since we often have a certain degree of control over the temperature of at least some of the ingredients of concrete, it is useful to consider the relative influence of changing their temperature. For instance, for a water/cement ratio of 0.5 and an aggregate/cement ratio of 5.6, a decrease of 1 °C (or 1 °F) in the temperature of fresh concrete can be obtained by lowering the temperature of either the cement by 9 °C (9 °F) *or* of the water by 3.6 °C (3.6 °F) *or* of the aggregate by 1.6 °C (1.6 °F). It can thus be seen that, because of its relatively small quantity in the mix, a greater temperature drop is required for cement than for the other ingredients; moreover, it is much easier to cool the water than the cement or the aggregate.

It is possible, furthermore, to use ice as part of the mixing water. This is even more effective because more heat is abstracted from the other ingredients to provide the latent heat of fusion of ice. In this case, the

165

temperature of the fresh concrete is given by

$$T = \frac{0.22(T_a W_a + T_c W_c) + T_w W_w + T_a W_{wa} - L W_i}{0.22(W_a + W_c) + W_w + W_{wa} + W_i} \tag{9.2}$$

where the terms are as in Eq. (9.1) except that the total mass of water added to the mix is the mass of fluid water W_w at temperature T_w plus the mass of ice W_i; L is the ratio of the latent heat of fusion of ice to the specific heat of water, and is equivalent to 80 °C (144 °F).[1]

Care is required when ice is used because it is essential that all the ice has melted completely before the completion of mixing.

Although it is less effective actively to cool the aggregate, a useful reduction in the placing temperature of concrete can be achieved simply and cheaply by shading the aggregate stockpiles from the direct rays of the sun, and by controlled sprinkling of the stockpiles so that heat is lost by evaporation. Other measures used are to bury the water pipes, paint all exposed pipes and tanks white, spray the formwork with water before commencing the placing of concrete, and to commence placing in the evening.

With regard to the choice of suitable mix proportions in order to reduce the effects of a high air temperature, the cement content should be as low as possible so that the total heat of hydration is low. To avoid workability problems, the aggregate type and grading should be chosen so that high absorption rates are avoided and the mix is cohesive; contaminants in the aggregate, such as sulphates, while always undesirable, are particularly harmful as they can cause a flash or a false set.

To reduce the loss of workability and also to increase the setting time, a set-retarding admixture can be used (see Chapter 8); this has the advantage of preventing the formation of cold joints in successive lifts. High-dosage levels of the admixture may be required and advice from an admixture specialist should be sought for the particular application.

After placing, evaporation of water from the mix has to be prevented. Evaporation rates greater than $0.5 \, \text{kg/m}^2$ ($0.1 \, \text{lb/ft}^2$) of the exposed concrete surface per hour have to be avoided in order to ensure satisfactory curing and to prevent plastic cracking. The rate of evaporation depends upon the air temperature, the concrete temperature, the relative humidity of the air, and the wind speed; values of rate of evaporation can be estimated from Fig. 9.2. The concrete should be protected from the sun as, otherwise, if a cold night follows, thermal cracking can occur due to the restraint to contraction on cooling from the original, unnecessarily high temperature. The extent of cracking is directly related to the difference in temperature between the concrete and the surrounding air (see Chapter 13).

In dry weather, wetting the concrete and allowing evaporation to take

---

[1] Latent heat of fusion of ice = 335 kJ/kg (144 BTU/lb).
Specific heat of water = 4.2 kJ/kg/°C (1 BTU/lb/°F).

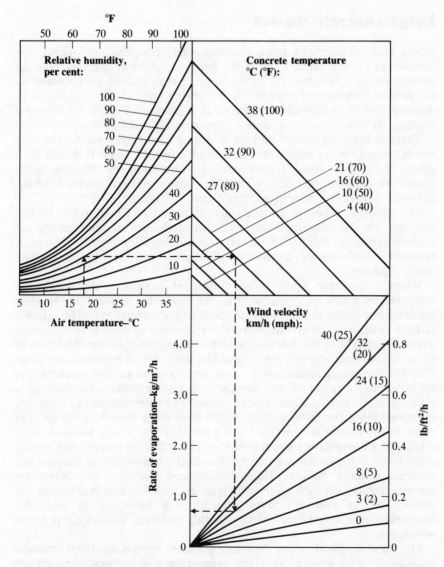

Fig. 9.2: Effect of concrete and air temperatures, relative humidity, and wind velocity on the rate of evaporation of surface moisture from concrete (Based on: ACI 305.R–77.)

place results in effective cooling as well as effective curing. Other methods of curing (see Chapter 10) are less effective. If plastic sheeting or membranes are used, they should be white so as to reflect the rays of the sun. Large exposed areas of concrete, such as highways and runways, are particularly vulnerable to this type of temperature problem, and the placing and curing of concrete in such cases should be carefully planned and executed.

167

Large concrete masses

When large volumes of plain (unreinforced) concrete are placed, for instance in gravity dams, there is the danger of *thermal cracking* because of restraint to contraction on cooling from a temperature peak caused by the heat of hydration of cement. Such cracking may take several weeks to develop. Quite independently, there is a danger of early-age thermal cracking in thinner sections, unless appropriately reinforced.

Thermal cracking should be clearly distinguished from *plastic cracking* which occurs on, or near, the surface of concrete while it is still in a plastic state when rapid evaporation of water from the concrete takes place. We may add that drying can also cause *shrinkage cracking,* which normally occurs at a later stage than thermal cracking.

The different types of cracking are discussed in Chapter 13. In this chapter, we are concerned only with the influence of temperature on thermal cracking, although there are other influencing factors: degree of restraint, coefficient of thermal expansion of concrete, and its tensile strain capacity.

When a concrete mass is not insulated from the atmosphere, a temperature gradient exists within the concrete because its interior becomes hot whilst the surface loses heat to the atmosphere. The interior is thus restrained from full thermal expansion, so that a compressive stress is induced in the interior, which is balanced by a tensile stress in the exterior. Both stresses are relieved to some extent by creep (see page 215) but the tensile stress may be sufficient to cause surface cracking. As the concrete starts to cool and contract, the tensile stress in the exterior is relieved and any surface cracks close and are therefore rendered harmless. Since the interior wants to contract more than the exterior, the strain in the former is restrained and a tensile stress is now induced, with a balancing compressive stress in the exterior. During this cooling phase, there is less relief of stress by creep than in the heating phase because the concrete is more mature. Thus, the induced tensile stress, caused by *internal restraint* on cooling, may be large enough to cause cracking in the interior of the concrete mass. Hence, it is necessary to limit the temperature differential or gradient within concrete if cracking is to be avoided.

On the other hand, when the entire concrete mass is insulated from the outside air or earth, so that the temperature is uniform throughout, cracking will occur only if the total mass is wholly or partly *externally* restrained from contracting during the cooling period. This form of restraint is termed *external restraint,* and to avoid cracking it is necessary to minimize the difference between the peak temperature of the concrete and the ambient temperature or to minimize the restraint. The tolerable temperature differential between the peak temperature and the final ambient temperature should be limited to about 20 °C (36 °F) when flint gravel aggregate is used, and 40 °C (72 °F) using certain limestone aggregates, but can be as high as 130 °C (234 °F) using some lightweight aggregates (see Chapter 18).

Several measures can be taken to minimize the temperature difference or gradient:

(a) Cool the ingredients of the mix by any of the methods given on page 165, so as to reduce the temperature of the fresh concrete to about 7 °C (45 °F). By this means, the difference between the peak temperature and the ambient temperature on cooling will be reduced.

(b) Cool the surface of the concrete, but *only* for sections less than about 500 mm (or 20 in.) thick, using formwork which offers little insulation, e.g. steel. Here, cooling the surface of the concrete reduces the temperature rise of the core without causing harmful temperature gradients and thus inducing internal restraint.

(c) Insulate the entire surface of the concrete (including the upper surface) for sections more than about 500 mm (or 20 in.) thick, using a suitable material for the formwork, so that the temperature gradients are minimized. The concrete will then be allowed to expand and contract freely, provided there is *no external restraint*.

(d) Select the mix ingredients carefully.

The choice of mix ingredients is, in part, dependent upon the other factors influencing cracking, besides the temperature. A suitable aggregate can help to reduce the coefficient of thermal expansion of concrete and increase its tensile strain capacity. For instance, concrete made with angular aggregate has a greater tensile strain capacity than concrete made with a rounded aggregate. Likewise, lightweight aggregate leads to a greater tensile strain capacity than normal weight aggregate. However, this advantage is offset, to some extent, by the requirement of a higher cement content when lightweight aggregate is used for the same strength and workability.

Generally, the use of low heat cement, pozzolan replacement, a low cement content, and use of water-reducing admixtures which make it possible, are beneficial in reducing the peak temperature. The choice of the type of cement is governed by the heat evolution characteristics which affect the temperature rise, viz. the *rate* at which the heat is evolved and the *total* heat. The latter is of course greater the higher the cement content per unit volume of concrete. In small sections, the rate of heat evolution is more important with regard to the temperature rise because heat is being steadily dissipated whereas, in massive sections, the temperature rise is more dependent on the total heat evolved because of greater self-insulation.

We can see thus that the temperature rise depends on a number of factors: cement type and quantity (or, strictly speaking, the type and quantity of all the cementitious materials), the section size, the insulating characteristics of the formwork, and the placing temperature of the concrete. With respect to the latter, we can note that the higher the placing temperature the faster the hydration of the cement and the greater the temperature rise.

In practice, the lowest temperature rise is given by a blend of sulphate resisting Portland (Type V) cement and ground granulated blast-furnace slag. The next best is a blend of ordinary Portland (Type I) cement and slag, and then part replacement of Portland cement by fly ash (PFA). In massive sections, the quantity of the cementitious material, i.e. cement plus slag or fly ash, is governed more by impermeability and durability requirements (maximum water/cement ratio) than by a specified 28-day compressive strength, which need not exceed 14 MPa (2000 psi). However, in structural reinforced concrete, a higher early strength may be critical so that ordinary Portland (Type I) cement alone and in larger quantities may have to be used; it is therefore necessary to adopt alternative measures to minimize the ill-effects of temperature rise.

We have referred earlier to the tolerable temperature differentials. The differential in a given case can be calculated from the knowledge of the thermal characteristics of the concrete and of its thermal insulation but, in practice, the temperature at various points should be monitored by thermocouples. It is then possible to adjust the insulation so as to keep within the limiting temperature differentials. The insulation must control the loss of heat by evaporation, as well as by conduction and radiation. To achieve the first, a plastic membrane or a curing compound should be used, whilst soft board will insulate against the other forms of heat loss; plastic coated quilts are useful in all respects.

The formwork striking times are important with regard to minimizing the temperature differentials. With thin sections, say less than 500 mm (or 20 in.), early formwork removal allows the concrete surface to cool more rapidly. However, for massive isolated sections, the insulation must remain in place until the whole section has cooled sufficiently, so that when the formwork is finally removed, the drop in surface temperature does not exceed *one-half* of the values given earlier, e.g. 10 °C (18 °F) for concrete made with flint gravel aggregate. The reason for the lower values of tolerable temperature differentials is that, when the insulation is removed, cooling is more rapid so that creep cannot help in increasing the tensile strain capacity of the concrete. For this reason, the formwork and insulation of large sections may have to remain in place for up to two weeks before the concrete has cooled to a safe level of temperature. However, if the section is subject to external restraint, this measure will not prevent cracking, and other remedial measures have to be considered. These involve the sequence of construction and the provision of movement joints, and are referred to in Chapter 13.

Cold-weather concreting

The problems of cold-weather concreting arise from the action of frost on *fresh* concrete (see page 287). If the concrete which has not yet *set* is allowed to freeze, the mixing water converts to ice and there is an increase in the overall volume of the concrete. Since there is now no water available for chemical reactions, the setting and hardening of the concrete are delayed, and, consequently, there is little cement paste that can be disrupted by the formation of ice. When at a later stage thawing takes place, the concrete will set and harden in its expanded state so that it will contain a large volume of pores and consequently have a low strength.

It is possible to revibrate the concrete when it thaws and thus re-compact it, but such a procedure is not generally recommended since it is difficult to ascertain exactly when the concrete has started to set.

If freezing occurs *after* the concrete has set, but before it has developed an appreciable strength, the expansion associated with the formation of ice causes disruption and an irreparable loss of strength. If, however, the concrete has acquired a sufficient strength before freezing, it can withstand the internal pressure generated by the formation of ice from the remaining mixing water. Its quantity is small because, at this stage, some of the mixing water will have combined with the cement in the process of hydration, and some will be located in the small gel pores and thus not be able to freeze. Unfortunately, it is not easy to establish the age at which the concrete is strong enough to resist freezing, although some rule-of-thumb data are available. Generally, the more advanced the hydration of the cement and the higher the strength of concrete the less vulnerable it is to frost damage.

In addition to being protected from frost damage at an early age, concrete has to be able to withstand any subsequent cycles of freezing and thawing in service, if such are likely. For concrete made at normal temperature, this is considered in Chapter 15. At this point, we are concerned with preventing the freezing of fresh concrete and protecting the concrete during initial hydration. To achieve this we must ensure that the placing temperature is high enough to prevent freezing of the mix water and that the concrete is thermally protected for a sufficient time to develop an adequate strength. Table 9.1 gives the recommended minimum concrete placing temperatures for various air temperatures and sizes of section when concreting in cold weather. We can see that the minimum tolerable concrete temperature, as *placed* and *maintained*, is lower for larger sections because they lose less heat.

From the same table, we can note that, when the air temperature is below 5 °C (40 °F), the concrete has to be mixed at a higher temperature in order to allow for heat losses during transportation and placing. Moreover, we must make sure that the fresh concrete is not deposited against a frozen surface. Furthermore, in order to avoid the possibility of thermal cracking in the first 24 hours after the end of protection, when the concrete cools to the ambient temperature, the maximum allowable temperature drop during those 24 hours must not exceed the values given in Table 9.1.

It may be noted that lightweight aggregate concrete retains more heat so that the minimum temperatures as placed and maintained may be lower.

The recommended periods of continuous protection for *air-entrained* concrete made with normal weight aggregate, placed and maintained at the temperatures given in Table 9.1, are shown in Table 9.2. Where freezing and thawing is likely to occur in service, air-entrained concrete should of course be used but, if the construction involves non-air-entrained concrete, the protection times of Table 9.2 should be at least

Table 9.1: Recommended concrete temperatures for cold weather concreting

Air temperature	Minimum dimension of section			
	smaller than 300 mm (12 in)	300–900 mm (12–36 in.)	900–1800 mm (36–72 in.)	above 1800 mm (72 in.)
	Minimum concrete temperature as *placed and maintained*			
Below 5 °C (40 °F)	13 °C (55 °F)	10 °C (50 °F)	7 °C (45 °F)	5 °C (40 °F)
	Minimum concrete temperature as *mixed* for indicated air temperature			
Above −1 °C (30 °F)	16 °C (60 °F)	13 °C (55 °F)	10 °C (50 °F)	7 °C (45 °F)
−18 to −1 °C (0 to 30 °F)	18 °C (65 °F)	16 °C (60 °F)	13 °C (55 °F)	10 °C (50 °F)
Below −18 °C (0 °F)	21 °C (70 °F)	18 °C (65 °F)	16 °C (60 °F)	13 °C (55 °F)
	Maximum concrete temperature *drop* permitted in first 24 hours after end of protection			
	28 °C (50 °F)	22 °C (40 °F)	17 °C (30 °F)	11 °C (20 °F)

Based on ACI 306R–78.

172

doubled because such concrete, particularly in a saturated state, is more vulnerable to damage by frost (see Chapter 15). The protection period given in Table 9.2 depends on the type and quantity of cement, on whether an accelerator is used, and on the service conditions. These protection times should ensure avoiding both early-age frost damage and later-age durability problems.

Table 9.2: **Recommended protection times for cold-weather concreting (using air-entrained concrete)**

Cement type, admixture, cement content	Protection time for preventing frost damage (days) for service category:				Protection time for safe level of strength (days) for service category:			
	1	2	3	4	1	2	3	4
Ordinary Portland (Type I) cement Modified (Type II) cement	2	3	3	3	2	3	6	see Table 9.3
Rapid-hardening Portland (Type III) cement, Accelerator, 20 per cent extra cement	1	2	2	2	1	2	4	see Table 9.3

Service categories: 1 – no load, no exposure
                     2 – no load, exposure
                     3 – partial load, exposure
                     4 – full load, exposure.
Based on ACI 306R–78

In the case where a high proportion of the design strength of structural concrete must be achieved before it is safe to remove forms and shoring, the protection times are those given in Table 9.3; these values are typical for 28-day strengths of 21 to 34 MPa (3000 to 5000 psi); for other service conditions and types of concrete, the protection times should be deduced from a pre-determined strength–maturity relation (see Chapter 10).

From Tables 9.2 and 9.3, it is apparent that to achieve a high rate of heat development (and, hence, an early temperature rise) rapid-hardening Portland (Type III) cement or an accelerating admixture should be used, preferably with a rich mix having a low water/cement ratio.

We referred earlier to the required minimum temperature of concrete at the time of placing. We should aim at a value between 7 and 21 °C (45 to 70 °F). Exceeding the upper value might have an adverse effect on the long-term strength. The temperature of the concrete at the time of placing is a function of the temperature of the mix ingredients and can be calculated by Eq. (9.1). If necessary, we can heat the appropriate ingredients. By analogy to what was mentioned in the section dealing with hot weather concreting, it is easier and more effective to heat the water, but it is inadvisable to exceed a temperature of 60 to 80 °C (140 to

Table 9.3: **Recommended protection times for fully loaded concrete exposed to cold weather**

Type of cement	Duration of protection (days)			
	Percentage of 28-day strength			
	50	65	85	95
	For concrete temperature of 10 °C (50 °F):			
Ordinary Portland (Type I) cement	6	11	21	29
Modified (Type II) cement	9	14	28	35
Rapid-hardening Portland (Type III) cement	3	5	16	26
	For concrete temperature of 21 °C (70 °F):			
Ordinary Portland (Type I) cement	4	8	16	23
Modified (Type II) cement	6	10	18	24
Rapid-hardening Portland (Type III) cement	3	4	12	20

Based on ACI 306R–78

180 °F) as flash set of cement may result; the difference in temperature between the water and the cement is relevant. Also, it is important to prevent the cement from coming into direct contact with hot water as agglomerations of cement (cement balls) may result, and for this reason the order of feeding the mix ingredients into the mixer must be suitably arranged.

If heating the water does not sufficiently raise the temperature of the concrete, the aggregate may be heated indirectly, i.e. by steam through coils, up to about 52 °C (125 °F). Direct heating with steam would lead to a variable moisture content of the aggregate. When the temperature of the aggregate is below 0 °C (32 °F), the absorbed moisture is in a frozen state. Therefore, not only the heat required to raise the temperature of the ice from the temperature of the aggregate T_a to 0 °C (32 °F), but also the heat required to change the ice into water (latent heat of fusion) have to be taken into account. In this case, the temperature of the freshly mixed concrete becomes

$$T = \frac{0.22(T_a W_a + T_c W_c) + T_w W_w + W_{wa}(0.5T_a - L)}{0.22(W_a + W_c) + W_w + W_{wa}} \qquad (9.3)$$

174

where 0.5 is the ratio of specific heat of ice to that of water, and L is as defined in Eq. (9.2).

After placing, an adequate temperature of the concrete is obtained by insulating it from the atmosphere and, if necessary, by constructing enclosures around the structure and providing a source of heat within the enclosure. The form of heating should be such that the concrete does not dry out rapidly, that no part of it is heated excessively, and that no high concentration of CO_2 (which would cause carbonation, see page 240) in the atmosphere results. For these reasons, exhaust steam is probably the best source of heat. Jacket-like steel forms with circulating hot water are sometimes used.

In important structures, the temperature of the concrete should be monitored. In deciding on the location of thermometers or thermo-couples, it should be remembered that corners and faces are particularly vulnerable to frost. Monitoring the temperature makes it possible to adjust the insulation or the heating so as to allow for changes in atmospheric conditions such as a wind accompanying a sudden lowering of the air temperature, a condition which aggravates the frost action. On the other hand, snow acts as an insulator and thus provides natural protection.

Bibliography

9.1 ACI-COMMITTEE 305R–89, Hot-weather concreting, Part 2, *ACI Manual of Concrete Practice*, 1990.

9.2 ACI-COMMITTEE 306R–88, Cold-weather concreting, Part 2, *ACI Manual of Concrete Practice*, 1990.

9.3 T. A. HARRISON, Early-age thermal crack control in concrete, *CIRIA Report* 91, pp. 48 (Construction Industry Research and Information Association 1981).

Problems

9.1 What particular precautions would you take when concreting: (i) in winter, and (ii) in hot weather?

9.2 What are the causes of thermal cracking of concrete walls?

9.3 What are the thermal problems in mass concrete?

9.4 What are the thermal problems in a very large pour of reinforced concrete?

9.5 What are the physical changes in concrete subjected to one cycle of freezing?

9.6 Why is insulation sometimes used in placing large concrete pours?

9.7 What are the effects of hot weather on fresh concrete?

9.8 What are the effects of hot weather on hardened concrete?

9.9 What special measures are necessary for concreting in hot weather?

9.10 Two cements have the same total heat of hydration but different rates of heat evolution. Compare their performance in mass concrete.

9.11 What is the action of frost on fresh concrete?

9.12 For cold weather concreting, what materials would you heat prior to putting in the mixer? Give your reasons. Are there any limitations on temperature?

9.13 What is the maximum rate of evaporation of water from a fresh concrete surface below which plastic cracking will not occur?

9.14 Should you place concrete in lifts or continuously? Give your reasons.

9.15 When would you place ice in the concrete mixer?

9.16 State the measures used to minimize temperature gradients.

9.17 Describe the methods of control of temperature in placing mass concrete.

9.18 How does the type of aggregate influence thermal cracking of concrete?

9.19 Why is lightweight aggregate superior to normal weight aggregate for reducing the risk of thermal cracking?

9.20 What type of cement would you choose for: (i) mass concrete, and (ii) reinforced concrete?

9.21 Discuss the influence of formwork striking times on the risk of thermal cracking.

9.22 How would you reduce: (i) internal restraint, and (ii) external restraint to temperature changes?

9.23 Why is the size of section relevant to the minimum temperature at placing necessary to prevent frost damage at early ages?

9.24 Give a suitable range of temperature of concrete at the time of placing in order to prevent frost damage.

9.25 How can the temperature of the concrete at the time of placing be raised and subsequently maintained?

9.26 A $1:1.8:4.5$ mix has a water/cement ratio of 0.6 and a cement content of $300 \, \text{kg/m}^3$ ($178 \, \text{lb/yd}^3$). The temperatures of the ingredients are as follows:

    cement: 18 °C (64 °F)
    aggregate: 30 °C (86 °F)
    water: 20 °C (68 °F)

Assuming that the aggregate has negligible absorption, calculate the temperature of the fresh concrete.

Answer: 26°C (79°F)

9.27 For the concrete used in question 9.26, how much ice is required to reduce the temperature of the fresh concrete to 16 °C (61 °F)? What is the mass of liquid water added to maintain the same water/cement ratio?

Answer: $67 \, \text{kg/m}^3$ ($112 \, \text{lb/yd}^3$)
              $113 \, \text{kg/m}^3$ ($191 \, \text{lb/yd}^3$)

176

10

Development of strength

In order to obtain good quality concrete, the placing of an appropriate mix must be followed by curing in a suitable environment during the early stages of hardening. *Curing* is the name given to procedures used for promoting the hydration of cement, and thus, the development of strength of concrete, the curing procedures being control of the temperature and of the moisture movement from and into the concrete. The latter affects not only strength but also durability. This chapter deals with the various methods of curing, both at normal and at elevated temperatures, the latter increasing the rate of the chemical reactions of hydration and of gain in strength. We should note, however, that early application of a higher temperature can adversely affect the longer-term strength. Consequently, the influence of temperature has to be carefully considered.

Normal curing

The object of curing at normal temperature is to keep concrete saturated, or as nearly saturated as possible, until the originally water-filled space in the fresh cement paste has been occupied to the desired extent by the products of hydration of cement. In the case of site concrete, active curing nearly always ceases long before the maximum possible hydration has taken place. The influence of moist curing on strength can be gauged from Fig. 10.1. Tensile and compressive strengths are affected in a similar manner. The failure to gain strength in consequence of inadequate curing, i.e. through loss of water by evaporation, is more pronounced in thinner elements and in richer mixes, but less so for lightweight aggregate concrete. The influence of curing conditions on strength is lower in the case of air-entrained than non-air-entrained concrete.

The necessity for curing arises from the fact that hydration of cement can take place only in water-filled capillaries. This is why loss of water by evaporation from the capillaries must be prevented. Furthermore, water lost internally by *self-desiccation* has to be replaced by water from outside, i.e. ingress of water into the concrete must take place.

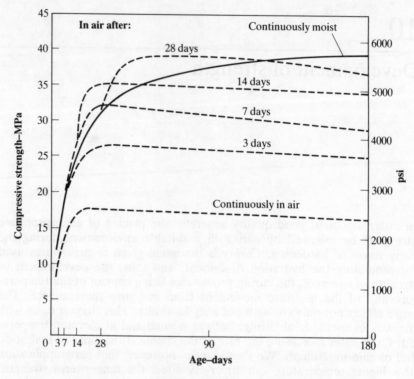

Fig. 10.1: Influence of moist curing on the strength of concrete with a
water/cement ratio of 0.50
(From: W. H. PRICE, Factors influencing concrete strength, *J.
Amer. Concr. Inst.*, 47, pp. 417–32 (Feb. 1951).)

Self-desiccation occurs in sealed concrete when the water/cement ratio is
less than about 0.5 (see Chapter 2), because the internal relative
humidity in the capillaries decreases below the minimum value necessary
for hydration to take place, viz. 80 per cent.

It must be stressed that for a satisfactory development of strength it is
not necessary for *all* the cement to hydrate, and indeed this is only rarely
achieved in practice. If, however, curing proceeds until the capillaries in
the hydrated cement paste have become segmented (see page 113), then
the concrete will be impermeable (as well as of satisfactory strength) and
this is vital for good durability. To achieve this condition, evaporation of
water from the concrete surface has to be prevented. Evaporation in the
early stages after placing depends on the temperature and relative
humidity of the surrounding air and on the velocity of wind which effects
a change of air over the surface of the concrete. As stated in Chapter 9,
evaporation rates greater than 0.5 kg/m² per h (0.1 lb/ft² per h) have to
be avoided (see Fig. 9.2).

Methods of curing

No more than an outline of the different means of curing will be given here as the actual procedures used vary widely, depending on the conditions on site and on the size, shape, and position of the concrete in question.

In the case of concrete members with a small surface/volume ratio, curing may be aided by oiling and wetting the forms before casting. The forms may be left in place for some time and, if of appropriate material, wetted during hardening. If they are removed at an early age, the concrete should be sprayed and wrapped with polythene sheets or other suitable covering.

Large horizontal surfaces of concrete, such as highway slabs, present a more serious problem. In order to prevent crazing of the surface on drying out, loss of water must be prevented even prior to setting. As the concrete is at that time mechanically weak it is necessary to suspend a covering above the concrete surface. This protection is required only in dry weather, but may also be useful to prevent rain marring the surface of fresh concrete.

Once the concrete has set, wet curing can be provided by keeping the concrete in contact with water. This may be achieved by spraying or flooding (ponding), or by covering the concrete with wet sand, earth, sawdust or straw. Periodically-wetted hessian or cotton mats can be used, or alternatively an absorbent covering with access to water can be placed over the concrete. A continuous supply of water is naturally more efficient than an intermittent one, and Fig. 10.2 compares the strength development of concrete cylinders whose top surface was flooded during the first 24 hours with that of cylinders covered with wet hessian. The difference is greatest at low water/cement ratios where self-desiccation operates rapidly.

Another means of curing is to seal the concrete surface by an impermeable membrane or by waterproof reinforced paper or by plastic sheets. A membrane, provided it is not punctured or damaged, will effectively prevent evaporation of water from the concrete but will not allow ingress of water to replenish that lost by self-desiccation. The membrane is formed by sealing compounds applied in liquid form by hand or by spraying after the free water has disappeared from the surface of the concrete but before the pores in the concrete dry out so that they can absorb the compound. The membrane may be clear, white or black. The opaque compounds have the effect of shading the concrete, and light colour leads to a lower absorption of heat from the sun, and hence to a smaller rise in the temperature of the concrete. The effectiveness (as measured by the strength of the concrete) of a white membrane and of white translucent sheets of polyethylene is the same. ASTM C 309–81 prescribes membrane curing compounds, and ASTM C 171–69 (re-approved 1980) prescribes plastic and reinforced paper sheet materials for curing. The tests for the efficiency of curing materials are prescribed by ASTM C 156–80a. To comply with the specifications for highway and bridge works, BS 8110: Part 1: 1985 requires a *curing efficiency* of 75 per

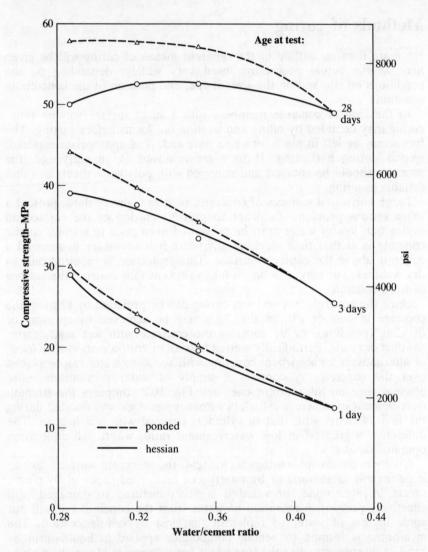

Fig. 10.2: Influence of curing conditions on strength of test cylinders
(From: P. KLIEGER, Early high strength concrete for prestressing,
Proc. of World Conference on Prestressed Concrete, pp. A5-1–A5-
14 (San Francisco, July 1957).)

cent for curing membranes. The curing efficiency is assessed by comparing the loss of moisture from a sealed specimen with the loss from an unsealed specimen made and cured under prescribed conditions.

Except when used on concrete with a high water/cement ratio, sealing membranes reduce the degree and rate of hydration compared with efficient wet curing. However, wet curing is often applied only intermittently so that in practice sealing may lead to better results than would otherwise be achieved. Reinforced paper, once removed, does not interfere with the adhesion of the next lift of concrete, but the effects of membranes in this respect have to be ascertained in each case. Plastic sheeting can cause discoloration or mottling because of non-uniform condensation of water on the underside of the sheet. To prevent this condition, and therefore loss of water, when plastic sheets are used they must rest tightly against the concrete surface.

The period of curing cannot be prescribed in a simple way but, if the temperature is above 10 °C (50 °F), ACI 308–81 lays down a minimum of 3 days for rapid-hardening Portland (Type III) cement, a minimum of 7 days for ordinary Portland (Type I) cement, and a minimum of 14 days for low-heat Portland (Type IV) cement. However, the temperature also affects the length of the required period of curing and BS 8110: Part 1: 1985 lays down the normal curing periods for different cements and exposure conditions as given in Table 10.1; when the temperature falls

Table 10.1: **Minimum period of protection required for different cements and curing conditions, as prescribed by BS 8110: Part 1: 1985**

Type of cement	Minimum period of curing and protection (days) for average surface temperature of concrete:		
	between 5 and 10 °C (41 and 50 °F)	above 10 °C (50 °F)	between* 5 and 25 °C (41 and 77 °F)
	(1) Damp/unprotected (relative humidity ≥80 per cent) protected from sun and wind		
All types	No special requirements		
	(2) Intermediate between conditions (1) and (3)		
Ordinary Portland (Type I) Sulphate-resisting (Type V) Rapid-hardening (Type III)	4	3	$\dfrac{60}{t+10}$
Other cements with or without slag of PFA	6	4	$\dfrac{80}{t+10}$
	(3) Dry/unprotected (relative humidity ≤50 per cent) unprotected from sun and wind		
Ordinary Portland (Type I) Sulphate-resisting (Type V) Rapid-hardening (Type III)	6	4	$\dfrac{80}{t+10}$
Other cements with or without slag or PFA	10	7	$\dfrac{140}{t+10}$

* The formulae given can be used to calculate the minimum period of protection; t is temperature (°C).

181

below 5 °C (41 °F), special precautions are necessary (see Chapter 9). ACI Standard 308–81 also gives extensive information on curing. Striking times for formwork are given in a British publication by the Construction Industry Research and Information Association (CIRIA) (see Bibliography).

High-strength concrete should be cured at an early age because partial hydration may make the capillaries discontinuous: on renewal of curing, water would not be able to enter the interior of the concrete and no further hydration would result. However, mixes with high water/cement ratios always retain a large volume of continuous capillaries so that curing can be effectively resumed later on. Nevertheless, it is advisable to commence curing as soon as possible because in practice early drying may lead to shrinkage and cracking (see Chapter 13).

Influence of temperature

Generally, the higher the temperature of the concrete at placement the greater the initial rate of strength development, but the lower the long-term strength. This is why it is important to reduce the temperature of fresh concrete when concreting in hot climates (see Chapter 9). The explanation is that a rapid initial hydration causes a non-uniform distribution of the cement gel with a poorer physical structure, which is probably more porous than the structure developed at normal temperatures. With a high initial temperature, there is insufficient time available for the products of hydration to diffuse away from the cement grains and for a uniform precipitation in the interstitial space. As a result, a concentration of hydration products is built up in the vicinity of the hydrating cement grains, a process which retards subsequent hydration and, thus, the development of longer-term strength.

The influence of the curing temperature on strength is illustrated in Fig. 10.3, which clearly indicates a higher initial strength development, but a lower 28-day strength, as the temperature increases. It should be noted that for the tests reported in this Figure the temperature was kept constant up to and including testing. However, when the concrete was cooled to 20 °C (68 °F) over a period of two hours prior to testing, only temperatures above 65 °C (150 °F) had a deleterious effect (Fig. 10.4). Thus, the temperature at the time of testing also appears to affect strength.

The results of Figs 10.3 and 10.4 are for neat ordinary Portland (Type I) cement compacts, but a similar influence of temperature occurs with concrete. Figure 10.5 shows that a higher temperature produces a higher strength during the first day, but for the ages of 3 to 28 days, the situation changes radically: for any given age, there is an optimum temperature which produces a maximum strength, but this optimum temperature decreases as the period of curing increases. With ordinary (Type I) or modified Portland (Type II) cement, the optimum temperature to produce a maximum 28-day strength is approximately 13 °C (55 °F). For

182

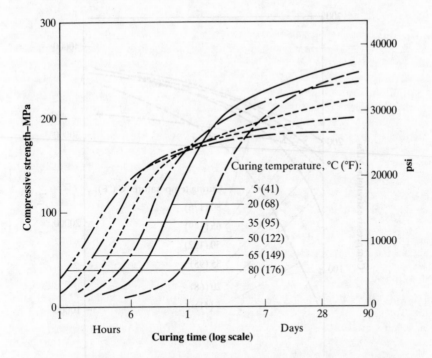

Curing temperature, °C (°F):

5 (41)
20 (68)
35 (95)
50 (122)
65 (149)
80 (176)

Hours   Curing time (log scale)   Days

Fig. 10.3: Relation between compressive strength and curing time of neat
cement paste compacts at different curing temperatures. The
temperature of the specimens was kept constant up to and including
the period of testing
(From: CEMENT AND CONCRETE ASSOCIATION, Research
and development – Research on materials, *Annual Report,* pp.
14–19 (Slough 1976).)

rapid-hardening Portland (Type III) cement, the corresponding optimum
temperature is lower. It is interesting to note that even concrete cast at
4 °C (40 °F) and stored at a temperature below the freezing point of water
is capable of hydration (Fig. 10.5). Furthermore, when the same concrete
was stored at 23 °C (73 °F) beyond 28 days, its strength at three months
exceeded that of similar concrete stored continuously at 23 °C (73 °F), as
shown in Fig. 9.1.

The observations so far pertain to concrete made in the laboratory, and
it seems that the behaviour on site in a hot climate may not be the same.
Here, there are some additional factors acting: ambient humidity, direct
radiation of the sun, wind velocity, and method of curing; these factors
were already mentioned in Chapter 9. It should be remembered, too,
that the quality of concrete depends on *its* temperature and not on that of
the surrounding atmosphere, so that the size of the member is also a
factor because of the heat of hydration of cement. Likewise, curing by
flooding in windy weather results in a loss of heat due to evaporation so

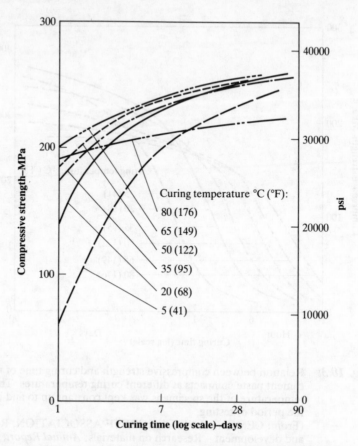

Fig. 10.4: Relation between compressive strength and curing time of neat cement paste compacts at different curing temperatures. The temperature of the specimens was moderated to 20 °C (68 °F) at a constant rate over a two-hour period prior to testing (water/cement ratio = 0.14; ordinary Portland (Type I) cement)
(From: CEMENT AND CONCRETE ASSOCIATION, Research and development – Research on materials, *Annual Report*, pp. 14–19 (Slough 1976).)

that the temperature of concrete is lower, and hence the strength greater, than when a sealing compound is used. Evaporation immediately after casting is also beneficial for strength of high water/cement ratio mixes because water is drawn out of the concrete while capillaries can still collapse, so that the effective water/cement ratio and porosity are decreased. If, however, evaporation is allowed to lead to the drying of the surface, plastic shrinkage and cracking may result.

In general terms, however, concrete cast and made in summer can be expected to have a lower strength than the same mix cast in winter.

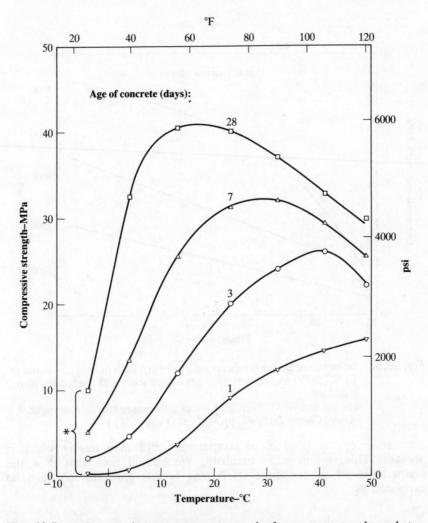

Fig. 10.5: Influence of temperature on strength of concrete cast and cured at
the temperature indicated
(Based on: P. KLIEGER, Effect of mixing and curing temperature
on concrete strength, *J. Amer. Concr. Inst.*, 54, pp. 1063–81 (June
1958).)
* Concrete cast at 4 °C (39 °F) and cured at −4 °C (25 °F) from the
age of 1 day

Maturity 'rule'

In the previous section, we considered the beneficial effect of temperature on the gain of strength, but pointed out the need for an initial curing
period at normal temperature. Figure 10.6 shows some typical data.

The temperature effect is cumulative and can be expressed as a

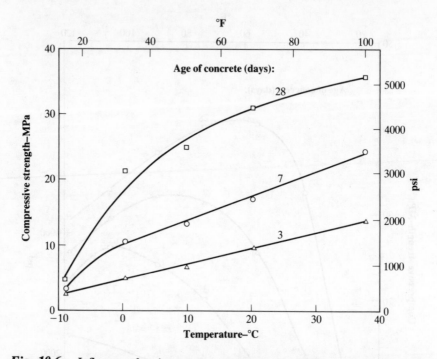

Fig. 10.6: Influence of curing temperature on the strength of concrete cured at
10 °C (50 °F) for the first 24 hours before storing at the temperature
indicated
(Based on: W. H. PRICE, Factors influencing concrete strength, *J.
Amer. Concr. Inst.*, 47, pp. 417–32 (Feb. 1951).)

summation of the product of temperature and time during which it
prevails. This is known as maturity. We should note that it is the
temperature of the concrete itself that is relevant. Maturity can be
expressed as

$$M = \sum T . \Delta t$$

where Δt is time interval (usually in days) during which the temperature
is T, and T is the temperature measured from a datum of -11 °C (12 °F),
which is the temperature below which strength development ceases.
Thus, for 30 °C (86 °F), $T = 41$ °C (74 °F).

Hence, the units of maturity are °C days (°F days) or °C h (°F h).
Figure 10.7 shows the data of Fig. 10.6 with the strength expressed as a
function of maturity. If maturity is plotted on a logarithmic scale, the
relation beyond the initial period is approximately linear (Fig. 10.8).

This maturity 'rule' can be of practical use in estimating the strength of
concrete. However, the relation between strength and maturity depends
on the actual cement used, on the water/cement ratio, and on whether
any loss of water takes place during curing. Moreover, the harmful effect

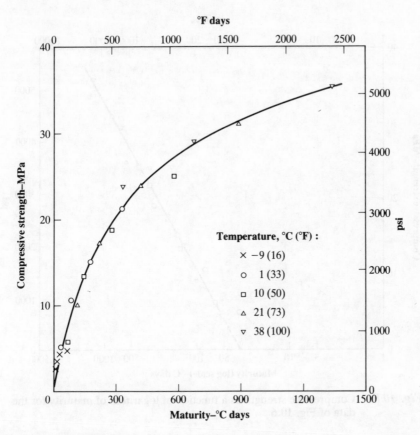

°F days

Fig. 10.7: Compressive strength as a function of maturity for the data of Fig. 10.6

of an early high temperature vitiates the maturity 'rule'. For these reasons, the maturity approach is not widely used and is useful only within a well-defined concreting system.

Steam curing

Since an increase in the curing temperature of concrete increases its rate of development of strength, the gain of strength can be accelerated by curing in steam. Steam at atmospheric pressure, i.e. when the temperature is below 100 °C (212 °F) is wet so that the process can be regarded as a special case of moist curing, and is known as steam curing. High-pressure steam curing, known as *autoclaving,* is an entirely different operation and is outside the scope of this book.

The primary object of steam curing is to obtain a sufficiently high early strength so that the concrete products may be handled soon after casting:

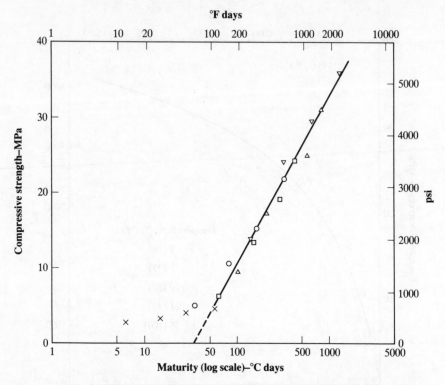

Fig. 10.8: Compressive strength as a function of logarithm of maturity for the data of Fig. 10.6

the forms can be removed, or the prestressing bed vacated, earlier than would be the case with normal moist curing, and less storage space is required: all these mean an economic advantage.

Because of the nature of operations required in steam curing, the process is mainly used with precast products. Steam curing is normally applied in special chambers or in tunnels through which the concrete members are transported on a conveyor belt. Alternatively, portable boxes or plastic covers can be placed over precast members, steam being supplied through flexible connections.

Owing to the adverse influence of temperature during the early stages of hardening on the later strength (see Fig. 10.9), a rapid rise in temperature must not be permitted. The adverse effect is more pronounced the higher the water/cement ratio of the mix, and is also more noticeable with rapid-hardening (Type III) than with ordinary Portland (Type I) cement. A delay in the application of steam curing is advantageous as far as later strength is concerned: a higher temperature requires a longer delay; in that case the strength–maturity relation is followed. However, in some cases, the later strength may be of lesser importance than the early strength requirement.

Practical curing cycles are chosen as a compromise between the early

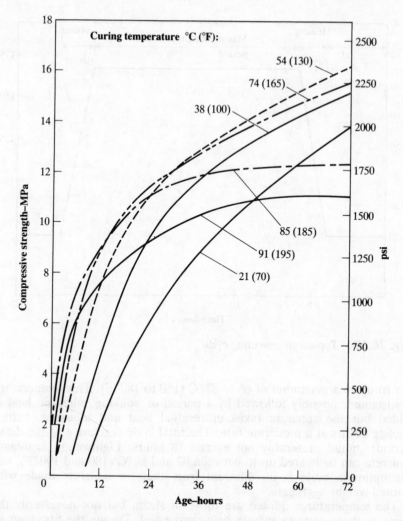

Fig. 10.9: Strength of concrete cured in steam at different temperatures (water/cement ratio = 0.50; steam curing applied immediately after casting)
(From: US BUREAU OF RECLAMATION, *Concrete Manual*, 8th Edn. (Denver, Colorado, 1975).)

and late strength requirements but are governed also by the time available (e.g. length of work shifts). Economic considerations determine whether the curing cycle should be suited to a given concrete mix or alternatively whether the mix ought to be chosen so as to fit a convenient cycle of steam curing. While details of an optimum curing cycle depend on the type of concrete product treated, a typical cycle would be that shown in Fig. 10.10. After a delay period (normal moist curing) of 3 to 5 hours, the temperature is raised at a rate of 22 to 33 °C per h (40 to 60 °F

189

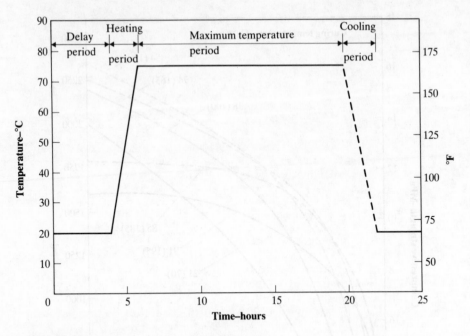

Fig. 10.10: Typical steam-curing cycle

per h) up to a maximum of 66 to 82 °C (150 to 180 °F). This temperature is sustained, possibly followed by a period of 'soaking' when no heat is added but the concrete takes in residual heat and moisture, before cooling occurs at a moderate rate. The total cycle (exclusive of the delay period) should preferably not exceed 18 hours. Lightweight aggregate concrete can be heated up to between 82 and 88 °C (180 and 190 °F), but the optimum cycle is no different from that for concrete made with normal weight aggregate.

The temperatures quoted are those of steam but not necessarily the same as those of the concrete being processed. During the first hour or two after placing in the curing chamber, the temperature of the concrete lags behind that of the air but later on, due to the heat of hydration of cement, the temperature of the concrete exceeds that of the air. Maximum use can be made of the heat stored in the chamber if the steam is shut off early and a prolonged curing period is allowed. A slow rate both of heating and of cooling is desirable as otherwise high temperature gradients within the concrete would cause internal stresses, possibly leading to cracking by thermal shock. This means that if the delay period is reduced then a slower rate of heating must be applied, not only to guard against thermal shock but also to ensure a satisfactory later strength.

Steam curing should never be used with high alumina cement because of the deleterious effect of hot-wet conditions on the strength of that cement (see page 34).

190

Bibliography

10.1 ACI COMMITTEE 308–81, Standard Practice for curing concrete, Part 2, *ACI Manual of Concrete Practice,* 1990.

10.2 ACI COMMITTEE 517.2R–87, Accelerated curing of concrete at atmospheric pressure – State of the Art, Part 5, *ACI Manual of Concrete Practice,* 1990.

10.3 T. A. HARRISON, Tables of minimum striking times for soffit and vertical formwork, *CIRIA Report* 67, pp. 23 (London, Construction Industry Research and Information Assoc., Oct. 1977).

Problems

10.1 What is the effect of temperature during the first 24 hours on the 28-day strength of concrete?

10.2 Compare the strength development of concrete stored moist from the time of de-moulding at 24 hours at 5 °C (41 °F) and 40 °C (104 °F).

10.3 What are the disadvantages of membrane curing compared with water curing?

10.4 Describe the positive features of membrane curing.

10.5 Describe a typical temperature cycle for steam curing.

10.6 What are the various temperature limits in a cycle of steam curing?

10.7 What is meant by self-desiccation and when does it occur?

10.8 What is meant by curing of concrete?

10.9 Why is curing important?

10.10 What are the limitations on the prediction of strength of concrete from its maturity?

10.11 Define maturity of concrete.

10.12 'The length of curing should be sufficient to produce impermeable concrete.' Discuss this statement.

10.13 Why does steam curing include a cooling period?

10.14 Explain the term curing efficiency.

10.15 What is meant by autoclaving concrete?

10.16 Compare the mechanism of hydration at high temperatures with that at normal temperature.

10.17 Give some advantages of steam curing.

10.18 What are the limitations on the validity of the maturity expression?

10.19 With what materials should steam curing never be used?

10.20 The relation between strength and maturity for a concrete is known to be as follows:

in SI units: $f_c = -33 + 21 \log_{10} M$
in US units: $f_c = -5570 + 3047 \log_{10} M$

Calculate the strength when the concrete is cured at 30 °C (86 °F) for 7 days. What temperature would be required to reach a strength of 30 MPa (4400 psi) at 28 days?

Answer: 18.6 MPa (2700 psi)
25°C (76°F)

11

Other strength properties

The title of this chapter refers to the strength of concrete under various types of loading other than static compression. In the design of structures, concrete is exploited so as not to rely on its tensile strength, which is low. Clearly, however, tensile stresses cannot be avoided. They are connected with shear and are generated by differential movements, such as shrinkage, which often result in cracking and impairment of durability. Consequently, we need to know how the tensile strength relates to compressive strength.

In some structures, repeated loading is applied, and here knowledge of fatigue strength is required. Strength under impact loading may also be of interest. In other circumstances, concrete surfaces are subjected to wear so that resistance to abrasion is of importance.

Finally, since most structural concrete contains steel reinforcement, the strength of bond between the two materials must be satisfactory in order to maintain the integrity of the structure.

Relation between tensile and compressive strengths

In the discussion of strength in Chapter 6, the theoretical compressive strength was stated to be eight times larger than the tensile strength. This implies a fixed relation between the two strengths. In fact, there is a close relation but not a direct proportionality: the ratio of the two strengths depends on the general *level* of strength of the concrete. Generally, the ratio of tensile to compressive strengths is lower the higher the compressive strength. Thus, for example, the tensile strength increases with age at a lower rate than the compressive strength. However, there are several other factors which affect the relation between the two strengths, the main ones being the method of testing the concrete in tension, the size of the specimen, the shape and surface texture of coarse aggregate, and the moisture condition of the concrete.

It is difficult to test concrete in *direct* (uniaxial) tension because of the problem of gripping the specimen satisfactorily (so that premature failure does not occur near to the end attachment) and because there must be no eccentricity of the applied load. Direct tensile test is therefore not standardized and rarely used. The standards ASTM C 78–84, ASTM C 496–71 (reapproved 1979) and BS 1881: 1983 all prescribe alternative methods of determining the tensile strength: in flexure (*modulus of rupture*) and in indirect tension (*splitting*) (see Chapter 16).

The different test methods yield numerically different results, ordered as follows: direct tension < splitting tension < flexural tension. The reasons for this are twofold. First, with the usual size of laboratory specimen, the volume of concrete subjected to tensile stress decreases in the order listed above and, statistically, there is a greater chance of a 'weak element' and therefore of failure in a larger volume than in a smaller volume. Second, both the splitting and flexural test methods involve non-uniform stress distributions which impede the propagation of a crack and, therefore, delay the ultimate failure; on the other hand, in the direct test, the stress distribution is uniform so that, once a crack has formed, it can propagate quickly through the section of the specimen. Figure 11.1 shows typical values of tensile strength as a function of compressive strength for the different methods of testing.

There is little influence of the type of the aggregate on the direct and splitting tensile strengths, but the flexural strength of concrete is greater when angular crushed aggregate is used than with rounded natural gravel. The explanation is that the improved bond of crushed aggregate holds the material together but is ineffective in direct or indirect tension. Since the compressive strength is little affected by the shape and surface texture of the aggregate, the ratio of flexural strength to compressive strength is greater for angular crushed aggregate, especially at higher compressive strengths.

The moisture condition of concrete influences the relation between the flexural and compressive strengths. Figure 11.1 compares continuously wet-stored concrete with concrete cured wet and then stored in a dry environment. Under these circumstances, the compressive strength of drying concrete is greater than when continuously wet-stored; the splitting and direct tensile strengths are not affected in a similar manner. However, the flexural strength of drying concrete is lower than that of wet concrete, probably because of the sensitivity of this test to the presence of shrinkage cracks.

A number of empirical formulae have been suggested to relate tensile (f_t) and compressive (f_c) strengths. Most are of the type

$$f_t = k f_c^n$$

where k and n are coefficients which depend on the main factors discussed earlier and, of course, on the shape of the compression specimen (cube or cylinder). There is little value in giving here specific values of the coefficients because they are affected by the properties of the mix used. However, the expression used by ACI is of interest:

$$f_t = k \sqrt{f_c}.$$

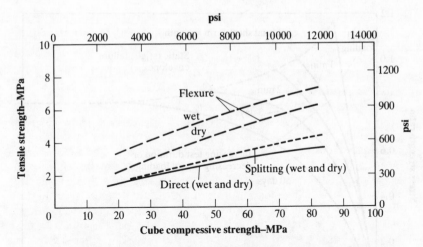

Fig. 11.1: Relation between tensile and compressive strengths of concrete
made with rounded coarse normal weight and lightweight aggregates
Flexural test: $100 \times 100 \times 500$ mm ($4 \times 4 \times 20$ in.) prisms,
Splitting test: 150×300 mm (6×12 in.) cylinders,
Direct test: 75×355 mm (3×14 in.) bobbins,
Compression test: 100 mm (4 in.) cubes.

Fatigue strength

Two types of failure in fatigue can take place in concrete. In the first,
failure occurs under a *sustained* load (or a slowly increasing load) near,
but below, the strength under an increasing load, as in a standard test;
this is known as *static fatigue* or *creep rupture*. The second type of failure
occurs under *cyclic* or repeated loading, and is known simply as fatigue.
In both instances, a time-dependent failure occurs only at stresses which
are greater than a certain threshold value but smaller than the short-term
static strength.

It may be appropriate at this stage to note that, in the standard
method, the compressive strength is determined in a test of short
duration, viz. 2 to 4 min. The duration of the test is important because
strength is dependent upon the rate of loading. This is why both
BS 1881: Part 116: 1983 and ASTM C 39–83b prescribe standard rates of
loading for the determination of compressive strength (see Chapter 16).

Figure 11.2 shows that, as the rate of loading decreases (or as the
duration of test increases), the application of a steadily increasing load
leads to a lower recorded strength than in the case of the standard test.
If, on the other hand, the load is applied extremely rapidly (or
instantaneously), a higher strength is recorded and the strain at failure
(*strain capacity*) is smaller. It follows that, at rapid rates of loading,
concrete appears more brittle in nature than under lower rates of loading
when creep (see Chapter 13) and microcracking increase the strain
capacity.

195

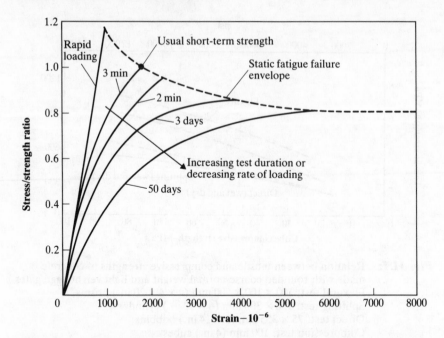

Fig. 11.2: Influence of test duration (or rate of loading) on strength and on strain capacity in compression
(Based on H. RÜSCH, Researches toward a general flexural theory for structural concrete, *ACI Journal,* Vol. 57, No. 1, pp. 1–28, (July 1960).)

Under low rates of loading, static fatigue occurs when the stress exceeds about 70 to 80 per cent of the short-term strength; this threshold value represents the onset of rapid development of microcracks, which eventually link and cause failure. Thus, when the stress exceeds the threshold value, concrete will fail after a period which is indicated by the failure envelope of Fig. 11.2.

A similar phenomenon takes place under a sustained load (see Fig. 11.3); here, a certain load is applied fairly quickly and then held constant. Above the same threshold of about 70 to 80 per cent of the short-term strength, the sustained load will eventually result in failure. At stresses below the threshold, failure will not occur and the concrete will continue to creep (see Chapter 12).

Static fatigue also occurs in tension at stresses greater than 70 to 80 per cent of the short-term strength but, of course, the tensile strain capacity is much lower than in compression.

Let us now consider the behaviour of concrete when a compressive stress alternates between zero and a certain fraction of the short-term static strength. Figure 11.4 shows that there is a change in the shape of the stress–strain curves under an increasing and decreasing load as the number of load cycles increases. Initially, the loading curve is concave toward the strain axis, then straight, and eventually concave toward the

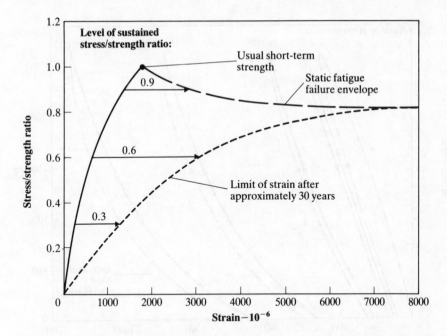

Fig. 11.3: Influence of sustained stress on strength and on strain capacity of
concrete in compression
(Based on H. RÜSCH, Researches toward a general flexural theory
for structural concrete, *ACI Journal*, Vol. 57, No. 1, pp. 1–28, (July
1960).)

stress axis. The extent of this latter concavity is reflected by an increase in
the elastic strain and, hence, by a decrease in the secant modulus of
elasticity (see Chapter 12), a feature which is an indication of how near
the concrete is to failure by fatigue.

The area enclosed by two successive curves on loading and unloading is
proportional to *hysteresis* and represents the irreversible energy of
deformation, i.e. energy due to crack formation or irreversible creep (see
Chapter 12). On first loading to a high stress, the hysteresis is large but
then decreases as the number of cycles increases. When fatigue failure is
approaching there is extensive cracking at the aggregate interfaces and,
consequently, hysteresis and non-elastic strain increase rapidly. As in
static fatigue, the non-elastic strain at fatigue failure and, hence, the
strain capacity are much larger than in short-term failure.

Let us now turn to fatigue. For a constant range of alternating stress,
the fatigue strength decreases as the number of cycles increases. This
statement is illustrated in Fig. 11.5 by the so-called *S–N curves*, where *S*
is the ratio of the maximum stress to the short-term static strength and *N*
is the number of cycles at failure. The maximum value of *S* below which
no failure occurs is known as *endurance limit*. Whereas mild steel has an
endurance limit of about 0.5, which means that when $S < 0.5$, *N* is
infinity, concrete does not appear to have a corresponding limit.

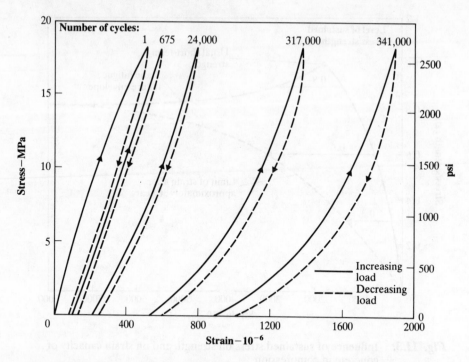

Fig. 11.4: Stress–strain relation of concrete under cyclic compressive loading

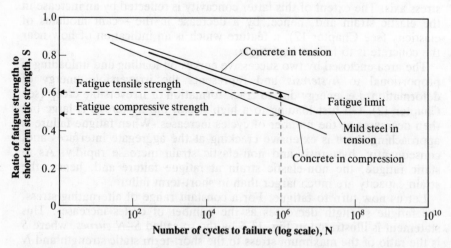

Fig. 11.5: Typical relations between the fatigue strength and the number of cycles for concrete and mild steel with a minimum stress of zero

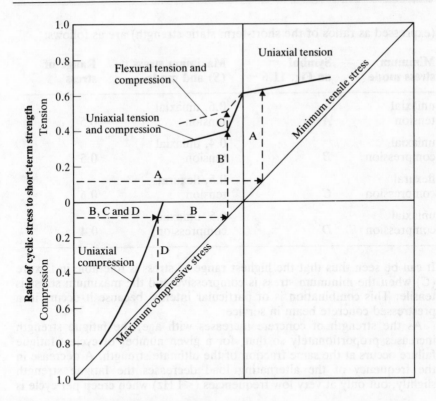

Fig. 11.6: Modified Goodman diagram for fatigue strength of concrete in uniaxial tension, flexure and compression after 1 million cycles; symbols indicate fatigue strength for a minimum load of 0.1 of the short-term static strength. Symbols are defined on page 200 (Based on H. A. W. CORNELISSEN, Fatigue of concrete in tension, *Heron,* Vol. 29, No. 4, pp. 68 (1984).)

Therefore, it is necessary to define the fatigue strength of concrete by considering a very large number of cycles, say, 1 million (see Fig. 11.5). In reality, the *S–N* curves for concrete have a very large scatter due to the uncertainty of the short-term strength of the actual fatigue specimen and due to the stochastic nature of fatigue. This means that, for a given cycle of stress, it is difficult precisely to determine the number of cycles to failure.

The effect of a change in the range of stress on the fatigue strength is represented by a *modified Goodman diagram,* as shown in Fig. 11.6. Here, the ordinate, measured from a line at 45° through the origin, indicates the range of stress to cause failure after 1 million cycles. For practical purposes, the lower load is the dead load and the upper load is that due to the dead plus live (transient) load. Figure 11.6 gives the fatigue strength for various combinations of compressive and tensile loads. For example, when the minimum stress is 0.1 of the short-term static strength, the corresponding maximum stress and range of stress (expressed as ratios of the short-term static strength) are as follows:

Minimum stress mode	Symbol on Fig. 11.6	Maximum stress (S) and mode	Range of stress
uniaxial tension	A	0.6, uniaxial tension	0.5
uniaxial compression	B	0.4, uniaxial tension	0.5
flexural compression	C	0.5, flexural tension	0.6
uniaxial compression	D	0.5, uniaxial compression	0.4

It can be seen thus that the highest range of stress is tolerable in flexure (C) when the minimum stress is compressive and the maximum stress is tensile. This combination is of particular interest because it occurs in a prestressed concrete beam in service.

As the strength of concrete increases with age, the fatigue strength increases proportionately so that, for a given number of cycles, fatigue failure occurs at the same fraction of the ultimate strength. A decrease in the frequency of the alternating load decreases the fatigue strength slightly, but only at very low frequencies (<1 Hz) when creep per cycle is

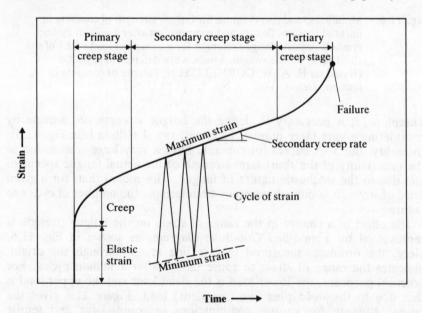

Fig. 11.7: Definition of secondary creep rate for assessing partial damage in fatigue

significant. The moisture condition of the concrete affects the fatigue strength only marginally except perhaps in the case of very dry concrete, which has a slightly higher fatigue strength than wet concrete.

So far, we have considered the fatigue strength of concrete subjected to a constant range of stress. In a practical situation, the range may vary, and it is the whole history of cyclic loading that is relevant. *Miner's rule* assumes that failure will occur if the total damage contribution M accumulated from a history of loadings is equal to unity. The individual damage contribution of one particular loading is equal to the ratio of the number of cycles, n_i, at a given cycle of stress i to the number of cycles to failure, N_i, at that stress. Thus,

$$M = \frac{n_1}{N_1} + \frac{n_2}{N_2} + \ldots + \frac{n_k}{N_k} = \sum_{i=1}^{k} \frac{n_i}{N_i} = 1.$$

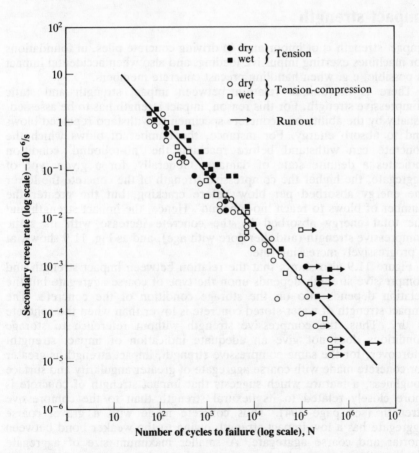

Fig. 11.8: Relation between the secondary creep rate and the number of cycles to failure
(From: H. A. W. CORNELISSEN, Fatigue of concrete in tension, *Heron,* Vol. 29, No. 4, pp. 68 (1984).)

For any cyclic stress i, the value of N_i is obtained from the appropriate $S-N$ curve. It is here that lies the weakness of Miner's rule: there is a large scatter associated with the $S-N$ curves, and, in consequence, Miner's rule is not very accurate.

An alternative approach is to relate the number of cycles to failure to the secondary creep rate (see Figs 11.7 and 11.8), which is a measure of the actual partial damage in fatigue. Although the scatter is much reduced, the time-dependent strain has to be monitored continuously. Application of Miner's rule on the basis of the secondary creep rate appears to be more satisfactory in that M is only slightly greater than unity and, therefore, $M = 1$ appears to be a safe failure criterion for design.

Impact strength

Impact strength is of importance in driving concrete piles, in foundations for machines exerting impulsive loading, and also when accidental impact is possible, e.g. when handling precast concrete members.

There is no unique relation between impact strength and static compressive strength. For this reason, impact strength has to be assessed, usually by the ability of a concrete specimen to withstand repeated blows and to absorb energy. For instance, the number of blows which the concrete can withstand before reaching the 'no-rebound' condition indicates a definite state of damage. Generally, for a given type of aggregate, the higher the compressive strength of the concrete the lower the energy absorbed per blow before cracking, but the greater the number of blows to reach 'no-rebound'. Hence, the impact strength and the total energy absorbed by the concrete increase with its static compressive strength (and therefore with age), and as Fig. 11.9 shows, at a progressively increasing rate.

Figure 11.9 also shows that the relation between impact strength and compressive strength depends upon the type of coarse aggregate but the relation depends also on the storage condition of the concrete. The impact strength of water-stored concrete is lower than when the concrete is dry. Thus, the compressive strength without reference to storage conditions does not give an adequate indication of impact strength. Moreover, for the same compressive strength, impact strength is greater for concrete made with coarse aggregate of greater angularity and surface roughness, a feature which suggests that impact strength of concrete is more closely related to its flexural strength than to the compressive strength (see page 194). Thus concrete made with a gravel coarse aggregate has a low impact strength owing to the weaker bond between mortar and coarse aggregate. A smaller maximum size of aggregate significantly improves the impact strength; so does the use of aggregate with a low modulus of elasticity and low Poisson's ratio. To provide a satisfactory impact strength, a cement content below $400 \, \text{kg/m}^3$ ($670 \, \text{lb/yd}^3$) is advantageous.

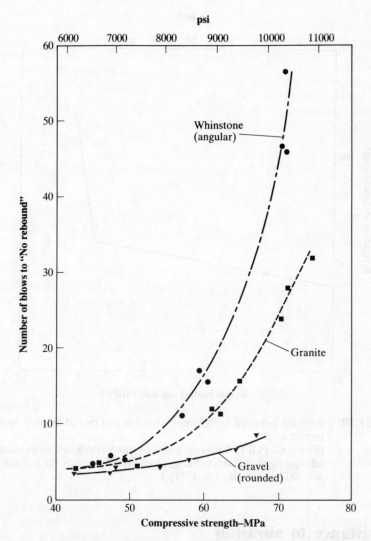

Fig. 11.9: Relation between compressive strength and number of blows to
'no-rebound' for concretes made with different aggregates and
ordinary Portland (Type I) cement, stored in water
(From: H. GREEN, Impact strength of concrete, *Proc. Inst. C. E.*,
28, pp. 383–96 (London, July, 1964); Building Research Establishment, Crown copyright.)

Impact loading can be considered as the application of a uniform stress
extremely rapidly, in which case the strength as measured is higher.
Figure 11.10 shows that the strength increases greatly when the rate of
application of stress exceeds about 500 GPa/s, reaching at 5 TPa/s more
than double the static compressive strength at normal rates of loading
(about 0.25 MPa/s).

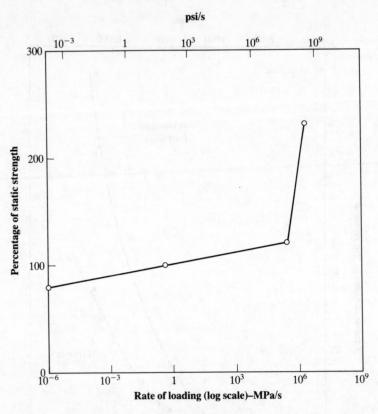

Fig. 11.10: Relation between compressive strength and rate of loading up to impact level
(From: C. POPP, Untersuchen über das Verhalten von Beton bei schlagartigen Beanspruchung, *Deutscher Ausschuss für Stahlbeton*, *No.* 281, pp. 66 (Berlin, 1977).)

Resistance to abrasion

Concrete surfaces can be subjected to various types of abrasive wear. For example, sliding or scraping can cause *attrition* and, in the case of hydraulic structures, the action of abrasive solids carried by water generally leads to *erosion* of concrete. For these reasons, it may be desirable to know the resistance of concrete to abrasion. However, this is difficult to assess as the damaging action varies depending on the cause of wear, and no one test procedure is satisfactory in evaluating the resistance of concrete to the various conditions of wear.

In all the tests, the depth of wear of a specimen is used as a measure of abrasion. ASTM C 779–82 prescribes three test procedures for laboratory or field use. In the *revolving disc test,* three flat surfaces revolve along a circular path at 0.2 Hz, each plate turning on its axis at 4.7 Hz; silicon

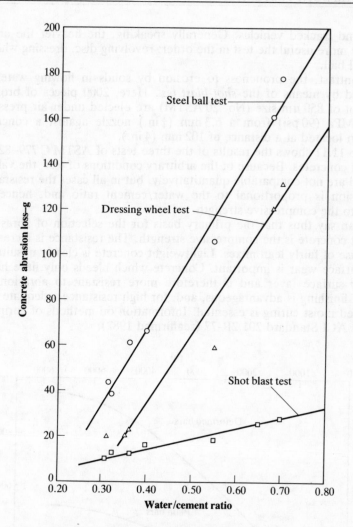

Fig. 11.11: Influence of the water/cement ratio of the mix on the abrasion loss of concrete for different tests
(From: F. L. SMITH, Effect of aggregate quality on resistance of concrete to abrasion, *ASTM Sp. Tech. Publicn. No. 205,* pp. 91–105 (1958).)

carbide is fed as an abrasive material between the plates and the concrete. In the *steel ball abrasion test,* a load is applied to a rotating head which is separated from the concrete by steel balls; the test is performed in circulating water in order to remove the eroded material. The *dressing wheel test* uses a drill press modified to apply a load to three sets of seven rotating dressing wheels which are in contact with the specimen; the driving head is rotated for 30 min at 0.93 Hz.

The tests prescribed by ASTM C 779–82 are useful in estimating the resistance of concrete to heavy foot traffic, to wheeled traffic, and to tyre chains and tracked vehicles. Generally speaking, the heavier the abrasion, the more useful the test in the order: revolving disc, dressing wheel, and steel ball.

By contrast, the proneness to erosion by solids in flowing water is measured by means of the *shot-blast test*. Here, 2000 pieces of broken steel shot of 850 μm size (No. 20 ASTM) are ejected under air pressure of 0.62 MPa (90 psi) from a 6.3 mm (¼ in.) nozzle against a concrete specimen located at a distance of 102 mm (4 in.).

Figure 11.11 shows the results of the three tests of ASTM C 779–82 on different concretes. Because of the arbitrary conditions of test, the values obtained are not comparable quantitatively, but in all cases the resistance to abrasion is proportional to the water/cement ratio and, hence, is related to the compressive strength.

We can say thus that the primary basis for the selection of abrasion-resistant concrete is the compressive strength. The resistance is increased by the use of fairly lean mixes. Lightweight concrete is clearly unsuitable when surface wear is important. Concrete which bleeds only little has a stronger surface layer and is therefore more resistant to abrasion. A delay in finishing is advantageous, and, for high resistance, adequate and prolonged moist curing is essential. Information on methods of curing is given in ACI Standard 201.2R-77 (reaffirmed 1982).

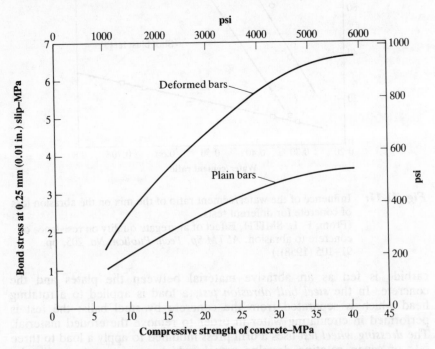

Fig. 11.12: Influence of the strength of concrete on bond determined by pull-out test
(From: W. H. PRICE, Factors influencing concrete strength, *J. Amer. Concr. Inst.*, 47, pp. 417–32 (Feb. 1951).)

Bond to reinforcement

The strength of bond between steel reinforcement and concrete arises primarily from friction and adhesion. Bond is affected by the properties both of steel and of concrete, and by the relative movements due to volume changes (e.g. shrinkage of concrete).

In general terms, bond strength is approximately proportional to the compressive strength of concrete up to about 20 MPa (3000 psi). For higher strengths of concrete, the increase in bond strength becomes progressively smaller and eventually negligible, as shown in Fig. 11.12.

The bond strength is not easily defined. ASTM C 234–71 (reapproved 1977) describes a pull-out test in which a 19 mm ($\frac{3}{4}$ in.) deformed bar is embedded in a 150 mm (6 in.) cube. The bar is pulled relative to the concrete until the bond fails, the concrete splits or a minimum slip of 2.5 mm occurs at the loaded end of the bar. The bond strength is then taken as the load on the bar at failure divided by the nominal embedded surface area of the bar.

Protective surface treatment of reinforcement may reduce the bond strength, probably because in treated steel the advantage of surface rust in bond is absent. However, the bond of galvanized reinforcement has been shown to be at least as good as that of ordinary steel bars and wires. The thickness of the galvanized coating is usually between 0.03 and 0.10 mm (0.001 and 0.004 in.), and steel is protected against corrosion even when cover to reinforcement is reduced by, perhaps, as much as 25 per cent. Moreover, galvanizing permits the use of lightweight aggregate concrete without increased cover.

Bibliography

11.1   ACI COMMITTEE 215R–74 (revised 1986), Considerations for design of concrete structures subjected to fatigue loading, Part 1: Materials and General Properties of Concrete, *ACI Manual of Concrete Practice*, 1984.

11.2   H. A. W. CORNELISSEN, Fatigue of concrete in tension, *Heron*, Vol. 29, No. 4, pp. 68 (1984).

11.3   A. M. NEVILLE, *Properties of Concrete* (London, Longman 1981).

Problems

11.1   Discuss the relation between impact strength and compressive strength of concrete.

11.2   Discuss the relation between compressive strength and tensile strength of concrete.

11.3 What precautions should you take to ensure a good resistance of concrete to abrasion?

11.4 How would you measure the resistance of concrete to abrasion?

11.5 What is fatigue or endurance limit?

11.6 What are the main factors influencing the fatigue strength of concrete?

11.7 What is the difference between fatigue and static fatigue?

11.8 For a given concrete, compare the direct tensile strength, flexural strength and splitting strength. Explain why these values are different.

11.9 Has the type of aggregate any effect on the tensile strength of concrete?

11.10 Has the moisture condition of the concrete any effect on its tensile strength?

11.11 What is creep rupture?

11.12 What is the effect of the rate of loading on strength of concrete?

11.13 What is hysteresis?

11.14 How is fatigue strength affected by the number of cycles?

11.15 Explain the terms: $S-N$ curve, Goodman diagram, and Miner's rule.

11.16 Explain when tertiary creep occurs.

12

Elasticity and creep

To be able to calculate the deformation and deflection of structural members, we have to know the relation between stress and strain. In common with most structural materials, concrete behaves nearly elastically when load is first applied. However, under sustained loading, concrete exhibits creep, i.e. the strain increases with time under a constant stress, even at very low stresses and under normal environmental conditions of temperature and humidity. Steel, on the other hand, creeps only at very high stresses at normal temperature, or even at low stresses at very high temperatures, and in both cases a time-dependent failure occurs. In contrast, in concretes subjected to a stress below about 60 to 70 per cent of the short-term strength there is no creep rupture or static fatigue (see Chapter 11). Like concrete, timber also creeps under normal environmental conditions.

The importance of creep in structural concrete lies mainly in the fact that the creep deformation is of the same order of magnitude as the elastic deformation. There are also other effects of creep, most of them detrimental, but some beneficial.

Elasticity

Let us first categorize the elastic behaviour of concrete in terms of the various types of elastic behaviour of engineering materials. The definition of *pure elasticity* is that strains appear and disappear immediately on application and removal of stress. The stress–strain curves of Fig. 12.1 illustrate two categories of pure elasticity: (a) is linear and elastic, and (b) is non-linear and elastic. Steel conforms approximately to case (a) whilst some plastics and timber follow case (b). Brittle materials, such as glass and most rocks, are described as linear and non-elastic (case (c)) because separate linear curves exist for the loading and unloading branches of the stress–strain diagram, and a permanent deformation exists after complete removal of load. The fourth category (case (d) of Fig. 12.1) can be described as non-linear and non-elastic behaviour, a permanent deformation existing after removal of load; the area enclosed by the loading and

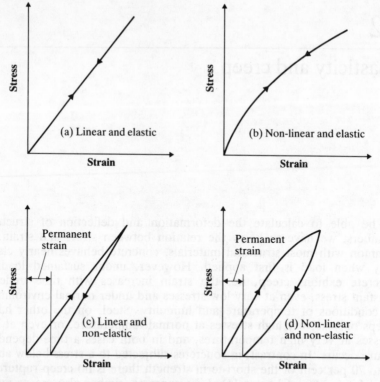

Fig. 12.1: Categories of stress–strain response

unloading curves represents the hysteresis (see page 197). This behaviour is typical of concrete in compression or tension loaded to moderate and high stresses but is not very pronounced at very low stresses.

The slope of the relation between stress and strain gives the modulus of elasticity, but the term Young's modulus can be applied strictly only to the linear categories of Fig. 12.1. However, we are concerned with determining the modulus of elasticity of concrete, and for this purpose let us consider Fig. 12.2, which is an enlarged version of Fig. 12.1(d). We can determine Young's modulus only for the initial part of the loading curve, but, when no straight portion of the curve is present, we can also measure the tangent to the curve at the origin. This is the *initial tangent modulus*. It is possible to find a *tangent modulus* at any point on the stress–strain curve, but this applies only to very small changes in load above or below the stress at which the tangent modulus is considered.

The magnitude of the observed strains and the curvature of the stress–strain relation depend to some extent on the rate of application of stress. When the load is applied extremely rapidly (<0.01 sec), recorded strains are greatly reduced and the curvature of the stress–strain curve becomes very small. An increase in loading time from 5 sec to about 2 min can increase the strain by up to 15 per cent, but within the range of 2 to about 10 min – a time normally required to test a specimen in an

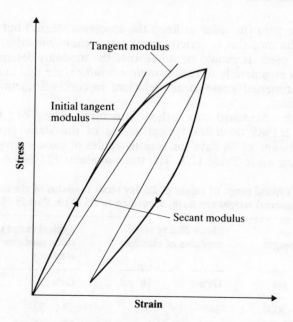

Fig. 12.2: Typical stress–strain curve for concrete
Note: In dry concrete a small concave part of the curve at the beginning of loading in compression is sometimes encountered due to the existence of fine shrinkage cracks

ordinary testing machine – the increase in strain and, hence, the degree of non-linear behaviour are very small.

The non-linearity in concrete at usual stresses is mainly due to creep; consequently, the demarcation between elastic and creep strains is difficult. For practical purposes, an arbitrary distinction is made: the deformation resulting from application of the design stress is considered elastic (initial elastic strain), and the subsequent increase in strain under the sustained design stress is regarded as creep. The modulus of 'elasticity' on loading defined in this way is the *secant modulus* of Fig. 12.2. There is no standard method of determining the secant modulus but usually it is measured at stresses ranging from 15 to 50 per cent of the short-term strength. Since the secant modulus is dependent on level of stress and on its rate of application, the stress and time taken to apply it should always be stated.

The determination of the initial tangent modulus is not easy but an approximate value can be obtained indirectly: the secant of the stress–strain curve on unloading (see Fig. 12.2) is often parallel to the initial tangent modulus, but this is not always the case. Also, the initial tangent modulus is approximately equal to the *dynamic modulus* (see page 213). Several cycles of loading and unloading reduce the subsequent creep, so that the stress–strain curve on subsequent loading exhibits only a small curvature; this method is prescribed by ASTM C 469–83 and BS 1881: Part 121: 1983. Since, in a testing machine, the stress or strain is not reduced to zero (in order to keep the specimen steady) but to some small value, the modulus is, strictly speaking, a *chord modulus,* but this term is rarely used. It should be noted that the modulus determined by these methods is generally termed a *static modulus* since it is determined from an experimental stress–strain relation, in contradistinction to the dynamic modulus.

The British Standard for the Structural Use of Concrete BS 8110: Part 2: 1985 tabulates typical values of the static modulus of elasticity at the age of 28 days for various values of cube strength at the same age, as shown in Table 12.1. The static modulus E_c (GPa or psi) can

Table 12.1: **Typical range of values of 28-day static modulus of elasticity for normal weight concrete, according to BS 8110: Part 2: 1985**

28-day cube compressive strength		Mean 28-day static modulus of elasticity		Typical range of 28-day static modulus of elasticity	
MPa	psi	GPa	10^6 psi	GPa	10^6 psi
20	3000	24	3.5	18 to 30	2.6 to 4.3
25	3500	25	3.6	19 to 31	2.7 to 4.5
30	4500	26	3.8	20 to 32	2.9 to 4.6
40	6000	28	4.1	22 to 34	3.2 to 4.9
50	7500	30	4.3	24 to 36	3.5 to 5.2
60	8500	32	4.6	26 to 38	3.8 to 5.5

be related to the cube compressive strength f_{cu} (MPa or psi) by the expression

$$\left. \begin{array}{l} \text{in SI units:}^1 \quad E_c = 9.1 f_{cu}^{0.33} \\[2mm] \text{in psi:}^1 \quad E_c = 0.255 f_{cu}^{0.33} \times 10^6 \end{array} \right\} \tag{12.1}$$

when the density of concrete ρ, is 2320 kg/m³ (145 lb/ft³), i.e. for typical normal weight concrete.

When the density ρ is between 1400 and 2320 kg/m³ (87 and 145 lb/ft³), the expression for static modulus is

$$\left. \begin{array}{l} \text{in SI units:}^1 \quad E_c = 1.7 \rho^2 f_{cu}^{0.33} \times 10^{-6} \\[2mm] \text{in psi:}^1 \quad E_c = 12.24 \rho^2 f_{cu}^{0.33} \end{array} \right\} \tag{12.2}$$

---

[1] In SI units, MPa is used for strength and stress, and GPa for modulus of elasticity. In psi units, both strength and modulus are in psi

Alternative expressions, recommended by BS 8110: Part 2: 1985, are given by Eqs (12.27) and (12.28).

The ACI Building Code 318–83 gives the following expression for the static modulus (GPa or psi) of normal weight concrete:

$$\left.\begin{array}{ll}\text{in SI units:}[1] & E_c = 4.70 f_{cyl}^{0.5} \\[2mm] \text{in psi:}[1] & E_c = 0.057 f_{cyl}^{0.5} \times 10^6\end{array}\right\} \tag{12.3}$$

where f_{cyl} is the cylinder compressive strength (MPa or psi).

When the density of concrete is betweeen 1500 and 2500 kg/m^3 (90 and 155 lb/ft^3), the static modulus is given by:

$$\left.\begin{array}{ll}\text{in SI units:}[1] & E_c = 43 \rho^{1.5} f_{cyl}^{0.5} \times 10^{-6} \\[2mm] \text{in psi:}[1] & E_c = 33 \rho^{1.5} f_{cyl}^{0.5}.\end{array}\right\} \tag{12.4}$$

The standards BS 1881: Part 209: 1990 and ASTM C 215–60 (reapproved 1976) prescribe the measurements of the dynamic modulus of elasticity using specimens similar to those employed in the determination of the flexural strength (see page 309), viz. beams 150 × 150 × 750 mm (6 × 6 × 30 in.) or 100 × 100 × 400 mm (4 × 4 × 20 in.). As shown in Fig. 12.3, the specimen is clamped at its middle section, and an electro-magnetic exciter unit is placed against one end face of the specimen and a pick-up against the other. The exciter is driven by a variable-frequency oscillator with a range of 100 to 10 000 Hz. Longitudinal vibrations propagated within the specimen are received by the pick-up, are amplified, and their amplitude is measured by an appropriate indicator. The frequency of excitation is varied until resonance is obtained at the fundamental (i.e. lowest) frequency of the specimen; this is indicated by the maximum deflection of the indicator. If this frequency is n Hz, L is the length of the specimen (mm or in.), and ρ its density (kg/m^3 or lb/ft^3), then the dynamic

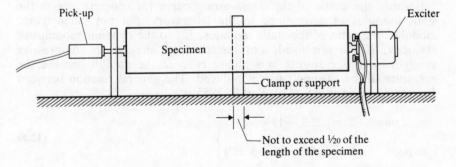

Fig. 12.3: Test arrangement for the determination of the dynamic modulus of elasticity (longitudinal vibration) given in BS 1881: Part 209: 1990

213

modulus of elasticity, E_d (GPa or psi) is

in SI units:[1] $E_d = 4n^2 L^2 \rho \times 10^{-15}$

in psi:[1] $\qquad E_d = 6n^2 L^2 \rho \times 10^{-6}.$

$$\left. \right\} \qquad \textbf{(12.5)}$$

In addition to the test based on the longitudinal resonance frequency, tests using the transverse (flexural) frequency and the torsional frequency can also be used.

Factors influencing the modulus of elasticity

Equations (12.1) to (12.4) could be seen to suggest unique relations between the modulus of elasticity and strength. However, these relations are valid only in general terms. For instance, the moisture condition of the specimen is a factor: a wet specimen has a modulus higher by 3 to 4 GPa (0.45 to 0.60×10^6 psi) than a dry one, while the recorded strength varies in the opposite sense. The properties of aggregate also influence the modulus of elasticity, although they do not significantly affect the compressive strength. Considering the basic two-phase composite models for concrete (see page 4), we see that the influence of the aggregate arises from the value of the modulus of the aggregate and its volumetric proportion. Thus, the higher the modulus of aggregate the higher the modulus of concrete, and, for aggregate having a higher modulus than the cement paste (which is usually the case), the greater the volume of aggregate the higher the modulus of concrete.

The relation between the modulus of elasticity of concrete and strength depends also on age: the modulus increases more rapidly than strength.

The modulus of elasticity of lightweight aggregate concrete is usually between 40 and 80 per cent of the modulus of normal weight concrete of the same strength, and, in fact, is similar to that of the cement paste. It is not surprising, therefore, that in the case of lightweight concrete, the mix proportions are of little influence on the modulus.

Because the shape of the stress–strain curve for concrete affects the static modulus as determined in the laboratory, but not the dynamic modulus, the ratio of the static modulus, E_c, to the dynamic modulus of elasticity, E_d, is not fixed. For instance, an increase in compressive strength or in age results in a higher ratio of the moduli because the curvature of the loading curve is reduced. The general relation between E_c, and E_d given in BS 8110: Part 2: 1985 is:

in SI units: $E_c = 1.25 E_d - 19$

in psi: $\qquad E_c = 1.25 E_d - (2.75 \times 10^6)$

$$\left. \right\} \qquad \textbf{(12.6)}$$

where E_c and E_d are expressed in GPa or psi, as appropriate. The relation does not apply to concretes containing more than 500 kg of cement per cubic metre (850 lb/yd³) of concrete or to lightweight aggregate concrete. For the latter, the following expression has been suggested:

$$\left. \begin{array}{ll} \text{in SI units:} & E_c = 1.04E_d - 4.1 \\ \text{in psi:} & E_c = 1.04E_d - (0.59 \times 10^6). \end{array} \right\} \tag{12.7}$$

When it is required to relate the dynamic modulus to strength, the static modulus may be estimated from Eq. (12.6) or (12.7) and substituted into the appropriate equation given earlier (Eqs (12.1) to (12.4)).

Poisson's ratio

The design and analysis of some types of structures require the knowledge of Poisson's ratio, viz. the ratio of the lateral strain accompanying an axial strain to the applied axial strain. The sign of the strains is ignored. We are usually interested in applied compression, and, therefore have axial contraction and lateral extension. Generally, Poisson's ratio, μ, for normal weight and lightweight concretes lies in the range of 0.15 to 0.20 when determined from strain measurements taken in the static modulus of elasticity tests (ASTM C 469–83 and BS 1881: Part 121: 1983).

An alternative method of determining Poisson's ratio is by dynamic means. Here, we measure the velocity of a pulse of ultrasonic waves and the fundamental resonant frequency of longitudinal vibration of a concrete beam specimen. The resonant frequency is found in the dynamics modulus test as prescribed by ASTM C 215–60 (reapproved 1976) and by BS 1881: Part 209: 1990 (see page 213), whilst the pulse velocity is obtained using the ultrasonic pulse apparatus prescribed by ASTM C 597–83 and by BS 1881: Part 203: 1986 (see page 321). The Poisson's ratio μ can then be calculated from the expression

$$\left(\frac{V}{2nL} \right)^2 = \frac{1 - \mu}{(1 + \mu)(1 - 2\mu)} \tag{12.8}$$

where V is the pulse velocity (mm/s or in./s), n is the resonant frequency (Hz), and L is the length of the beam (mm or in.). The value of Poisson's ratio determined dynamically is somewhat greater than from static tests, typically ranging from 0.2 to 0.24.

Creep

In the previous section it was seen that, in concrete, the relation between stress and strain is, strictly speaking, a function of time. Here, we are concerned with creep at stresses well below the threshold value at which static fatigue or, in the case of cyclic loading, fatigue occurs (see Chapter 11).

Creep is defined as the increase in strain under a sustained constant stress after taking into account other time-dependent deformations not associated with stress, viz. shrinkage, swelling and thermal deformations. Thus, creep is reckoned from the initial elastic strain as given by the secant modulus of elasticity (see page 211) at the age of loading.[2] Figures 12.4 and 12.5 illustrate the situation.

Let us consider the following examples of concrete loaded to a compressive stress σ_0 at the age t_0 and subjected to the same stress σ_0 until some time t $(t > t_0)$. In all these examples, it is assumed that the

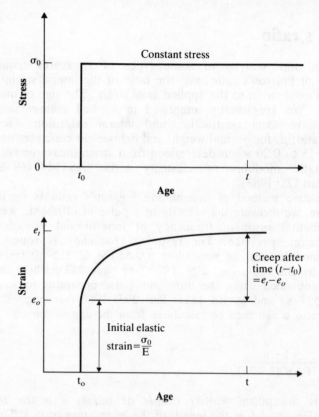

Fig. 12.4: Definition of creep under a constant stress σ_0; E is the secant modulus of elasticity at age t_0

[2] Strictly speaking, creep should be reckoned from the elastic strain at the time when creep is determined, since the elastic strain decreases with age due to the increase in the modulus of elasticity with age; this effect is ignored in the definition of creep.

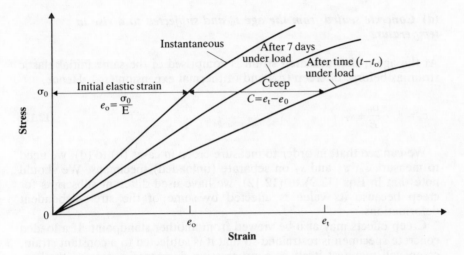

Fig. 12.5: Schematic stress–strain curves for concrete; on application of load, after 7 days under load, and after $(t - t_0)$ days under load; symbols as in Fig. 12.4

concrete has been cured in water until age t_0 and subsequently stored in various environments. The secant modulus of elasticity at the age t_0 was determined and is referred to as E. Hence, the elastic strain is σ_0/E.

(a) *Concrete sealed from the environment from the age t_0*

At the age t, the measured strain (ε_a) is comprised of initial elastic strain (σ_0/E) and creep (c_a). Hence,

$$c_a = \varepsilon_a - \frac{\sigma_0}{E}. \tag{12.9}$$

(b) *Concrete allowed to dry from the age t_0*

At the age t, the measured strain (ε_b) is comprised of the same initial elastic strain as before and of creep (c_b) and shrinkage (s_h). Since shrinkage is a contraction, we have

$$c_b = \varepsilon_b - \frac{\sigma_0}{E} - s_h. \tag{12.10}$$

(c) *Concrete stored in water from the age t_0*

At the age t, the measured strain (ε_c) is comprised of the same initial elastic strain as before and creep (c_c) and swelling (s_w). Swelling is, by definition, an expansion, so that

$$c_c = \varepsilon_c - \frac{\sigma_0}{E} + s_w. \tag{12.11}$$

217

(d) *Concrete sealed from the age t_0 and subjected to a rise in temperature*

At the age t, the measured strain is comprised of the same initial elastic strain as before, of creep (c_d) and of thermal expansion (s_T). Hence,

$$c_d = \varepsilon_d - \frac{\sigma_0}{E} + s_T. \tag{12.12}$$

We can see that, in order to measure creep in cases (b) to (d), we need to measure s_h, s_w and s_T on separate (unloaded) specimens. We should note that in Eqs (12.9) to (12.12), we have used different subscripts for creep because its value is affected by some of the stress-dependent deformations.

Creep effects may also be viewed from another standpoint. If a loaded concrete specimen is restrained so that it is subjected to a constant strain, creep will manifest itself as a progressive decrease in stress with time. This is termed *relaxation* and is shown in Fig. 12.6.

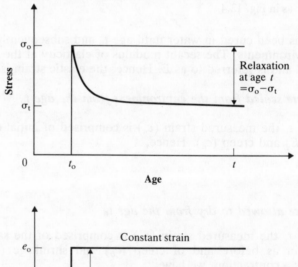

Fig. 12.6: Definition of relaxation for concrete subjected initially to stress σ_0 and then kept at a constant strain; E is the secant modulus of elasticity at age t_0

If a sustained load is removed after some time, the strain decreases immediately by an amount equal to the elastic strain. This strain is generally smaller than the initial elastic strain because of the increase in the modulus of elasticity with age. The *instantaneous recovery* is followed by a gradual decrease in strain, called *creep recovery* (Fig. 12.7). The shape of the creep recovery curve is similar to the creep curve but recovery approaches its maximum value much more rapidly. The creep recovery is always smaller than the preceding creep so that there is a *residual deformation* (even after a period under load of one day only).

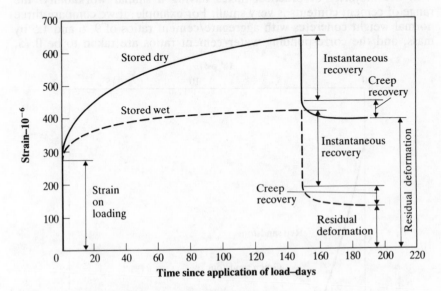

Fig. 12.7: Creep and creep recovery of concrete stored in water and in air from the age of 28 days, subjected to a stress of 9 MPa (1300 psi) and then unloaded; mix proportions 1 : 1.7 : 3.5 by mass; water/cement ratio of 0.5; specimen size 75 × 255 mm (3 × 10 in.) cylinder; cured in water

Creep is therefore not a completely reversible phenomenon, and the residual deformation can be viewed as irreversible creep which contributes to the hysteresis occurring in a short-term cycle of load (see page 197). A knowledge of creep recovery is of interest in connection with estimating stresses when relaxation occurs, e.g. in prestressed concrete.

Factors influencing creep

In normal weight aggregate concrete, the source of creep is the hardened cement paste since the aggregate is not liable to creep at the level of stress existing in concrete. Because the aggregate is stiffer than the cement paste, the main role of aggregate is to restrain the creep in the cement paste, the effect depending upon the elastic modulus of aggregate and its volumetric proportion. Hence, the stiffer the aggregate the lower the creep (Fig. 12.8), and the higher the volume of aggregate the lower the creep. The latter influence is shown in Fig. 12.9 in terms of cement paste content, which is complementary to the aggregate content by volume (as fractions of the total volume of hardened concrete); in other words, if the aggregate content by volume is g per cent, the cement paste content is $(100 - g)$ per cent.

In the majority of practical mixes having a similar workability, the range of cement contents is very small. For example, if we compare three normal weight concretes with aggregate/cement ratios of 9, 6 and 4.5 by mass, and the corresponding water/cement ratios are taken to be 0.75,

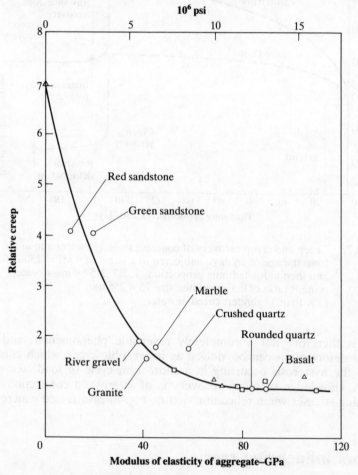

Fig. 12.8: Effect of modulus of elasticity of aggregate on relative creep of concrete (equal to 1 for an aggregate with a modulus of 69 GPa (10^7 psi))

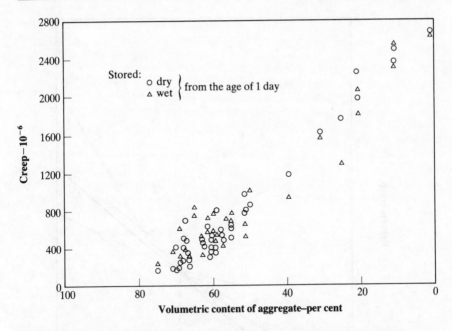

Fig. 12.9: Effect of volumetric content of aggregate on creep of concrete, corrected for variations in the water/cement ratio

0.55 and 0.40 by mass, we find the cement paste content to be 24, 27 and 29 per cent,[3] respectively. Hence, the creep of these concretes might be expected to differ little, but, in fact, this is not the case because there is another significant influence on creep, viz. the water/cement ratio.

It will be recalled from Chapter 6 that the water/cement ratio is the main factor influencing the porosity and, hence, the strength of concrete, so that a lower water/cement ratio results in a higher strength. Now, for a constant cement paste content, the effect of a decrease in water/cement ratio is to decrease creep (see Fig. 12.10), and therefore it can be expected that creep and strength are related. Indeed, within a wide range of mixes, creep is inversely proportional to the strength of concrete at the age of application of the load. Moreover, for a given type of concrete, we can expect creep to decrease as the age at application of load increases (see Fig. 12.11) because, of course, strength increases with age.

One of the most important external factors influencing creep is the relative humidity of the air surrounding the concrete. Generally, for a given concrete, creep is higher the lower the relative humidity, as illustrated in Fig. 12.12 for specimens cured at a relative humidity of 100 per cent, then loaded and exposed to different humidities. Thus, even

---

[3] Specific gravity of aggregate assumed to be 2.6.

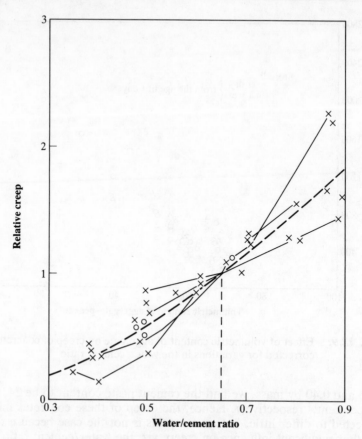

Fig. 12.10: Data of several investigators adjusted for the volumetric content of cement paste (to a value of 0.20), with creep expressed relative to the creep at a water/cement ratio of 0.65
(From: O. WAGNER, Das Kriechen unbewehrten Betons, *Deutscher Ausschuss für Stahlbeton, No.* 131, pp. 74, Berlin, 1958.)

though shrinkage has been taken into account in determining creep (see page 217), there is still an influence of drying on creep. This influence of relative humidity is much smaller or absent, in the case of specimens which have been allowed to dry prior to application of load so that hygral equilibrium with the surrounding medium exists under load; in this case, creep is much reduced. However, such a practice is not normally recommended as a means of reducing creep, especially for young concrete, because inadequate curing will lead to a low tensile strength and possibly to shrinkage-induced cracking (see page 255).

The influence of relative humidity on creep and on shrinkage is similar (see Chapter 13), and both deformations are also dependent on the size of the concrete member. When drying occurs at a constant relative humidity, creep is smaller in a larger specimen; this size effect is

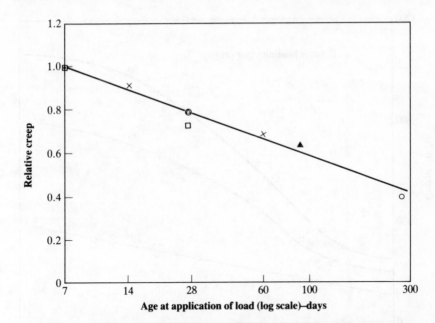

Fig. 12.11: Influence of age at application of load on creep of concrete relative to creep of concrete loaded at 7 days, for tests of different investigators: concrete stored at a relative humidity of approximately 75 per cent
(From: R. L'HERMITE, What do we know about plastic deformation and creep of concrete? *RILEM Bulletin,* No. 1, pp. 21–5, Paris, March 1959.)

expressed in terms of the volume/surface ratio of the concrete member (see Fig. 12.13). If no drying occurs, as in mass concrete, creep is smaller and is independent of size because there is no additional effect of drying on creep. The effects which have just been discussed can be viewed in terms of the values of creep given by Eqs (12.9) to (12.11): $c_b > c_c$ and $c_a \approx c_c$.

The influence of temperature on creep has become of increased interest in connection with the use of concrete in nuclear pressure vessels, but the problem is of significance also in other types of structures, e.g. bridges. The time at which the temperature of concrete rises relative to the time at which the load is applied affects the creep–temperature relation. If saturated concrete (simulated mass concrete) is heated and loaded at the same time, creep is greater than when concrete is heated during the curing period, prior to application of load. Figure 12.14 illustrates these two conditions. Creep is smaller when concrete is cured at a high temperature because strength is higher than when concrete is cured at normal temperature before heating and loading.

If unsealed concrete is subjected to a high temperature at the same time as, or just prior to, the application of load, there is a rapid increase in creep as the temperature increases to approximately 50 °C (about

223

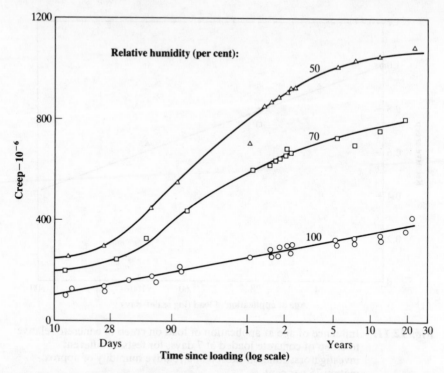

Fig. 12.12: Creep of concrete cured in fog for 28 days, then loaded and stored
at different relative humidities
(From: G. E. TROXELL, J. M. RAPHAEL and R. E. DAVIS,
Long-time creep and shrinkage tests of plain and reinforced
concrete, *Proc. ASTM.,* 58, pp. 1101–20 (1958).)

120 °F), then a decrease in creep down to about 120 °C (about 250 °F),
followed by another increase in creep to at least 400 °C (about 750 °F)
(see Fig. 12.15). The initial increase in creep is due to a rapid expulsion
of evaporable water; when all of that water has been removed, creep is
greatly reduced and becomes equal to that of pre-dried (desiccated)
concrete.

Figure 12.16 shows creep at low temperatures as a proportion of creep
at 20 °C (68 °F). At temperatures below 20 °C (68 °F), creep decreases
until the formation of ice which causes an increase in creep but below the
ice point creep again decreases.

In the discussion so far we have compared the influence of the various
factors on creep on the basis of equality of stress. Of course, creep is
affected by stress, and normally creep is assumed to be directly
proportional to the applied stress up to about 40 per cent of the
short-term strength, i.e. within the range of working or design stresses.
Hence, we can use the term: *specific creep,* viz. creep per unit of stress.
Above 40 to 50 per cent of the short-term strength, microcracking
contributes to creep so that the creep–stress relation becomes non-linear,
creep increasing at an increasing rate (see page 196).

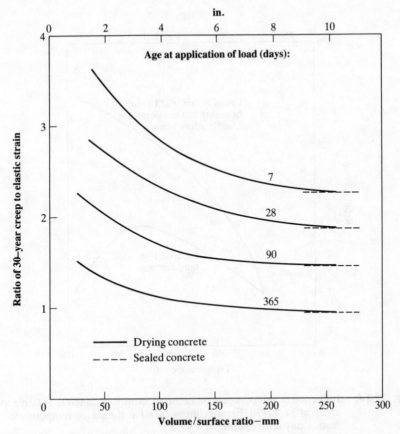

Fig. 12.13: Influence of volume/surface ratio on the ratio of creep to elastic strain for sealed concrete and for drying concrete stored at a relative humidity of 60 per cent

Creep is affected by the type of cement in so far as it influences the strength of the concrete at the time of application of load. On the basis of equality of the stress/strength ratio, most Portland cements lead to sensibly the same creep. On the other hand, on the basis of equality of stress, the specific creep increases (in the order of type of cement) as follows: high-alumina cement, rapid-hardening (Type III) and ordinary Portland (Type I). The order of magnitude of creep of Portland blast-furnace (Type IS), low-heat Portland (Type IV) and Portland-pozzolan (Type IP and P) cements is less clear, and so is the influence of partial replacement of cement by blast-furnace slag or by fly-ash (PFA) as the effect depends upon the storage environment. For example, when compared with ordinary Portland (Type I) cement, for sealed concrete, creep decreases with an increase in the level of replacement of slag or of fly-ash but, when there is concurrent drying, creep is sometimes higher. When such concrete is to be used it is recommended that tests be undertaken to assess creep.

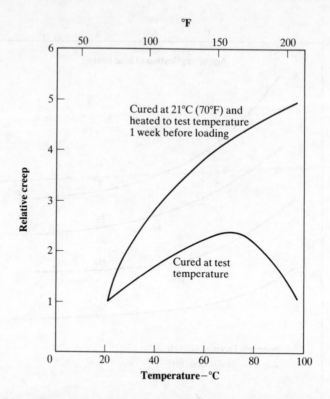

Fig. 12.14: Influence of temperature on creep of saturated concrete relative to creep at 21 °C (70 °F); specimens cured at the stated temperature from 1 day until loading at 1 year
(Based on: K. W. NASSER and A. M. NEVILLE, Creep of old concrete at normal and elevated temperatures, *ACI Journal*, 64, pp. 97–103, 1967.)

A similar recommendation applies when admixtures are used. For sealed concrete, neither plasticizers nor superplasticizers affect creep in both water-reduced and flowing concretes, but under drying conditions the effect of these admixtures on creep is uncertain.

Magnitude of creep

In the preceding sections, we gave relatively little information about the magnitude of creep. The reason for this apparent omission is that the presence of several factors influencing creep makes it impossible to quote reliable typical values. Nevertheless, it may be useful to give some indication.

Figure 12.7 shows the development of creep with time for a 1:1.7:3.5 mix with a water/cement ratio of 0.5, made with quartzitic rounded

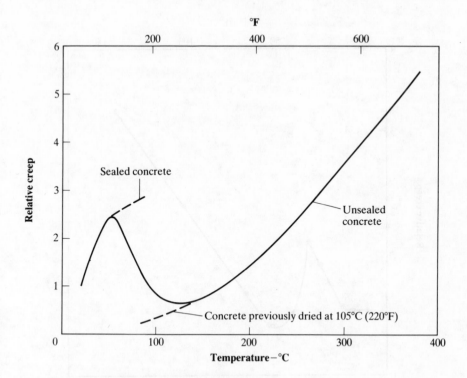

Fig. 12.15: Influence of temperature on creep of unsealed concrete relative to creep at 20 °C (68 °F); specimens moist-cured for 1 year and then heated to the test temperature 15 days before loading
(Based on: J. C. MARÉCHAL, Le fluage du béton en fonction de la température, *Materials and Structures*, 2, No. 8, pp. 111–15, Paris, 1969.)

gravel aggregate. The initial parabolic shape of the curve, gradually flattening out, is always present. For practical purposes, we are usually interested in creep after several months or years, or even in the ultimate (or limiting) value of creep. We know that the increase in creep beyond 20 years under load (within the range of working stresses) is small, and, as a guide, we can assume that:

about 25 per cent of the 20-year creep occurs in 2 weeks;
about 50 per cent of the 20-year creep occurs in 3 months;
about 75 per cent of the 20-year creep occurs in 1 year.

Several methods of estimating creep are available (see Bibliography) but, with unknown materials, it may be necessary to determine creep of concrete by experiment. ASTM C 512–82 (reapproved 1983) describes a test method. In practice, a test can rarely be run for longer than several months, and extrapolation is useful. Typical equations relating creep

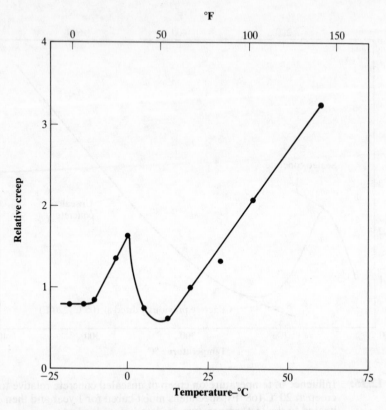

Fig. 12.16: Creep of sealed concrete at low temperatures relative to creep at 20 °C (68 °F)
(From: R. JOHANSEN and C. H. BEST, Creep of concrete with and without ice in the system, *RILEM Bulletin*, No. 16, pp. 47–57, Paris, Sept. 1962.)

after any time under load, c_t, to creep after 28 days under load, c_{28}, are:

for sealed or saturated concrete: $c_t = c_{28} \times 0.5t^{0.2}$ **(12.13)**

for drying concrete: $c_t = c_{28}[-6.19 + 2.15 \log_e t]^{1/2.64}$ **(12.14)**

where t = time under load (days) > 28 days.

The above expressions are sensibly independent of mix proportions, type of aggregate, size of specimen, and age at loading.

Prediction of creep

The following methods are appropriate to normal weight concrete subjected to a constant stress and stored under normal constant environmental conditions. For other methods, other loadings and storage environments, the reader is referred to the Bibliography.

ACI 209R–82 expresses the *creep coefficient* $\phi(t, t_0)$ as a function of time:

$$\phi(t, t_0) = \frac{(t - t_0)^{0.6}}{10 + (t - t_0)^{0.6}} \times \phi_\infty(t_0) \tag{12.15}$$

where the creep coefficient is the ratio of specific creep $c(t, t_0)$ at age t due to a unit stress applied at the age t_0 to the initial elastic strain under a unit stress applied at the age t_0; age is measured in days. Since the initial elastic strain under a unit stress is equal to the reciprocal of the modulus of elasticity $E_c(t_0)$, we can write

$$\phi(t, t_0) = c(t, t_0) \times E_c(t_0). \tag{12.16}$$

In Eq. (12.15), $(t - t_0)$ is the time since application of load and $\phi_\infty(t, t_0)$ is the ultimate creep coefficient, which is given by

$$\phi_\infty(t, t_0) = 2.35 k_1 k_2 k_3 k_4 k_5 k_6. \tag{12.17}$$

For ages at application of load greater than 7 days for moist curing, or greater than 1 to 3 days for steam curing, the coefficient k_1 is estimated from:

for moist curing: $k_1 = 1.25 t_0^{-0.118}$ (12.18a)

for steam curing: $k_1 = 1.13 t_0^{-0.095}$. (12.18b)

The coefficient k_2 is dependent upon the relative humidity h (per cent):

$k_2 = 1.27 - 0.006 h$ (for $h \geqslant 40$). (12.19)

The coefficient k_3 allows for member size in terms of the volume/surface ratio, V/S, which is defined as the ratio of the cross-sectional area to the perimeter exposed to drying. For values of V/S smaller than 37.5 mm (1.5 in.), k_3 is given in Table 12.2. When V/S is between 37.5

Table 12.2: **Values of coefficient k_3 to allow for member size in the ACI method of predicting creep**

Volume/surface ratio		Coefficient
mm	in.	k_3
12.5	0.50	1.30
19	0.75	1.17
25	1.00	1.11
31	1.25	1.04
37.5	1.50	1.00

and 95 mm (1.5 and 3.7 in.), k_3 is given by:

for $(t - t_0) \leqslant 1$ year:

in SI units: $\quad k_3 = 1.14 - 0.00364 \dfrac{V}{S}$

$$\left. \right\} \tag{12.20a}$$

in US units: $\quad k_3 = 1.14 - 0.09246 \dfrac{V}{S}$

for $(t - t_0) > 1$ year:

in SI units: $\quad k_3 = 1.10 - 0.00268 \dfrac{V}{S}$

$$\left. \right\} \tag{12.20b}$$

in US units: $\quad k_3 = 1.10 - 0.06807 \dfrac{V}{S}$

When $V/S \geqslant 95$ mm (3.7 in.):

in SI units: $\quad k_3 = \frac{2}{3}[1 + 1.13\, e^{-0.0212(V/S)}]$

$$\left. \right\} \tag{12.21}$$

in US units: $\quad k_3 = \frac{2}{3}[1 + 1.13\, e^{-0.538(V/S)}]$

The coefficients to allow for the composition of the concrete are k_4, k_5 and k_6. Coefficient k_4 is given by:

in SI units: $\quad k_4 = 0.82 + 0.00264s$

$$\tag{12.22}$$

in US units: $\quad k_4 = 0.82 + 0.06706s$

where s = slump (mm or in.) of fresh concrete.

Coefficient k_5 depends on the fine aggregate/total aggregate ratio, A_f/A, in per cent and is given by

$$k_5 = 0.88 + 0.0024 \frac{A_f}{A}. \tag{12.23}$$

Coefficient k_6 depends on the air content a (per cent):

$$k_6 = 0.46 + 0.09a \geqslant 1. \tag{12.24}$$

The elastic strain-plus-creep deformation under a unit stress is termed the *creep function* Φ, which is given by:

$$\Phi(t, t_0) = \frac{1}{E_c(t_0)}[1 + \phi(t, t_0)] \tag{12.25}$$

230

where $E_c(t_0)$ is related to the compressive strength of test cylinders by Eq. (12.4). If the strength at age t_0 is not known, it can be found from the following relation:

$$f_{cyl}(t_0) = \frac{t_0}{X + Y \times t_0} \times f_{cyl28} \qquad (12.26)$$

where f_{cyl28} is the strength at 28 days, and X and Y are given in Table 12.3.

Table 12.3: Values of the constants X and Y in Eq. (12.26) using the ACI method of predicting creep

Type of cement	Curing condition	Constants of Eq. (12.26)	
		X	Y
Ordinary Portland (Type 1)	Moist	4.00	0.85
	Steam	1.00	0.95
Rapid-hardening Portland (Type III)	Moist	2.30	0.92
	Steam	0.70	0.98

In the UK, BS 8110: Part 2: 1985 recommends a method of estimating ultimate creep. For concrete with an average, high quality dense aggregate, the modulus of elasticity $E_c(t_0)$ is related to the compressive strength of cubes, $f_{cu}(t_0)$, as follows:

$$E_c(t_0) = E_{c28}\left[0.4 + 0.6\frac{f_{cu}(t_0)}{f_{cu28}}\right]. \qquad (12.27)$$

The modulus of elasticity at 28 days, E_{c28}, is obtained from the cube strength at 28 days f_{cu28} by the following expressions:

$$\left.\begin{array}{ll} \text{in GPa:} & E_{c28} = 20 + 0.2f_{cu28} \\ \text{in psi:} & E_{c28} = 2.9 \times 10^6 + 200f_{cu28}. \end{array}\right\} \qquad (12.28)$$

For lightweight aggregate concrete of density ρ, the modulus of elasticity given by the foregoing equations should be multiplied by $(\rho/2400)^2$ in SI units and by $(\rho/150)^2$ in psi units.

The strength ratio term of Eq. (12.27) is best obtained by measurement; however, the values of Table 12.4 may be used.

For very long time under load, the ultimate creep function Φ_∞ is given

Table 12.4: **Values of the strength ratio $\dfrac{f_{cu}(t_0)}{f_{cu28}}$ in Eq. (12.27) using the UK method of predicting ultimate creep**

Age t_0	Strength ratio
7	0.70
28	1.00
90	1.17
365	1.25

by:

$$\Phi_\infty = \frac{1}{E_c(t_0)}(1 + \phi_\infty) \tag{12.29}$$

where ϕ_∞ is the ultimate creep coefficient which is obtained from Fig. 12.17.

Given the ambient relative humidity, age at application of load, and volume/surface ratio, the ultimate creep function can be calculated from Eq. (12.29). If there is no moisture exchange, i.e. the concrete is sealed or we are dealing with mass concrete, creep is assumed to be equivalent to that of concrete with a volume/surface ratio greater than 200 mm (8 in.) at 100 per cent relative humidity.

An improvement in the accuracy of prediction of creep may be obtained by undertaking short-term tests and then by extrapolation by the use of Eqs (12.13) and (12.14). This approach is also recommended when untried aggregates or admixtures are contemplated.

Effects of creep

Creep of concrete increases the deflection of reinforced concrete beams and, in some cases, may be a critical consideration in design. In reinforced concrete columns, creep results in a gradual transfer of load from the concrete to the reinforcement. Once the steel yields, any increase in load is taken by the concrete, so that the full strength of both the steel and the concrete is developed before failure takes place – a fact recognized by design formulae. However, in eccentrically loaded, very slender columns, creep increases the deflection and can lead to buckling. In statically indeterminate structures, creep may relieve (by relaxation) the stress concentrations induced by shrinkage, temperature changes or movement of support. In all concrete structures, creep reduces internal stresses due to non-uniform or restrained shrinkage so that there is a reduction in cracking (see Chapter 13).

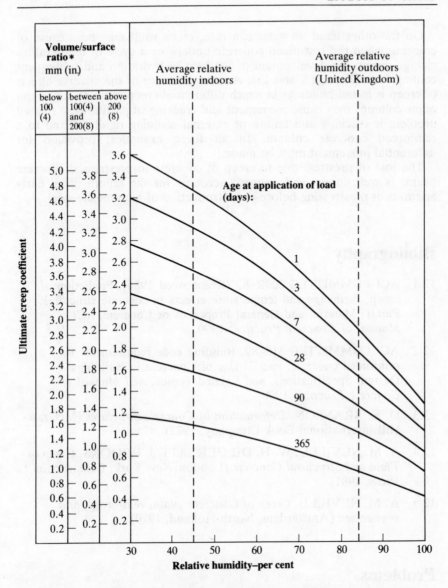

Fig. 12.17: Data for estimating the ultimate creep coefficient for use in Eq. (12.29)
(From: BS 8110: Part 2: 1985)
* Sometimes the term 'effective section thickness' is used to represent the size of a member; effective section thickness = 2 × volume/surface ratio

233

On the other hand, in mass concrete, creep itself may be a cause of cracking when the restrained concrete undergoes a cycle of temperature change due to the development of the heat of hydration and subsequent cooling (see Chapters 9 and 13). Another instance of the adverse effects of creep is in tall buildings in which differential creep between inner and outer columns may cause movement and cracking of partitions; a related problem is cracking and failure of external cladding rigidly affixed to a reinforced concrete column. In all these examples, provision for differential movement must be made.

The loss of prestress due to creep of concrete in prestressed concrete beams is well known and, indeed, accounts for the failure of all early attempts at prestressing before the introduction of high tensile steel.

Bibliography

12.1 ACI COMMITTEE 209R–82 (Reapproved 1986), Prediction of creep, shrinkage and temperature effects in concrete structures, Part 1: Material and General Properties of Concrete, *ACI Manual of Concrete Practice*, 1990.

12.2 ACI COMMITTEE 318–89, Building code requirements for reinforced concrete, Part 3: Use of Concrete in Buildings – Design, Specifications, and Related Topics, *ACI Manual of Concrete Practice*, 1990.

12.3 D. E. BRANSON, *Deformation of Concrete Structures* (McGraw-Hill International Book Company, 1977).

12.4 A. M. NEVILLE, W. H. DILGER and J. J. BROOKS, *Creep of Plain and Structural Concrete* (London/New York, Construction Press, 1981).

12.5 A. M. NEVILLE, *Creep of Concrete: plain, reinforced, and prestressed* (Amsterdam, North-Holland, 1970).

Problems

12.1 How would you estimate the creep of concrete made with an unknown aggregate compared with a known mix?

12.2 Comment on the magnitude of creep of concrete made with different cements.

12.3 Describe the role of aggregate in creep of concrete.

12.4 How does the Poisson's ratio of concrete vary with an increase in stress?

12.5 Draw a stress–strain curve for concrete loaded at a constant rate of strain.

12.6 Discuss the main factors affecting the creep of concrete.

12.7 What is the significance of the area within the hysteresis loop in the stress–strain curve on loading and unloading?

12.8 Compare the creep of mass concrete and concrete exposed to dry air.

12.9 What is Poisson's ratio?

12.10 What is the difference between the dynamic and static moduli of elasticity of concrete?

12.11 What is a secant modulus of elasticity?

12.12 What is a tangent modulus of elasticity?

12.13 What is the initial tangent modulus of elasticity?

12.14 Which modulus of elasticity would you use to determine the deformational response of concrete to small variations in stress?

12.15 What is the relation between the modulus of elasticity of concrete and strength?

12.16 How does the relation between the modulus of elasticity of concrete and strength vary with age?

12.17 What is the influence of the properties of aggregate on the modulus of elasticity of concrete?

12.18 What is the significance of the shape of the descending part of the stress–strain curve for concrete?

12.19 What is the significance of the area under the stress–strain curve?

12.20 What is meant by specific creep?

12.21 How would you assess creep of concrete containing PFA, or slag or a superplasticizer?

12.22 Define creep of concrete.

12.23 Discuss the beneficial and harmful effects of creep of concrete.

12.24 Would concrete having a zero creep be beneficial?

12.25 Calculate the static modulus of elasticity of normal weight concrete which has a compressive strength of 30 MPa (4400 psi) using the British and American expressions.

> *Answer:* British: 28.0 GPa (4.1×10^6 psi)
> US: 25.7 GPa (3.8×10^6 psi)

12.26 Use the ACI method to estimate the 30-year specific creep of concrete given the following information:

Age at application of load	14 days
Curing condition	moist
Storage environment	relative humidity of 70 per cent
Volume/surface ratio	50 mm (2 in.)
Slump	75 mm (3 in.)
Fine aggregate/total aggregate ratio	30 per cent
Air content	2 per cent
14-day cylinder compressive strength	30 MPa (4400 psi)
Density of concrete	2400 kg/m^3 (150 lb/ft^3)

Answer: 59.6×10^{-6} per MPa
$(0.41 \times 10^{-6}$ per psi)

12.27 Use the BS 8110: Part 2: 1985 method to estimate the 30-year specific creep of the concrete given in question 12.26, assuming the cube strength is 35 MPa (6000 psi).

Answer: 106×10^{-6} per MPa $(0.73 \times 10^{-6}$ per psi)

13

Deformation and cracking independent of load

In addition to deformation caused by the applied stress, volume changes due to shrinkage and temperature variation are of considerable importance because in practice these movements are usually partly or wholly restrained, and therefore they induce stress. Thus, although we categorize shrinkage (or swelling) and thermal changes as independent of stress, the real life situation is, unfortunately, not so simple. The main danger is the presence of tensile stress induced by some form of restraint to these movements because, of course, concrete is very weak in tension and prone to cracking. Cracks must be avoided or controlled and minimized because they impair the durability and structural integrity, and are also aesthetically undesirable.

Shrinkage and swelling

Shrinkage is caused by loss of water by evaporation or by hydration of cement, and also by carbonation. The reduction in volume, i.e. volumetric strain, is equal to 3 times the linear contraction, and in practice we measure shrinkage simply as a linear strain. Its units are thus mm per mm (in. per in.) usually expressed in 10^{-6}.

While the cement paste is plastic it undergoes a volumetric contraction whose magnitude is of the order of 1 per cent of the absolute volume of dry cement. This contraction is known as *plastic shrinkage*. It is caused by the loss of water by evaporation from the surface of concrete or by suction by dry concrete below. The contraction induces tensile stress in the surface layers because they are restrained by the non-shrinking inner concrete, and, since the concrete is very weak in its plastic state, plastic cracking at the surface can readily occur (see page 255).

Plastic shrinkage is greater the greater the rate of evaporation of water, which in turn depends on the air temperature, the concrete temperature, the relative humidity of the air and wind speed. According to ACI 305.R–77 (revised 1982) evaporation rates greater than $0.5\,kg/h/m^2$ ($0.1\,lb/h/ft^2$) of the exposed concrete surface have to be avoided in order to prevent plastic cracking (see page 166). Clearly, a complete prevention of evaporation immediately after casting reduces plastic shrinkage. Because it is the loss of water from the cement paste that is responsible for plastic shrinkage, it is greater the larger the cement content of the mix (Fig. 13.1), or lower the larger the aggregate content (by volume).

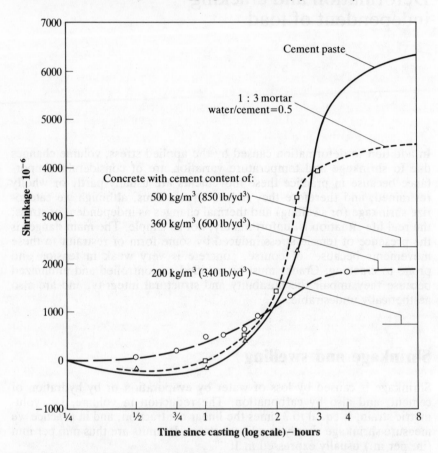

Fig. 13.1: Influence of cement content of the mix on plastic shrinkage in air at 20 °C (68 °F) and 50 per cent relative humidity with wind velocity of 1.0 m/s (2.25 mph)
(Based on: R. L'HERMITE, Volume changes of concrete, *Proc. Int. Symp. on the Chemistry of Cement*, Washington D.C., 1960, pp. 659–94.)

Even when no moisture movement to or from the set concrete is possible *autogenous shrinkage* occurs. This is caused by loss of water used up in hydration and, except in massive concrete structures, is not distinguished from shrinkage of hardened concrete due to loss of water to the outside. Autogenous shrinkage is very small, typically 50×10^{-6} to 100×10^{-6}.

If there is a continuous supply of water to the concrete during hydration, concrete expands due to absorption of water by the cement gel; this process is known as *swelling*. In concrete made with normal weight aggregate, swelling is ten to twenty times smaller than shrinkage. On the other hand, swelling of lightweight concretes can be as large as 20 to 80 per cent of the shrinkage of hardened concrete after 10 years.

Drying shrinkage

Withdrawal of water from hardened concrete stored in unsaturated air causes drying shrinkage. A part of this movement is irreversible and should be distinguished from the reversible part or *moisture movement*. Figure 13.2(a) shows that if concrete which has been allowed to dry in air of a given relative humidity is subsequently placed in water (or at a higher humidity) it will swell due to absorption of water by the cement paste. Not all the initial drying shrinkage is, however, recovered even after prolonged storage in water. For the usual range of concretes, the reversible moisture movement (or *wetting expansion*) represents about 40 to 70 per cent of the drying shrinkage, but this depends on the age before the onset of first drying. For instance, if concrete is cured so that it is fully hydrated before being exposed to drying, then the reversible moisture movement will form a greater proportion of the drying shrinkage. On the other hand, if drying is accompanied by extensive carbonation (see page 240) the cement paste is no longer capable of moisture movement so that the residual or irreversible shrinkage is larger.

The pattern of moisture movement under alternating wetting and drying – a common occurrence in practice – is shown in Fig. 13.2(b). The magnitude of this cyclic moisture movement clearly depends upon the duration of the wetting and drying periods but it is important to note that drying is very much slower than wetting. Thus, the consequence of prolonged dry weather can be reversed by a short period of rain. The movement depends also upon the range of relative humidity and on the composition of the concrete, as well as on the degree of hydration at the onset of initial drying. Generally, lightweight concrete has a higher moisture movement than concrete made with normal weight aggregate.

The irreversible part of shrinkage is associated with the formation of additional physical and chemical bonds in the cement gel when adsorbed water has been removed. The general pattern of behaviour is as follows. When concrete dries, first of all, there is the loss of free water, i.e. water in the capillaries which is not physically bound. This process induces internal relative humidity gradients within the cement paste structure so that, with time, water molecules are transferred from the large surface area of the calcium silicate hydrates into the empty capillaries and then out of the concrete. In consequence, the cement paste contracts but the reduction in volume is not equal to the volume of water removed because the initial loss of free water does not cause a significant volumetric contraction of the paste and because of internal restraint to consolidation by the calcium silicate hydrate structure.

239

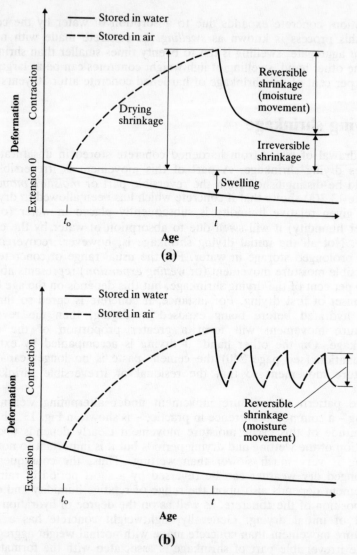

Fig. 13.2: Moisture movement in concrete: (a) concrete which has dried from age t_0 until age t and was then re-saturated, and (b) concrete which has dried from age t_0 until age t and was then subjected to cycles of drying and wetting

Carbonation shrinkage

In addition to shrinkage upon drying, concrete undergoes carbonation shrinkage. Many experimental data include both types of shrinkage but their mechanism is different. By carbonation we mean the reaction of CO_2 with the hydrated cement. The gas CO_2 is of course present in the atmosphere: about 0.03 per cent by volume in rural air; 0.1 per cent, or even more, in an unventilated laboratory, and generally up to 0.3 per cent in large cities. In the presence of moisture, CO_2 forms carbonic acid, which reacts with $Ca(OH)_2$ to form $CaCO_3$; other cement compounds are also decomposed. A concomitant of the process of carbonation is a contraction of concrete known as carbonation shrinkage.

Carbonation proceeds from the surface of the concrete inwards but does so extremely slowly. The actual rate of carbonation depends on the permeability of the concrete, its moisture content, and on the CO_2 content and relative humidity of the ambient medium. Since the permeability of concrete is governed by the water/cement ratio and the effectiveness of curing, concrete with a high water/cement ratio and inadequately cured will be more prone to carbonation, i.e. there will be a greater depth of carbonation. The extent of carbonation can be easily determined by treating a freshly broken surface with phenolphthalein – the free $Ca(OH)_2$ is coloured pink while the carbonated portion is uncoloured.

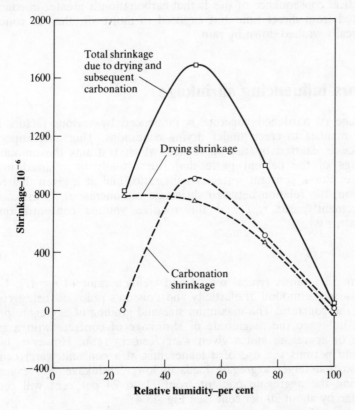

Fig. 13.3: Drying shrinkage and carbonation shrinkage of mortar at different relative humidities
(Based on: G. J. VERBECK, Carbonation of hydrated Portland cement, *ASTM Sp. Tech. Publicn. No. 205*, pp. 17–36 (1958).)

241

Carbonation of concrete (made with ordinary Portland (Type I) cement) results in a slightly increased strength and a reduced permeability, possibly because water which is released by the decomposition of $Ca(OH)_2$ on carbonation, aids the process of hydration and $CaCO_3$ is deposited in the voids within the cement paste. However, much more importantly, carbonation neutralizes the alkaline nature of the hydrated cement paste and thus the protection of steel from corrosion is vitiated. Consequently, if the full depth of cover to reinforcement is carbonated and moisture and oxygen can ingress, corrosion of steel and possibly cracking (see page 280) will result.

Figure 13.3 shows the drying shrinkage of mortar specimens dried in CO_2-free air at different relative humidities, and also the total shrinkage after subsequent carbonation. We can see that carbonation increases shrinkage at intermediate humidities but not at 100 per cent or 25 per cent. In the latter case, there is insufficient water in the pores within the cement paste for CO_2 to form carbonic acid. On the other hand, when the pores are full of water the diffusion of CO_2 into the paste is very slow. A practical consequence of this is that carbonation is greater in concrete protected from direct rain, but exposed to moist air, than in concrete periodically washed down by rain.

Factors influencing shrinkage

Shrinkage of hardened concrete is influenced by various factors in a similar manner to creep under drying conditions. The most important influence is exerted by the aggregate, which restrains the amount of shrinkage of the cement paste that can actually be realized in the concrete. For a constant water/cement ratio, and at a given degree of hydration, the relation between shrinkage of concrete s_{hc}, shrinkage of neat cement paste s_{hp}, and the relative volume concentration of aggregate g is

$$s_{hc} = s_{hp}(1 - g)^n. \tag{13.1}$$

Figure 13.4 shows typical results and yields a value of $n = 1.7$, but n depends on the moduli of elasticity and Poisson's ratios of the aggregate and of the concrete. The maximum size and grading of aggregate *per se* do not influence the magnitude of shrinkage of concrete with a given volume of aggregate and a given water/cement ratio. However, larger aggregate permits the use of a leaner mix at a constant water/cement ratio, so that larger aggregate leads to lower shrinkage. For example, increasing the aggregate content from 71 to 74 per cent will reduce shrinkage by about 20 per cent (see Fig. 13.4).

The type of aggregate, or strictly speaking its modulus of elasticity, influences shrinkage of concrete so that lightweight concrete exhibits a higher shrinkage than concrete made with normal weight aggregate; a change in the modulus of elasticity of aggregate is reflected by a change

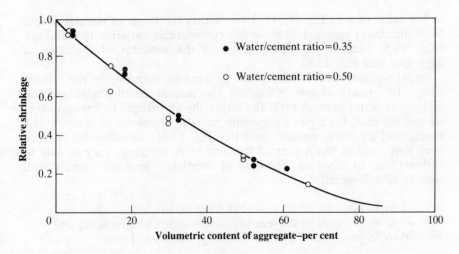

Fig. 13.4: Influence of volumetric content of aggregate in concrete (by volume) on the ratio of the shrinkage of concrete to the shrinkage of neat cement paste
(Based on: G. PICKETT, Effect of aggregate on shrinkage of concrete and hypotheses concerning shrinkage, *J. Amer. Concr., Inst.*, 52, pp. 581–90 (Jan. 1956).)

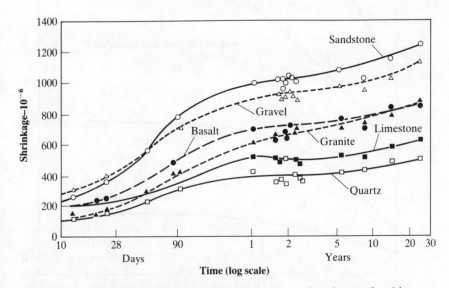

Fig. 13.5: Shrinkage of concretes of fixed mix proportions but made with different aggregates, and stored in air at 21 °C (70 °F) and a relative humidity of 50 per cent
Time reckoned since end of wet curing at the age of 28 days
(From: G. E. TROXELL, J. M. RAPHAEL and R. E. DAVIS, Long-time creep and shrinkage tests of plain and reinforced concrete, *Proc. ASTM.*, 58, pp. 1101–20 (1958).)

243

in the value of n in Eq. (13.1). Even within the range of normal weight (non-shrinking) aggregates, there is a considerable variation in shrinkage (Fig. 13.5) because of the variation in the modulus of elasticity of aggregate (see Fig. 12.8).

So far we have said nothing about the intrinsic shrinkage of the cement paste. Its quality clearly influences the magnitude of shrinkage: the higher the water/cement ratio the larger the shrinkage. In consequence, we can say that, for a given aggregate content, shrinkage of concrete is a function of the water/cement ratio (see Fig. 13.6). Shrinkage takes place over long periods but a part of the long-term shrinkage may be due to carbonation. In any case, the rate of shrinkage decreases rapidly with time so that, generally:

(a)   14 to 34 per cent of 20-year shrinkage occurs in 2 weeks;
(b)   40 to 80 per cent of 20-year shrinkage occurs in 3 months; and
(c)   66 to 85 per cent of 20-year shrinkage occurs in 1 year.

The relative humidity of the air surrounding the concrete greatly affects the magnitude of shrinkage, as shown in Fig. 13.7. In the shrinkage test prescribed in BS 1881: Part 5: 1970, the specimens are dried for a

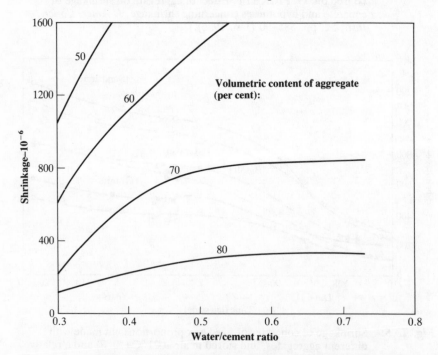

Fig. 13.6:   Influence of water/cement ratio and aggregate content on shrinkage
(Based on: S. T. A. ODMAN, Effects of variations in volume,
surface area exposed to drying, and composition of concrete on
shrinkage, *RILEM/CEMBUREAU Int. Colloquium on the
Shrinkage of Hydraulic Concretes,* 1, 20 pp. (Madrid, 1968).)

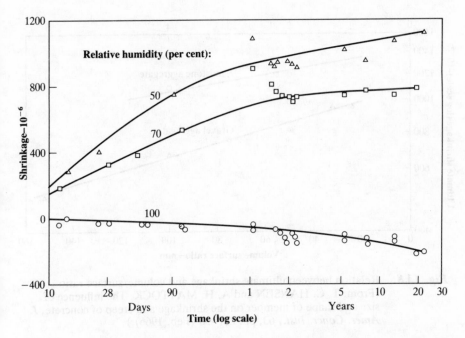

Fig. 13.7: Relation between shrinkage and time for concretes stored at different relative humidities
Time reckoned since end of wet curing at the age of 28 days
(From: G. E. TROXELL, J. M. RAPHAEL and R. E. DAVIS, Long-time creep and shrinkage tests of plain and reinforced concrete, *Proc. ASTM.*, 58, pp. 1101–20 (1958).)

specified period under prescribed conditions of temperature and humidity. The shrinkage occurring under these accelerated conditions is of the same order as that after a long exposure to air with a relative humidity of approximately 65 per cent, the latter being representative of the average of indoor (45 per cent) and outdoor (85 per cent) conditions in the UK. In the US, ASTM C 157–80 prescribes a temperature of 23 °C (73 °F) and a relative humidity of 53 per cent for the determination of shrinkage.

The magnitude of shrinkage can be determined using a measuring frame fitted with a micrometer gauge or a dial gauge reading strain to 10×10^{-6}, or by means of an extensometer or of fixed strain gauges.

The actual shrinkage of a given concrete member is affected by its size and shape. However, the influence of shape is small so that shrinkage can be expressed as a function of the ratio volume/exposed surface. Figure 13.8 shows that there is a linear relation between the logarithm of ultimate shrinkage and the volume/surface ratio.

The lower shrinkage of large members is due to the fact that only the outer part of the concrete is drying and its shrinkage is restrained by the non-shrinking core. In practice, then, we have *differential* or *restrained shrinkage*. In consequence, no test measures true shrinkage as an intrinsic

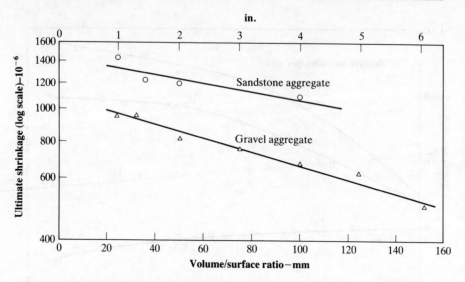

Fig. 13.8: Relation between ultimate shrinkage and volume/surface ratio (From: T. C. HANSEN and A. H. MATTOCK, The influence of size and shape of member on the shrinkage and creep of concrete, *J. Amer. Concr. Inst.*, 63, pp. 267–90 (Feb. 1966).)

property of concrete, so that the specimen size should always be reported.

Prediction of drying shrinkage and swelling

According to ACI 209.R-82, shrinkage $s_h(t, \tau_0)$ at time t (days), measured from the start of drying at τ_0 (days), is expressed as follows:

for moist curing:

$$s_h(t, \tau_0) = \frac{t - \tau_0}{35 + (t - \tau_0)} s_{h\infty}$$

for steam curing: $\quad$ (13.2a)

$$s_h(t, \tau_0) = \frac{t - \tau_0}{55 + (t - \tau_0)} s_{h\infty}$$

(13.2b)

where $s_{h\infty}$ = ultimate shrinkage, and

$$s_{h\infty} = 780 \times 10^{-6} k_1' k_2' k_3' k_4' k_5' k_6' k_7'.$$

(13.3)

For curing times different from 7 days for moist-cured concrete, the age coefficient k_1' is given in Table 13.1; for steam curing with a period of 1 to 3 days, $k_1' = 1$.

Table 13.1: Shrinkage coefficient k_1' for use in Eq. (13.3)

Period of moist curing (days)	Shrinkage coefficient k_1'
1	1.2
3	1.1
7	1.0
14	0.93
28	0.86
90	0.75

The humidity coefficient k_2' is

$$k_2' = 1.40 - 0.010h \quad (40 \leqslant h \leqslant 80)$$

$$k_2' = 3.00 - 0.30h \quad (80 \leqslant h \leqslant 100)$$

(13.4)

where h = relative humidity (per cent).

Since $k_2' = 0$ when $h = 100$ per cent, the ACI method does not predict swelling.

Coefficient k_3' allows for the size of the member in terms of the volume/surface ratio V/S (see page 245). For values of $V/S < 37.5$ mm (1.75 in.) k_3' is given in Table 13.2. When V/S is between 37.5 and 95 mm (1.75 and 3.75 in.):

for $(t - \tau_0) \leqslant 1$ year

$$\left.\begin{array}{l} \text{in SI units: } k_3' = 1.23 - 0.006 \dfrac{V}{S} \\[2ex] \text{in US units: } k_3' = 1.23 - 0.152 \dfrac{V}{S} \end{array}\right\}$$

(13.5(a))

Table 13.2: Shrinkage coefficient k_3' for use in Eq. (13.3)

Volume/surface ratio, V/S		Coefficient k_3'
mm	in.	
12.5	0.50	1.35
19	0.75	1.25
25	1.00	1.17
31	1.25	1.08
37.5	1.50	1.00

for $(t - \tau_0) \geq 1$ year

in SI units: $\quad k_3' = 1.17 - 0.006 \dfrac{V}{S}$

$$(13.5(b))$$

in US units: $\quad k_3' = 1.17 - 0.152 \dfrac{V}{S}$

When $V/S \geq 95$ mm (3.75 in.):

in SI units: $\quad k_3' = 1.2 \, e^{-0.00473(V/S)}$

$$(13.6)$$

in US units: $\quad k_3' = 1.2 \, e^{-0.120(V/S)}$

The coefficients which allow for the composition of the concrete are:

in SI units: $\quad k_4' = 0.89 + 0.00264 \, s$

$$(13.7)$$

in US units: $\quad k_4' = 0.89 + 0.067 \, s$

where s = slump of fresh concrete (mm or in.), and

$$k_5' = 0.30 + 0.014 \, \frac{A_f}{A}, \quad \left(\frac{A_f}{A} \leq 50\right)$$

$$(13.8)$$

$$k_5' = 0.90 + 0.002 \frac{A_f}{A}, \quad \left(\frac{A_f}{A} > 50\right)$$

where A_f/A = fine aggregate/total aggregate ratio by mass (per cent). Also,

in SI units: $\quad k_6' = 0.75 + 0.00061\gamma$

$$(13.9)$$

in US units: $\quad k_6' = 0.75 + 0.00036\gamma$

where γ = cement content (kg/m³ or lb/yd³), and

$$k_7' = 0.95 + 0.008A \qquad (13.10)$$

where A = air content (per cent).

In the UK, BS 8110: Part 2: 1985 gives values of shrinkage and swelling after periods of exposure of 6 months and 30 years (see Fig. 13.9) for various relative humidities of storage and volume/surface ratios. The data apply to concretes made with high-quality, dense, non-shrinking aggregates and to concretes with an effective original water content of 8 per cent of the original mass of concrete. (This value corresponds to approximately 190 litres per cubic metre of concrete.) For concretes with other water contents, the shrinkage of Fig. 13.9 is adjusted in proportion to the actual water content.

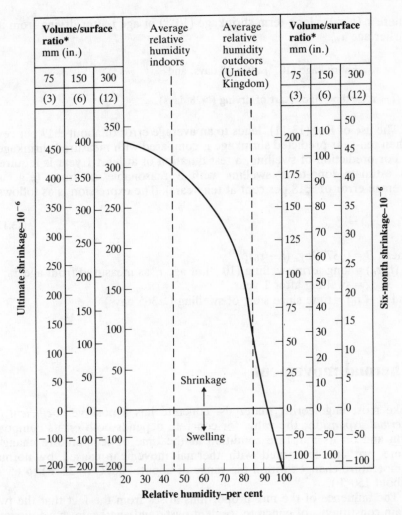

Volume/surface ratio* mm (in.)			Average relative humidity indoors	Average relative humidity outdoors (United Kingdom)	Volume/surface ratio* mm (in.)		
75	150	300			75	150	300
(3)	(6)	(12)			(3)	(6)	(12)

Fig. 13.9: Prediction of shrinkage and swelling of high quality dense aggregate concretes
(From: BS 8110: Part 2: 1985.)
* Sometimes the term 'effective section thickness' is used to represent the size of a member; effective section thickness = 2 × volume/surface ratio

An improvement in the accuracy of prediction of shrinkage is obtained by undertaking short-term tests of 28-day duration and then extrapolating to obtain long-term values. The following expression is applicable for both normal weight and lightweight concretes, stored in any drying environment at normal temperature:

$$s_h(t, \tau_0) = s_{h28} + 100[3.61 \log_e (t - \tau_0) - 12.05]^{\frac{1}{2}} \tag{13.11}$$

where $s_h(t, \tau_0) = $ long-term shrinkage (10^{-6}) at age t after drying from an earlier age τ_0,

$s_{h28} = $ shrinkage (10^{-6}) after 28 days, and

$(t - \tau_0) = $ time since start of drying $(>28 \text{ days})$.

The use of Eq. (13.11), leads to an average error of about ±17 per cent when ten-year predicted shrinkage is compared with measured shrinkage.

For prediction of swelling, a test duration of at least 1 year is required to estimate long-term swelling with a reasonable accuracy (e.g. an average error of ±18 per cent at ten years). The expression is as follows:

$$s_w(t, \tau_0) = s_{w365}^B \tag{13.12}$$

where $B = 0.377[\log_e (t - \tau_0)]^{0.55}$
$s_w(t, \tau_0) = $ long-term swelling (10^{-6}) at age t, as measured from age τ_0,
$s_{w365} = $ swelling after 1 year,
and $(t - \tau_0) = $ time since start of swelling $(>365 \text{ days})$.

Thermal movement

Like most engineering materials, concrete has a positive *coefficient of thermal expansion*; the value for concrete depends both on its composition and on its moisture condition at the time of temperature change. Here, we are concerned with thermal movement caused by normal temperature changes within the range of about $-30\,°C$ $(-22\,°F)$ to $65\,°C$ (about $150\,°F$).

The influence of the mix proportions arises from the fact that the two main constituents of concrete, cement paste and aggregate, have dissimilar thermal coefficients. The coefficient for concrete is affected by these two values and also by the volumetric proportions and elastic properties of the two constituents. In fact, the role of the aggregate here is similar to that in shrinkage and creep, i.e. the aggregate restrains the thermal movement of the cement paste, which has a higher thermal coefficient. The coefficient of thermal expansion of concrete, α_c, is related to the thermal coefficients of aggregate, α_g, and of cement paste, α_p, as follows:

$$\alpha_c = \alpha_p - \frac{2g(\alpha_p - \alpha_g)}{1 + \dfrac{k_p}{k_g} + g\left[1 - \dfrac{k_p}{k_g}\right]} \tag{13.13}$$

where $g = $ volumetric content of aggregate,

and k_p/k_g = stiffness ratio of cement paste to aggregate, approximately equal to the ratio of their moduli of elasticity.

Equation (13.13) is represented in Fig. 13.10, from which it is apparent that, for a given type of aggregate, an increase in its volume concentration reduces α_c while, for a given volume concentration, a lower thermal coefficient of aggregate also reduces α_c; the influence of the stiffness ratio is small.

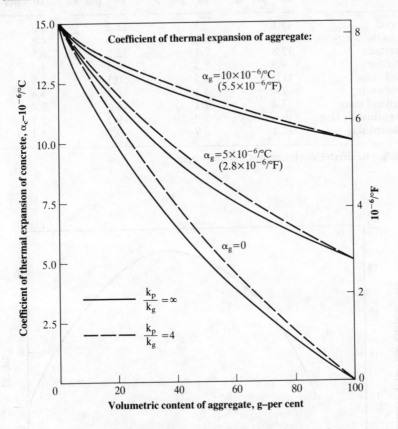

Fig. 13.10:   Influence of volumetric content of aggregate and of aggregate type on linear coefficient of thermal expansion of concrete, using Eq. (13.13); $\alpha_p = 15 \times 10^{-6}/°C$
(Based on: D. W. HOBBS, The dependence of the bulk modulus, Young's modulus, creep, shrinkage and thermal expansion of concrete upon aggregate volume concentration, *Materials and Construction*, Vol. 4, No. 20, pp. 107–14, 1971.)

Typical values of the coefficient of thermal expansion of aggregate are given in Table 13.4. Values for the cement paste vary between 11×10^{-6} and 20×10^{-6} per °C (6×10^{-6} and 11×10^{-6} per °F) depending on the moisture condition. This dependence is due to the fact that the thermal coefficient of cement paste has two components: the *true (kinetic) thermal*

Table 13.3: **Coefficient of thermal expansion of 1:6 concretes made with different aggregates**

Type of aggregate	Linear coefficient of thermal expansion			
	Air-cured concrete		Water-cured concrete	
	10^{-6} per °C	10^{-6} per °F	10^{-6} per °C	10^{-6} per °F
Gravel	13.1	7.3	12.2	6.8
Granite	9.5	5.3	8.6	4.8
Quartzite	12.8	7.1	12.2	6.8
Dolerite	9.5	5.3	8.5	4.7
Sandstone	11.7	6.5	10.1	5.6
Limestone	7.4	4.1	6.1	3.4
Portland stone	7.4	4.1	6.1	3.4
Blast-furnace slag	10.6	5.9	9.2	5.1
Foamed slag	12.1	6.7	9.2	5.1

Building Research Establishment, Crown copyright

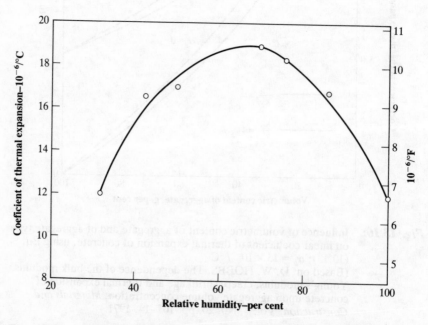

Fig. 13.11: Relation between ambient relative humidity and the linear coefficient of thermal expansion of neat cement paste cured normally
(From: S. L. MEYERS, How temperature and moisture changes may affect the durability of concrete, *Rock Products*, pp. 153–7 (Chicago, Aug. 1951).)

coefficient, which is caused by the molecular movement of the paste, and the *hygrothermal expansion coefficient.* The latter arises from an increase in the internal relative humidity (water vapour pressure) as the temperature increases, with a consequent expansion of the cement paste. No hygrothermal expansion is possible when the paste is totally dry or when it is saturated since there can be no increase in water vapour pressure. However, Fig. 13.11 shows that hygrothermal expansion occurs at intermediate moisture contents, and for a young paste has a maximum at a relative humidity of 70 per cent. For an older paste, the maximum hygrothermal expansion is smaller and occurs at a lower internal relative humidity. In concrete, the hygrothermal effect is naturally smaller.

Table 13.3 gives the values of the thermal coefficient for 1:6 concretes cured in air at a relative humidity of 64 per cent and also for saturated (water-cured) concretes made with different types of aggregate.

Temperature near freezing results in a minimum value of the coefficient of thermal expansion; at still lower temperatures, the coefficient is higher again. Figure 13.12 shows the values for saturated concrete tested in saturated air. In concrete slightly dried after a period of initial curing, then stored and tested at a relative humidity of 90 per cent, the decrease in the thermal coefficient is absent. The behaviour of saturated concrete is of interest because of its vulnerability to freezing and thawing (see Chapter 15).

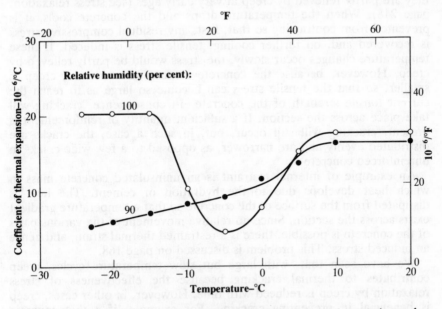

Fig. 13.12: Relation between the linear coefficient of thermal expansion and temperature of concrete specimens stored and tested at the age of 55 days under different conditions of humidity
(From: F. H. WITTMANN and J. LUKAS, Experimental study of thermal expansion of hardened cement paste, *Materials and Structures,* 7, No. 40, pp. 247–52 (Paris, July–Aug. 1974).)

Effects of restraint and cracking

Since stress and strain occur together, any restraint of movement introduces a stress corresponding to the restrained strain.[1] If this stress and the restrained strain are allowed to develop to such an extent that they exceed the strength or strain capacity of concrete, then cracks will take place. Later on, we shall discuss cracking in terms of stress and strength rather than of restrained strain and strain capacity, although either concept can be used.

Restraint can induce both compression and tension but in the majority of cases it is tension which causes problems. There are two forms of restraint: *external* and *internal*. The former exists when movement of a section in a concrete member is fully or partially prevented by external, rigid, or part-rigid, adjacent members or by foundations. Internal restraint exists when there are temperature and moisture gradients within the section. Combinations of external and internal restraints are possible.

To illustrate the effect of external restraint let us consider a section of a completely insulated concrete member whose ends are fully restrained; the section is subjected to a cycle of temperature. As the temperature increases, the concrete is prevented from expanding so that compressive stresses develop uniformly across the section. These stresses are usually small compared with the compressive strength of concrete; moreover, they are partly relieved by creep at very early ages (see stress relaxation, page 218). When the temperature drops and the concrete cools, it is prevented from contracting so that, first, any residual compressive stress is recovered and, on further cooling, tensile stress is induced. If these temperature changes occur slowly, the stress would be partly relieved by creep. However, because the concrete is now more mature, creep is smaller, so that the tensile stress can become so large as to reach the current tensile strength of the concrete. In consequence, cracking will take place across the section. If a sufficient quantity of reinforcement is present, cracking will still occur, but, in such a case, the cracks are distributed evenly and are narrower, as opposed to a few wide cracks in unreinforced concrete.

An example of internal restraint is an uninsulated concrete mass in which heat develops due to the hydration of cement. The heat is dissipated from the surface of the concrete so that a temperature gradient exists across the section. Since no relative movement of the various parts of the concrete is possible, there is a restrained thermal strain, and hence an induced stress. This problem is discussed on page 168.

We have seen that, with large but slow temperature cycles, creep contributes to thermal cracking because the effectiveness of stress relaxation by creep is reduced with time. However, in other cases, creep is beneficial in preventing cracking. For example, if a thin concrete

---

[1] Restrained strain is the difference between 'free' strain and measured strain.

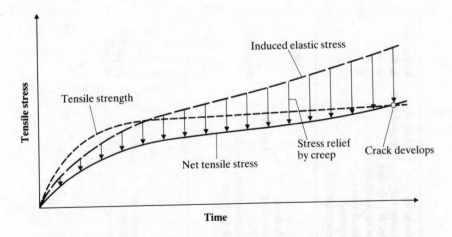

Fig. 13.13: Schematic pattern of crack development when tensile stress due to restrained shrinkage is relieved by creep

member is externally restrained so that contraction due to shrinkage is prevented, the induced uniform elastic tensile stress is relieved by creep (see Fig. 13.13). In cases of thicker members, with no external restraint, but where a moisture gradient exists, shrinkage of the surface layers is restrained by the core so that a tensile stress exists at the outside and a compressive stress inside the concrete. Creep again relieves the stresses, but if the tensile stress exceeds the current strength, surface shrinkage cracking will occur.

Types of cracking

We have referred on several occasions to cracking, and it may be useful to review the various types of cracks. We are not concerned here with cracks caused by an excessive applied load but only with those intrinsic to concrete. These are of three types: *plastic cracks, early-age thermal cracks,* and *drying shrinkage cracks.* Actually, there exist also other types of non-structural cracks; these are listed in Table 13.4 and shown schematically in Fig. 13.14.

Plastic cracks develop before the concrete has hardened (i.e. between 1 and 8 hours after placing) and are in the form of *plastic shrinkage cracks* and *plastic settlement cracks.* The cause and prevention of the former was considered on page 237; the latter develop when settlement of concrete on bleeding is uneven due to the presence of obstructions. These can be in the form of large reinforcing bars or even of unequal depth of the concrete which is monolithically placed. To reduce the incidence of plastic settlement cracks we can use an air-entraining admixture (see Chapter 15) so as to reduce bleeding and also increase the cover to the top steel. Plastic settlement cracks can be eliminated by revibration of

Table 13.4: Classification of intrinsic cracks

Type of cracking	Symbol in Fig. 13.14	Subdivision	Most common location	Primary cause (excluding restraint)	Secondary causes/factors	Remedy (assuming basic redesign is impossible); in all cases reduce restraint	Time of appearance	Reference in this book
Plastic settlement	A	Over reinforcement	Deep sections	Excess bleeding	Rapid early drying conditions	Reduce bleeding (air entrainment) or revibrate	10 min to 3 h	page 255
	B	Arching	Top of columns					
	C	Change of depth	Trough and waffle slabs					
Plastic shrinkage	D	Diagonal	Roads and slabs	Rapid early drying	Low rate of bleeding	Improve early curing	30 min to 6 h	pages 237 and 255
	E	Random	Reinforced concrete slabs					
	F	Over reinforcement	Reinforced concrete slabs	Ditto plus steel near surface				

G Early thermal contraction	External restraint	Thick walls	Excess heat generation	Rapid cooling	Reduce heat and/or insulate	One day to two or three weeks	pages 168 and 254
H	Internal restraint	Thick slabs	Excess temperature gradients				
I Long-term drying shrinkage		Thin slabs and walls	Inefficient joints	Excess shrinkage Inefficient curing	Reduce water content Improve curing	Several weeks or months	page 255
J	Against formwork	'Fair faced' concrete	Impermeable formwork	Rich mixes Poor curing	Improve curing and finishing	One to seven days, sometimes much later	page 258
K Crazing	Floated concrete	Slabs	Over-trowelling				
L Corrosion of reinforcement	Natural	Columns and beams	Lack of cover	Poor quality concrete	Eliminate causes listed	More than two years	page 275
M Alkali-aggregate reaction		Damp locations	Reactive aggregate plus high-alkali cement	Eliminate causes	More than five years listed		page 273

From: Concrete Society Technical Report 22.

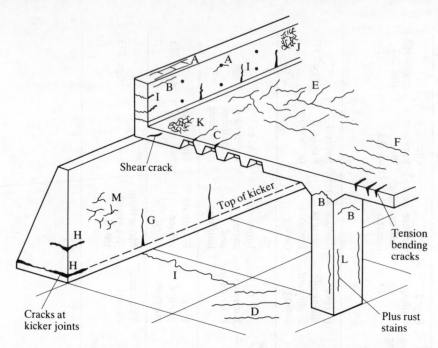

Fig. 13.14: Schematic representation of the various types of cracking which can occur in concrete
(From: CONCRETE SOCIETY, Non-structural cracks in concrete, *Technical Report,* No. 22, pp. 38 (London, 1982).)

concrete at a suitable time; this is at the latest possible time when a vibrating poker can be inserted into the concrete and withdrawn without leaving a significant trace.

The locations of early-age thermal cracks are shown in Fig. 13.14; their causes and method of prevention are discussed in Chapter 9.

As mentioned on page 255, drying shrinkage cracks in large sections are induced by tensile stresses due to internal restraint caused by *differential shrinkage* between the surface and the interior of the concrete. Drying shrinkage cracks, which take weeks or months to develop, occur also because of external restraint to movement provided by another part of the structure or by the subgrade. Drying shrinkage cracking is best reduced by reducing shrinkage (see page 239). Moreover, adequate curing is essential so as to increase the tensile strength of the concrete, together with the elimination of external restraints by the provision of movement joints. The width of shrinkage cracks can be controlled by reinforcement placed as near to the surface as possible, bearing in mind the requirements of cover. Other types of cracking are caused by corrosion of reinforcement and by alkali–aggregate reaction; these are discussed in Chapter 14.

A related form of drying shrinkage cracking is *surface crazing* on walls and slabs, which takes place when the surface layer of the concrete has a higher water content than the interior concrete (see Table 13.4 and Fig. 13.14). Surface crazing usually occurs earlier than drying shrinkage cracking.

The causes, evaluation and repair of non-structural cracks in concrete are fully considered by ACI 224.R–80 and by the Concrete Society Technical Report No. 22.

Bibliography

13.1 ACI COMMITTEE 209.R–82 (Reapproved 1986), Prediction of creep, shrinkage and temperature effects in concrete structures, Part 1: Materials and general properties of concrete, *ACI Manual of Concrete Practice,* 1990.

13.2 ACI COMMITTEE 224.R–89, Control of cracking in concrete structures, Part 3: Uses of concrete in buildings – design, specifications, and related topics, *ACI Manual of Concrete Practice,* 1990.

13.3 ACI 305.R–89, Hot weather concreting, Part 2: Construction practices and inspection pavements, *ACI Manual of Concrete Practice,* 1990.

13.4 CONCRETE SOCIETY, Non-structural cracks in concrete, *Technical Report,* No. 22, pp. 38, (London, 1982).

13.5 T. A. HARRISON, Early-age thermal crack control in concrete, *CIRIA Report* 91, pp. 48, (London, Construction Industry Research and Information Assoc., 1981).

13.6 D. W. HOBBS, The dependence of the bulk modulus, Young's modulus, creep, shrinkage and thermal expansion of concrete upon aggregate volume concentration, *Materials and Construction,* Vol. 4, No. 20, pp. 107–14, 1971.

Problems

13.1 What is the cause of plastic settlement?
13.2 What is crazing?
13.3 What are pop-outs?
13.4 Describe the various causes of cracking in concrete.
13.5 Discuss the influence of mix proportions of concrete on shrinkage.
13.6 Describe the mechanism of drying shrinkage of concrete.
13.7 Compare the carbonation of concrete exposed to intermittent rain and protected from rain.
13.8 What is the effect of wind on fresh concrete?
13.9 What are the main reactions in carbonation of concrete?
13.10 What is autogenous healing of concrete?
13.11 Discuss the main factors affecting the shrinkage of concrete.

13.12 How can unsuitable mix proportions of concrete lead to non-structural cracking?

13.13 How can curing procedures lead to non-structural cracking?

13.14 How can restraint of movement of a member lead to shrinkage cracking?

13.15 Explain what is meant by restrained shrinkage.

13.16 Describe the phenomenon of shrinkage of cement paste.

13.17 Discuss the influence of aggregate on the shrinkage of concrete made with a given cement paste.

13.18 What is autogenous shrinkage?

13.19 What is carbonation shrinkage?

13.20 Describe a test to determine the depth of carbonation of concrete.

13.21 At what rate does the carbonation of concrete progress?

13.22 Explain what is meant by differential shrinkage.

13.23 Explain plastic shrinkage cracking.

13.24 When is the maximum cement content of concrete specified?

13.25 State how you would assess the drying shrinkage of: (i) normal weight concrete, and (ii) lightweight concrete.

13.26 What are the consequences of creep of concrete with respect to cracking?

13.27 How would you assess the drying shrinkage of concrete containing: (i) PFA, (ii) slag, and (iii) a superplasticizer?

13.28 Can the coefficient of thermal expansion of concrete be estimated from the thermal expansion of the two main constituents: cement paste and aggregate? Discuss.

13.29 Explain the terms: true kinetic thermal coefficient, and hygrothermal expansion coefficient.

13.30 Discuss the influence of low temperature on the coefficient of thermal expansion of concrete.

13.31 What is restrained strain?

13.32 Give examples of external restraint and internal restraint of shrinkage.

13.33 How can drying shrinkage cracking be reduced?

13.34 Use the ACI method to predict the ultimate drying shrinkage of concrete, given the following information:

Length of moist curing	14 days
Storage conditions	relative humidity of 70 per cent
Volume/surface ratio	50 mm (2 in.)
Slump	75 mm (3 in.)
Fine aggregate/total aggregate ratio	30 per cent
Cement content	300 kg/m^3 (505 lb/yd^3)
Air content	2 per cent

Answer: 312×10^{-6}

13.35 Use the BS 8110: Part 2: 1985 method to estimate the 30-year

shrinkage of the concrete given in question 13.34; assume the
original water content is 8 per cent.

Answer: 320×10^{-6}

13.36 Calculate the shrinkage required to cause cracking in a fully
restrained plain concrete member, given that the tensile strength is
3 MPa (450 psi) and the modulus of elasticity is 30 GPa (4.4×10^6 psi); assume the concrete to be brittle and have zero creep.

Answer: 100×10^{-6}

13.37 If the concrete of question 13.36 undergoes creep so that the
effective modulus of elasticity is 20 GPa (2.9×10^6 psi), what would
the value of shrinkage be to cause cracking?

Answer: 150×10^{-6}

14

Permeability and durability

The durability of concrete is one of its most important properties because it is essential that concrete should be capable of withstanding the conditions for which it has been designed throughout the life of a structure.

Lack of durability can be caused by external agents arising from the environment or by internal agents within the concrete. Causes can be categorized as physical, mechanical and chemical. Physical causes arise from the action of frost (see Chapter 15) and from differences between the thermal properties of aggregate and of the cement paste (see Chapter 13), while mechanical causes are associated mainly with abrasion (see Chapter 11).

In this chapter, we are concerned with chemical causes: attack by sulphates, acids, sea water, and also by chlorides, which induce electro-chemical corrosion of steel reinforcement. Since this attack takes place *within* the concrete mass, the attacking agent must be able to penetrate throughout the concrete, which therefore has to be permeable. Permeability is, therefore, of critical interest. The attack is aided by the internal transport of agents by diffusion due to internal gradients of moisture and temperature and by osmosis.

Permeability

Permeability is the ease with which liquids or gases can travel through concrete. This property is of interest in relation to the water-tightness of liquid-retaining structures and to chemical attack.

Although there are no prescribed tests by ASTM and BS, the permeability of concrete can be measured by means of a simple laboratory test but the results are mainly comparative. In such a test, the sides of a concrete specimen are sealed and water under pressure is applied to the top surface only. When steady state conditions have been reached (and this may take about 10 days) the quantity of water flowing through a given thickness of concrete in a given time is measured. The water permeability is expressed as a *coefficient of permeability, k,* given by Darcy's equation

$$\frac{1}{A}\frac{dq}{dt} = k\frac{\Delta h}{L}$$

(14.1)

where $\frac{dq}{dt}$ is the rate of flow of water,

A is the cross-sectional area of the sample,
Δh is the drop in hydraulic head through the sample, and
L is the thickness of the sample.

The coefficient k is expressed in m/s or ft/s.

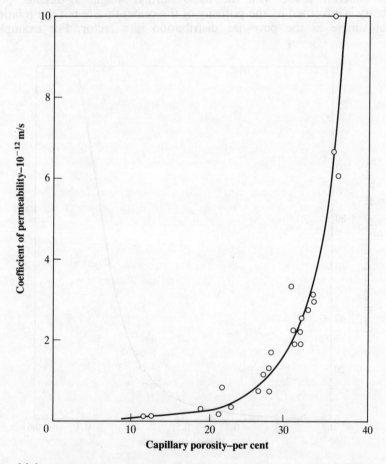

Fig. 14.1: Relation between permeability and capillary porosity of cement paste
(From: T. C. POWERS, Structure and physical properties of hardened Portland cement paste, *J. Amer. Ceramic Soc.,* 41, pp. 1–6 (Jan. 1958).)

There is a further test, prescribed by BS 1881: Part 5: 1970, for the determination of the *initial surface absorption,* which is defined as the rate of flow of water into concrete per unit area, after a given time, under a constant applied load, and at a given temperature. This test gives information about the very thin 'skin' of the concrete only.

Permeability of concrete to air or other gases is of interest in structures such as sewage tanks and gas purifiers, and in pressure vessels in nuclear reactors. Equation (14.1) is applicable, but in the case of air permeability the steady condition is reached in a matter of hours as opposed to days. We should note, however, that there is no unique relation between air and water permeabilities for any concrete, although they are both mainly dependent on the water/cement ratio and the age of the concrete.

For concrete made with the usual normal weight aggregate, permeability is governed by the porosity of the cement paste but the relation is not simple as the pore-size distribution is a factor. For example,

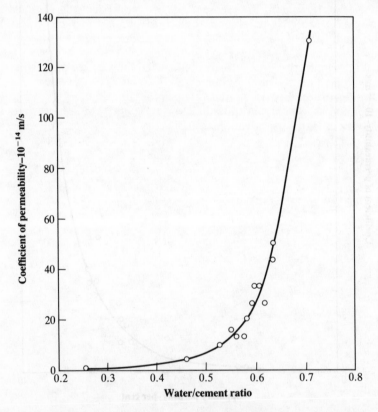

Fig. 14.2: Relation between permeability and water/cement ratio for mature cement pastes (93 per cent of cement hydrated)
(From: T. C. POWERS, L. E. COPELAND, J. C. HAYES and H. M. MANN, Permeability of Portland cement paste, *J. Amer. Concr. Inst.,* 51, pp. 285–98 (Nov. 1954).)

although the porosity of the cement gel is 28 per cent, its permeability is very low, viz. 7×10^{-16} m/s (2.3×10^{-15} ft/s), because of the extremely fine texture of the gel and the very small size of the gel pores. The permeability of hydrated cement paste as a whole is greater because of the presence of larger capillary pores, and, in fact, its permeability is generally a function of capillary porosity (see Fig. 14.1). Since capillary porosity is governed by the water/cement ratio and by the degree of hydration (see Chapter 6), the permeability of cement paste is also mainly dependent on those parameters. Figure 14.2 shows that, for a given degree of hydration, permeability is lower for pastes with lower water/cement ratios, especially below a water/cement ratio of about 0.6, at which the capillaries become segmented or discontinuous (see page 113). For a given water/cement ratio, the permeability decreases as the cement continues to hydrate and fills some of the original water space (see Fig. 14.3), the reduction in permeability being faster the lower the water/cement ratio.

The large influence of segmenting of capillaries on permeability illustrates the fact that permeability is not a simple function of porosity. It is possible for two porous bodies to have similar porosities but different permeabilities, as shown in Fig. 14.4. In fact, only one large passage connecting capillary pores will result in a large permeability, while the porosity will remain virtually unchanged.

From the durability viewpoint, it may be important to achieve low permeability as quickly as possible. Consequently, a mix with a low water/cement ratio is advantageous because the stage at which the

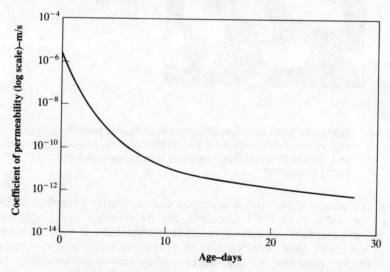

Fig. 14.3: Reduction in permeability of cement paste with the progress of hydration; water/cement ratio = 0.7
(Based on: T. C. POWERS, L. E. COPELAND, J. C. HAYES and H. M. MANN, Permeability of Portland cement paste, *J. Amer. Concr. Inst.*, 51, pp. 285–98 (Nov. 1954).)

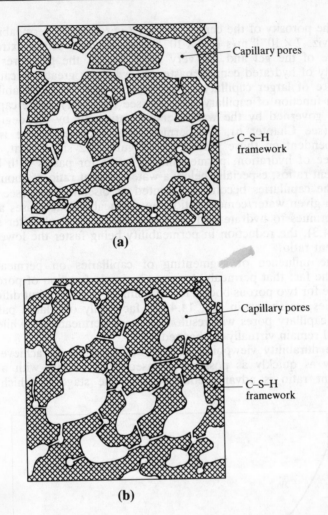

Fig. 14.4: Schematic representation of materials of similar porosity but: (a) high permeability – capillary pores interconnected by large passages, and (b) low permeability – capillary pores segmented and only partly connected.

capillaries become segmented is achieved after a shorter period of moist curing (see Table 6.1). ACI Standard 301–72 (revised 1975) suggests that, to be watertight, structural concrete should have a water/cement ratio of not more than 0.48 for exposure to fresh water and not more than 0.44 for exposure to sea water. A maximum permeability of 1.5×10^{-11} m/s (4.8×10^{-11} ft/s) is often recommended.

So far we have considered the permeability of cement paste which has been moist cured. The permeability of *concrete* is generally of the same order when it is made with normal weight aggregates which have a permeability similar to that of the cement paste, but the use of a more porous aggregate will increase the permeability of concrete. Interruption of moist curing by a period of drying will also cause an increase in permeability because of the creation of water passages by minute shrinkage cracks around aggregate particles, especially the large ones.

Permeability of steam-cured concrete is generally higher than that of moist-cured concrete and, except for concrete subjected to a long curing temperature cycle, supplemental fog curing may be required to achieve an acceptably low permeability.

While a low water/cement ratio is essential for the concrete to have a low permeability, it is not by itself sufficient. The concrete must be dense, and therefore a well-graded aggregate has to be used. This argument can be illustrated by reference to no-fines concrete (see Chapter 18), which can have a low water/cement ratio but a high permeability through passages outside the cement paste, as in the case of porous pipes.

Sulphate attack

Concrete attacked by sulphates has a characteristic whitish appearance, damage usually starting at the edges and corners and followed by cracking and spalling of the concrete. The reason for this appearance is that the essence of sulphate attack is the formation of calcium sulphate (gypsum) and calcium sulphoaluminate (ettringite), both products occupying a greater volume than the compounds which they replace so that expansion and disruption of hardened concrete take place.

It will be recalled from Chapter 2 that gypsum is added to the cement clinker in order to prevent flash set by the hydration of the tricalcium aluminate (C_3A). Gypsum quickly reacts with C_3A to produce ettringite which is harmless because, at this stage, the concrete is still in a semi-plastic state so that expansion can be accommodated.

A similar reaction takes place when hardened concrete is exposed to sulphates from external sources. A typical sulphate solution is the groundwater of some clays which contain sodium, calcium or magnesium sulphates. The sulphates react with both $Ca(OH)_2$ and the hydrated C_3A to form gypsum and ettringite, respectively.

Magnesium sulphate has a more damaging effect than other sulphates because it leads to the decomposition of the hydrated calcium silicates as well as of $Ca(OH)_2$ and of hydrated C_3A; hydrated magnesium silicate is eventually formed and it has no binding properties.

The extent of sulphate attack depends on its concentration and on the permeability of the concrete, i.e. on the ease with which sulphate can travel through the pore system. If the concrete is very permeable, so that water can percolate right through its thickness, $Ca(OH)_2$ will be leached out. Evaporation at the 'far' surface of the concrete leaves behind deposits of calcium carbonate, formed by the reaction of $Ca(OH)_2$ with carbon dioxide; this deposit, of whitish appearance, is known as *efflorescence*. Efflorescence is generally not harmful. However, extensive leaching of $Ca(OH)_2$ will increase porosity so that concrete becomes

267

Table 14.1. Requirements of BS 8110:Part 1:1985 for concrete exposed to sulphate attack

Class	Concentration of sulphates expressed as SO_3			Type of cement	Minimum content of cementitious material in kg/m³ (lb/yd³) for nominal maximum size of aggregate				Maximum free water/ cementitious material ratio
	In soil		In ground-water						
	Total SO_3 per cent	SO_3 in 2:1 water:soil extract g/litre	g/litre		40 mm (1½ in.)	20 mm (¾ in.)	14 mm (⅝ in.)	10 mm (⅜ in.)	
1	less than 0.2	less than 1.0	less than 0.3	Ordinary Portland (Type I); Rapid-hardening Portland (Type III) – combined with or without PFA or slag. Low-heat Portland (Type IV) Sulphate-resisting Portland (Type V)	—	—	—	—	
2(a)	0.2 to 0.5	1.0 to 1.9	0.3 to 1.2	As for class I	300* (500)	330 (560)	350 (590)	370 (620)	0.5
2(b)				Ordinary Portland (Type I); Rapid-hardening Portland (Type III) – combined with 25 to 40 per cent of PFA, or combined with 70 to 90 per cent of slag.	280* (470)	310 (520)	330 (560)	350 (590)	0.55

Class				Type of cement					Max. free water/cement ratio
2(c)				Sulphate-resisting Portland (Type V); Supersulphated	250* (420)	280 (490)	300 (500)	320 (540)	0.55
3(a)	0.5 to 1.0	1.9 to 3.1	1.2 to 2.5	As for Class 2(b)	350 (590)	380 (640)	400 (670)	430 (710)	0.45
3(b)				As for Class 2(c)	300 (500)	330 (560)	350 (590)	370 (620)	0.50
4	1.0 to 2.0	3.1 to 5.6	2.5 to 5.0	As for Class 2(c)	340 (570)	370 (620)	390 (660)	410 (690)	0.45
5	more than 2	more than 5.6	more than 5.0	Sulphate-resisting Portland (Type V); Supersulphated – adequate protective coatings such as sheet polyethylene or polychloroprene, or surface coatings based on asphalt, chlorinated rubber, epoxy or polyurethane materials	340 (570)	370 (620)	390 (660)	410 (690)	0.45

* Values to be at least 300 kg/m³ (500 lb/yd³) for prestressed concrete.

Table 14.2: **Requirements of ACI 318–83 for concrete exposed to sulphate attack**

Sulphate exposure	Water-soluble sulphate (SO_4) in soil per cent by mass	Sulphate (SO_4) in water ppm	Type of cement	Normal weight aggregate concrete Maximum free water/cement ratio*	Lightweight aggregate concrete Minimum compressive strength in MPa (psi)
Negligible	0.00–0.10	0–150	–	–	–
Moderate (seawater)	0.10–0.20	150–1500	Modified (Type II, Portland-pozzolan (Type IP(MS)), Portland blast-furnace (Type IS(MS))	0.50	26 (3750)
Severe	0.20–2.00	1500–10 000	Sulphate-resisting Portland (Type V)	0.45	29 (4250)
Very severe	over 2.00	over 10 000	Sulphate-resisting Portland (Type V) plus pozzolan (fly ash or other suitable material)	0.45	29 (4250)

* A lower water/cement ratio may be required for protection against corrosion of embedded items or from freezing and thawing (Table 14.6).

progressively weaker and more prone to chemical attack. Crystallization of other salts also causes efflorescence.

Salts attack concrete only when present in solution, and not in solid form. The strength of the solution is expressed as concentration, for instance, as the number of parts by mass of sulphur trioxide (SO_3) per million parts of water (ppm). A concentration of 1000 ppm is considered to be moderately severe, and 2000 ppm very severe, especially if magnesium sulphate is the predominant constituent.

Since it is C_3A that is attacked by sulphates, the vulnerability of concrete to sulphate attack can be reduced by the use of cement low in C_3A, viz. sulphate-resisting (Type V) cement. Improved resistance is obtained also by the use of Portland blast-furnace (Type IS) cement and of Portland-pozzolan (Type IP) cement; the exact mechanism by which these cements are beneficial is uncertain.[1] However, it must be emphasized that the type of cement is of secondary importance, or even of none, unless the concrete is dense and has a low permeability, i.e. a low water/cement ratio. The water/cement ratio is the vital factor but a high cement content facilitates full compaction at low water/cement ratios. Typical requirements for concrete exposed to sulphate attack are given in Tables 14.1 and 14.2.

Attack by sea water

Sea water contains sulphates and could be expected to attack concrete in a similar manner to that described in the previous section but, because chlorides are also present, sea-water attack does not generally cause expansion of the concrete. The explanation lies in the fact that gypsum and ettringite are more soluble in a chloride solution than in water, which means that they can be more easily leached out by the sea water. In consequence, there is no disruption but only a very slow increase in porosity and, hence a decrease in strength.

On the other hand, expansion can take place as a result of the pressure exerted by the crystallization of salts in the pores of the concrete. Crystallization occurs above the high-water level at the point of evaporation of water. Since, however, the salt solution rises in the concrete by capillary action, the attack takes place only when water can penetrate into the concrete so that permeability of the concrete is again of great importance.

Concrete between tide marks, subjected to alternating wetting and drying, is severely attacked, while permanently immersed concrete is attacked least. However, the attack by sea water is slowed down by the blocking of pores in the concrete due to the deposition of magnesium

---

[1] Concretes made with these cements require prolonged moist curing (see Table 10.1).

hydroxide which is formed, together with gypsum, by the reaction of magnesium sulphate with $Ca(OH)_2$.

In some cases, the action of sea water on concrete is accompanied by the destructive action of frost, of wave impact and of abrasion. Additional damage can be caused by rupture of concrete surrounding reinforcement which has corroded due to electro-chemical action set up by absorption of salts by the concrete (see page 274).

Sea-water attack can be prevented by the same measures which are used to prevent sulphate attack but, here, the type of cement is of little importance compared to the requirement of low permeability. In reinforced concrete, adequate cover to reinforcement is essential – at least 50 to 75 mm (2 to 3 in.). A cement content of 350 kg/m³ (600 lb/yd³) above the water mark and 300 kg/m³ (500 lb/yd³) below it, and a water/cement ratio of not more than 0.40 to 0.45 are recommended. A well-compacted concrete and good workmanship, especially in the construction joints, are of vital importance.

Acid attack

No Portland cement is resistant to attack by acids. In damp conditions, sulphur dioxide (SO_2) and carbon dioxide (CO_2), as well as some other fumes present in the atmosphere, form acids which attack concrete by dissolving and removing a part, of the hydrated cement paste and leave a soft and very weak mass. This form of attack is encountered in various industrial conditions, such as chimneys, and in some agricultural conditions, such as floors of dairies.

In practice, the degree of attack increases as acidity increases; attack occurs at values of pH below about 6.5, a pH of less than 4.5 leading to severe attack. The rate of attack also depends on the ability of hydrogen ions to be diffused through the cement gel(C–S–H) after $Ca(OH)_2$ has been dissolved and leached out.

As mentioned in Chapter 5, concrete is also attacked by water containing free carbon dioxide in concentrations of at least 15 to 60 ppm; such acidic waters are moorland water and flowing pure water formed by melting ice or by condensation. Peaty water with carbon dioxide levels in excess of 60 ppm is particularly aggressive – it can have a pH as low as 4.4.

Although alkaline in nature, domestic sewage causes deterioration of sewers, especially at fairly high temperatures, when sulphur compounds in the sewage are reduced by anaerobic bacteria to H_2S. This is not a destructive agent in itself, but it is dissolved in moisture films on the exposed surface of the concrete and undergoes oxidation by anaerobic bacteria, finally producing sulphuric acid. The attack occurs, therefore, above the level of flow of the sewage. The cement is gradually dissolved and progressive deterioration of concrete takes place.

Attack of $Ca(OH)_2$ can be prevented or reduced by fixing it. This is achieved by treatment with diluted water glass (sodium silicate) to form calcium silicates in the pores. Surface treatment with coal-tar pitch, rubber or bituminous paints, epoxy resins, and other agents has also been used successfully. The degree of protection achieved by the different treatments varies, but in all cases it is essential that the protective coat adheres well to the concrete and remains undamaged by mechanical agencies, so that access for inspection and renewal of the coating is generally necessary. Detailed information on surface coatings is given in some of the publications listed in the Bibliography.

Alkali–aggregate reaction

Concrete can be damaged by a chemical reaction between the active silica constituents of the aggregate and the alkalis in the cement; this process is known as alkali–silica reaction. The reactive forms of silica are opal (amorphous), chalcedony (cryptocrystalline fibrous), and tridymite (crystalline). These materials occur in several types of rocks: opaline or chalcedonic cherts, siliceous limestones, rhyolites and rhyolitic tuffs, dacite and dacite tuffs, andesite and andesite tuffs, and phyllites. Since lightweight aggregates are often composed of predominantly amorphous silicates, they appear to have the potential for being reactive with alkali in cement. However, there is no evidence of damage of lightweight aggregate concrete caused by alkali–aggregate reaction.

Aggregates containing reactive silica are found mostly in the western part of the US, and to a much smaller extent in the Midlands and South-West of the UK. They are found also in numerous other countries.

The reaction starts with the attack of the siliceous minerals in the aggregate by the alkaline hydroxides derived from the alkalis (Na_2O and K_2O) in the cement. The alkali–silicate gel formed attracts water by absorption or by osmosis and thus tends to increase in volume. Since the gel is confined by the surrounding cement paste, internal pressures result and eventually lead to expansion, cracking and disruption of the cement paste (pop-outs and spalling) and to *map cracking* of the concrete (see Fig. 13.14 and Table 13.4). Expansion of the cement paste appears to be due to the hydraulic pressure generated by osmosis, but expansion can also be caused by the swelling pressure of the still solid products of the alkali–silica reaction. For this reason, it is believed that it is the swelling of the hard aggregate particles that is most harmful to concrete. The speed with which the reaction occurs is controlled by the size of the siliceous particles: fine particles (20 to 30 μm) lead to expansion within four to eight weeks while larger ones do so only after some years. It is generally the very late occurrence of damage due to the alkali–aggregate reaction, often after more than five years, that is a source of worry and uncertainty.

Other factors influencing the progress of the alkali–aggregate reaction are the porosity of the aggregate, the quantity of the alkalis in the cement, the availability of water in the paste and the permeability of the cement paste. The reaction takes place mainly in the exterior of the concrete under permanently wet conditions, or when there is alternating wetting and drying, and at higher temperatures (in the range: 10 to 38 °C (50 to 100 °F)); consequently, avoidance of these environments is recommended.

Although it is known that certain types of aggregate tend to be reactive, there is no simple way of determining whether a given aggregate will cause excessive expansion due to reaction with the alkalis in cement. For potentially safe aggregates, service records have generally to be relied upon but as little as 0.5 per cent of vulnerable aggregate can cause damage.

In the US, a petrographic examination of aggregates is prescribed by ASTM C 295–79, which indicates the amount of reactive minerals but these are not easily recognized especially when there is no previous experience of the aggregate. There exists also a chemical method (ASTM C 289–81) but again this method is not reliable. Probably the most suitable test is the mortar-bar test (ASTM C 227–81). Here, the suspected aggregate, crushed if necessary and made up to a prescribed grading, is used to make special sand-cement mortar bars, using a cement with an equivalent alkali content of not less than 0.6 per cent. The bars are stored over water at 38 °C (100 °F), at which temperature the expansion is more rapid and usually higher than at higher or lower temperatures. The reaction is also accelerated by the use of a fairly high water/cement ratio. The aggregate under test is considered harmful if it expands more than 0.05 per cent after 3 months or more than 0.1 per cent after 6 months.

The ASTM mortar-bar test has not been found to be suitable for UK aggregates. Here, tests on concrete specimens are generally thought to be more appropriate but such tests have yet to be devised. To minimize the risk of alkali–silica reaction the following precautions are recommended in the UK:

(a) Prevent contact between the concrete and external source of moisture.

(b) Use Portland cements with an alkali content of not more than 0.6 per cent expressed as Na_2O. (This is the sum of the actual Na_2O content plus 0.658 times the K_2O content of the cement.)

(c) Use a blend of ordinary Portland (Type I) cement and ground granulated blast-furnace slag, with a minimum of 50 per cent of slag.

(d) Use a blend of ordinary Portland (Type I) cement and PFA with a minimum of 25 per cent of PFA (fly ash), provided that the alkali content of the concrete supplied by the Portland cement *component* is less than $3.0 \, kg/m^3$ ($5.0 \, lb/yd^3$). The alkali content of concrete is calculated by multiplying the alkali content of Portland cement (expressed as a fraction) by the maximum expected Portland cement content.

(e) Limit the alkali content of the concrete to $3.0 \, kg/m^3$ ($5.0 \, lb/yd^3$) which is now the alkali content of the composite cement (expressed as a fraction) times the maximum expected content of the cementitious material.

274

(f) Use a combination of aggregates which is judged to be potentially safe.

We should note that neither slag nor PFA is assumed to contribute reactive alkalis to the concrete, although those materials have fairly high levels of alkalis. However, most of these alkalis are probably contained in the glassy structures of the slag or PFA and take no part in the reaction with aggregate. Moreover, the silica in PFA, paradoxically, attenuates the harmful effects of the alkali–silica reaction. The reaction still takes place, but the finely-divided siliceous material in PFA forms, preferentially, an innocuous product. In other words, there is a pessimum[2] content of reactive silica in the concrete above which little damage occurs.

In the US, ACI 201.2R–77 (reaffirmed 1982) recommends the use of a low alkali cement (not more than 0.6 per cent) and the use of a suitable pozzolanic material, as prescribed by ASTM C 618–84. If potentially deleterious cement–aggregate combinations cannot be avoided, the use of a pozzolan and of at least 30 per cent (by mass) of limestone coarse aggregate is recommended. It is important, however, that the resulting concrete should not exhibit an increase in shrinkage or a reduced resistance to frost damage (with air entrainment, of course). To put it more generally, it is vital to guard against introducing new undesirable features of concrete while curing other ills.

Another type of deleterious alkali–aggregate reaction is that between some dolomitic limestone aggregates and alkalis in cement; this is the *alkali–carbonate reaction*. Important differences between the alkali–silica and alkali–carbonate reactions are the absence of significant quantities of alkali–carbonate gel, the expansive reactions being nearly always associated with the presence of clay, and the uncertainty about the effect of pozzolan in controlling the reaction.

Alkali–carbonate reaction is rare and has not been found in the UK. Test methods developed in the US include petrographic examination to identify dolomitic limestones with the characteristic texture and composition in which relatively large crystals are scattered in a finer-grained matrix of calcite and clay, measurement of the length change of rock samples immersed in a solution of sodium hydroxide (ASTM C 586–69 (reapproved 1981)), and measurement of the length change of concrete specimens containing the suspect rock as aggregate.

Corrosion of reinforcement

The strongly alkaline nature of $Ca(OH)_2$ (pH of about 13) prevents the corrosion of the steel reinforcement by the formation of a thin protective

---

[2] Pessimum is opposite of optimum, i.e. a content of silica such that less damage occurs at higher and lower contents of silica.

film of iron oxide on the metal surface; this protection is known as *passivity*. However, if the concrete is permeable to the extent that carbonation reaches the concrete in contact with the steel or soluble chlorides can penetrate right up to the reinforcement, *and* water and oxygen are present, then corrosion of reinforcement will take place. The passive iron oxide layer is destroyed when the pH falls below about 11.0 and carbonation lowers the pH to about 9. The formation of rust results in an increase in volume compared with the original steel so that swelling pressures will cause cracking and spalling of the concrete (see Fig. 13.14 and Table 13.14).

The harmful effect of carbonation was discussed in Chapter 13, while the deleterious action of chlorides from aggregate, from added calcium chloride or from external sources (de-icing salts and marine environment) was referred to in Chapter 8. Corrosion of steel occurs because of electro-chemical action which is usually encountered when two dissimilar metals are in electrical contact in the presence of moisture and oxygen. However, the same process takes place in steel alone because of differences in the electro-chemical potential on the surface which forms anodic and cathodic regions, connected by the electrolyte in the form of the salt solution in the hydrated cement. The positively charged ferrous ions Fe^{++} at the anode pass into solution while the negatively charged free electrons e^- pass along the steel into the cathode where they are absorbed by the constituents of the electrolyte and combine with water and oxygen to form hydroxyl ions $(OH)^-$. These then combine with the ferrous ions to form ferric hydroxide and this is converted by further oxidation to rust (see Fig. 14.5(a)).

Thus, we can write:

$$Fe \rightarrow Fe^{++} + 2e^- \text{ (anodic reaction)}$$
$$4e^- + O_2 + 2H_2O \rightarrow 4(OH)^- \text{ (cathodic reaction)}$$
$$Fe^{++} + 2(OH)^- \rightarrow Fe(OH)_2 \text{ (ferrous hydroxide)}$$
$$4Fe(OH)_2 + 2H_2O + O_2 \rightarrow 4Fe(OH)_3 \text{ (ferric hydroxide).}$$

It should be emphasized that these are schematic descriptions.

We see that oxygen is consumed, but water is regenerated and is needed only for the process to continue. Thus there is no corrosion in a completely dry atmosphere, probably below a relative humidity of 40 per cent. Nor is there much corrosion in concrete fully immersed in water, except when water can entrain air. It has been suggested that the optimum relative humidity for corrosion is 70 to 80 per cent. At higher relative humidities, the diffusion of oxygen is considerably reduced and also the environmental conditions are more uniform along the steel.

Chloride ions present in the cement paste surrounding the reinforcement react at anodic sites to form hydrochloric acid which destroys the passive protective film on the steel. The surface of the steel then becomes activated locally to form the anode, with the passive surface forming the cathode; the ensuing corrosion is in the form of localized pitting. In the presence of chlorides, the schematic reactions are (see Fig. 14.5(b)):

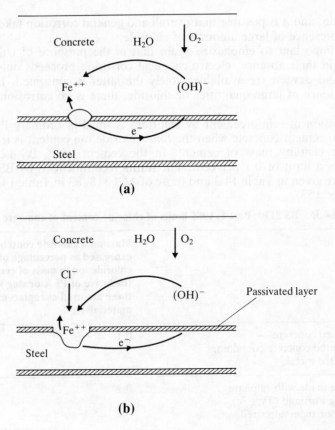

Fig. 14.5: Schematic representation of electro-chemical corrosion: (a) electro-chemical process, and (b) electro-chemical corrosion in the presence of chlorides

$$Fe^{++} + 2Cl^- \rightarrow FeCl_2$$
$$FeCl_2 + 2H_2O \rightarrow Fe(OH)_2 + 2HCl$$

Thus, Cl^- is regenerated. The other reactions, and especially the cathodic reaction, are as in the absence of chlorides.

We should note that the rust contains no chloride, although ferric chloride is formed at an intermediate stage.

Because of the acidic environment in the pit, once it has formed, the pit remains active and increases in depth. Pitting corrosion takes place at a certain potential, called the pitting potential. This potential is higher in dry concrete than at high humidities. As soon as a pit has started to form, the potential of the steel in the neighbourhood drops, so that no new pit is formed for some time. Eventually, there may be a large-scale spread of corrosion, and it is possible that overall and general corrosion takes place in the presence of large amounts of chloride.

It is important to emphasize again that in the presence of chlorides, just as in their absence, electro-chemical corrosion proceeds only when water and oxygen are available. Solely the latter is consumed. Even in the presence of large quantities of chloride, there is no corrosion of dry concrete.

Corrosion of reinforcement by chlorides is small in ordinary Portland (Type I) cement concrete when the *total* chloride ion content is less than 0.4 per cent by mass of cement. In the cement itself, BS 12: 1989 specifies a limit of 0.1 per cent. The limits recommended by BS 8110: 1985 are given in Table 14.3 and those of ACI 318–83 in Table 14.4 (also see page 153).

Table 14.3:   BS 8110: Part 1: 1985 limits of chloride content of concrete

Type	Maximum chloride content expressed as percentage of chloride ion in mass of cement (inclusive of PFA or slag when these are used as replacement materials)
Prestressed concrete Heat-cured concrete containing embedded metal	0.1
Concrete made with sulphate-resisting Portland (Type V) cement or supersulphated cement	0.2
Concrete containing embedded metal and made with the following cements: ordinary Portland (Type I), rapid-hardening Portland (Type III), Portland blast-furnace (Type IS), low-heat Portland (Type IV), low-heat Portland blast-furnace (Types I and IP), or combinations with slag or PFA	0.4

It is only the soluble chlorides that are relevant to corrosion of steel, other chlorides being fixed in the products of hydration. For example, the presence of C_3A may be beneficial in this respect since this reacts with chlorides to form calcium chloroaluminate. For this reason, the use of sulphate-resisting (Type V) cement, which has a low C_3A content, may increase the risk of corrosion induced by chlorides. However, the fixing may not be permanent and, moreover, simultaneous carbonation destroys the ability of the hydrated cement to fix the chloride so that corrosion may take place at a lower chloride content. Sulphate attack also leads to the liberation of chlorides into solution and thus aggravates the corrosion.

278

Table 14.4: ACI 318–83 limits of chloride content of concrete

Type	Maximum water soluble chloride ion (Cl⁻) in concrete by mass of cement
Prestressed concrete	0.06
Reinforced concrete exposed to chloride in service	0.15
Reinforced concrete which will be dry or protected from moisture in service	1.00
Other reinforced concrete construction	0.30

Table 14.5: Requirements of BS 8110: Part 1: 1985 to ensure durability under specified exposure conditions of reinforced and prestressed concrete made with normal weight aggregate

Condition of exposure (see Table 15.1)	Nominal cover of concrete in mm (in.)				
Mild	25 (1)	20($\frac{3}{4}$)	20($\frac{3}{4}$)	20($\frac{3}{4}$)	20($\frac{3}{4}$)
Moderate	–	35(1$\frac{1}{2}$)	30(1$\frac{1}{4}$)	25(1)	20($\frac{3}{4}$)
Severe	–	–	40(1$\frac{1}{2}$)	30(1$\frac{1}{4}$)	25(1)
Very severe	–	–	50(2)†	40(1$\frac{1}{2}$)†	30(1$\frac{1}{4}$)
Extreme	–	–	–	60(2$\frac{1}{2}$)†	50(2)
Maximum free water/ cementitious material‡ ratio	0.65	0.60	0.55	0.50	0.45
Minimum content of cementitious material‡ in kg/m³ (lb/yd³)	275 (460)	300 (510)	325 (550)	350 (590)	400 (670)
Minimum grade* MPa (psi)	30 (4400)	35 (5100)	40 (5800)	45 (6500)	50 (7300)

* Grade is characteristic cube strength (see page 330).
Note: This table applies when the maximum size of aggregate is 20 mm ($\frac{3}{4}$ in.). When it is 10 mm ($\frac{3}{8}$ in.) and 14 mm ($\frac{5}{8}$ in.), respectively, the content of cementitious material should be increased by 40 kg/m³ and 20 kg/m³ (70 lb/yd³ and 35 lb/yd³); conversely, for maximum size of aggregate of 40 mm (1$\frac{1}{2}$ in.) the content of cementitious material can be reduced by 30 kg/m³ (50 lb/yd³) but BS 8110: Part 1: 1985 imposes some overall limitations. Specifically, prestressed concrete must contain at least 300 kg/m³ (500 lb/yd³) of cementitious material.
† For exposure to freezing and thawing, air entrainment should be used.
‡ Inclusive of any ground granulated blast-furnace slag or PFA.

Table 14.6: **Requirements of ACI 318–83 for water/cement ratio and strength for special exposure conditions**

Exposure condition	Maximum water/cement ratio, normal density aggregate concrete	Minimum design strength* in MPa (psi), low density aggregate concrete
Concrete intended to be watertight:		
(a) exposed to fresh water	0.50	25 (3630)
(b) exposed to brackish or sea water	0.45	30 (4350)
Concrete exposed to freezing and thawing in a moist condition:		
(a) kerbs, gutters, guardrails or thin sections	0.45	30 (4350)
(b) other elements	0.50	25 (3630)
(c) in presence of de-icing chemicals	0.45	30 (4350)
For corrosion protection of reinforced concrete exposed to de-icing salts, brackish water, sea water or spray from these sources	0.40†	33 (4790)†

* See page 330.
† If minimum cover required by Table 14.7 is increased by 10 mm ($\frac{1}{2}$ in.), water/cement ratio may be increased to 0.45 for normal density concrete or design strength reduced to 30 MPa (4350 psi) for low density concrete.

The use of slag cement or Portland-pozzolan cement appears to be beneficial in restricting the mobility of chloride ions within the hydrated cement paste.

Since sea water contains chlorides, it is not advisable to use sea water as mixing or curing water; this was mentioned on page 75. In prestressed concrete, the consequences of the use of sea water as mixing water would be far more serious than in the case of reinforced concrete because of the danger of corrosion of prestressing wires, generally of small diameter.

The importance of cover to reinforcement was mentioned in the discussion of sea-water attack. The same precaution is needed to help to prevent corrosion of steel due to carbonation. By specifying a suitable concrete mix for a given environment, it is possible to ensure that the rate of advance of carbonation declines within a short time to a value smaller than 1 mm (0.04 in.) per year. Provided an adequate depth of cover is present, the passivity of the steel reinforcement should then be preserved for the design life of the structure. Table 14.5 gives the requirements of BS 8110: Part 1: 1985 for durability of reinforced and prestressed concretes exposed to various conditions. The corresponding requirements of ACI 318–83 are given in Tables 14.6 and 14.7 (see also page 368). It should be noted that air entrainment for freezing and thawing conditions and for exposure to de-icing salts is mandatory in the US and advisable in the UK (see Chapter 15).

Table 14.7: Requirements of ACI 318–83 for minimum cover for protection of reinforcement

Exposure condition	Minimum cover in mm (in.)		
	Reinforced concrete cast in situ	Precast concrete	Prestressed concrete
Concrete cast against, or permanently exposed to, earth	70 (3)	–	70 (3)
Concrete exposed to earth or weather:			
wall panels	40–50 $(1\frac{1}{2}-2)$	20–40 $(\frac{3}{4}-1\frac{1}{2})$	30 $(1\frac{1}{4})$
slabs and joists	40–50 $(1\frac{1}{2}-2)$	–	30 $(1\frac{1}{4})$
other members	40–50 $(1\frac{1}{2}-2)$	30–50 $(1\frac{1}{4}-2)$	40 $(1\frac{1}{2})$
Concrete not exposed to weather or in contact with earth:			
slabs, walls, joists	20–40 $(1-1\frac{1}{2})$	15–30 $(\frac{3}{4}-1\frac{1}{4})$	20 (1)
beams, columns	40 $(1\frac{1}{2})$	10–40 $(\frac{1}{2}-1\frac{1}{2})$	20–40 $(1-1\frac{1}{2})$
shells, folded plate members	15–20 $(\frac{3}{4}-1)$	10–15 $(\frac{1}{2}-\frac{3}{4})$	10 $(\frac{1}{2})$
non-prestressed reinforcement	–	–	20 (1)
Concrete exposed to de-icing salts, brackish water, sea water or spray from these sources:			
walls and slabs	50 (2)	40 $(1\frac{1}{2})$	–
other members	60 $(2\frac{1}{2})$	50 (2)	–

Note: Ranges of cover quoted depend on the size of steel used.

Prevention of pitting in non-carbonated concrete can be achieved by the application of a moderate level of cathodic protection or by restriction of the availability of oxygen. Cathodic protection involves connecting the reinforcement bars electrically, using an inactive anode and passing an opposing current to that generated in electro-chemical corrosion, thus preventing the latter.

Bibliography

Alkali–aggregate reaction

14.1 ACI COMMITTEE 201.2R–77 (Reapproved 1982), Guide to durable concrete, Part 1: Materials and General Properties of Concrete, *ACI Manual of Concrete Practice*, 1990.

14.2 BUILDING RESEARCH ESTABLISHMENT, Alkali aggregate reactions in concrete, *Digest No.* 258, pp. 8, (London, HMSO, May 1982).

14.3 CONCRETE SOCIETY, Minimising the risk of alkali-silica reaction, *Guidance Notes, Report of a Working Party*, (London, Sept. 1983).

Corrosion of reinforcement

14.4 ACI COMMITTEE 318–89, Building code requirements for reinforced concrete, Part 3: Use of Concrete in Buildings – Design, Specifications and Related Topics, *ACI Manual of Concrete Practice*, 1990.

14.5 A. M. NEVILLE, Corrosion of Reinforcement, *Concrete*, pp. 48–50 (London, June 1983).

14.6 C. L. PAGE, *Corrosion and protection of reinforcing steel in concrete*, 13th Annual Convention of Inst. Conc. Tech., University of Loughborough, 9 pp, March 1985.

Sea-water attack

14.7 ACI COMMITTEE 201, Guide to durable concrete, *Journal of American Concrete Institute*, 74, No. 12, pp. 573–609, (1977).

14.8 ACI COMMITTEE 515, A guide to the use of waterproofing, dampproofing, protective, and decorative barrier systems for concrete, *Concrete International*, Vol. 1, No. 11, pp. 41–81, (Nov. 1979).

14.9 ACI COMMITTEE 515, Guide for the protection of concrete against chemical attack by means of coatings and other corrosion-resistant materials, *Journal of American Concrete Institute*, Proceedings Vol. 53, No. 12, pp. 1305–91, (Dec. 1966).

14.10 ACI COMMITTEE 515.1R–79 (Revised 1985), A guide to the use of waterproofing, dampproofing, protective and decorative barrier systems for concrete, Part 5: Masonry, precast concrete, special processes, *ACI Manual of Concrete Practice, 1990.*

14.11 R. F. M. BAKKER, *Diffusion within and into concrete,* 13th Annual Convention of Inst. Conc. Tech., University of Loughborough, pp. 21, March 1985.

14.12 CEMENT AND CONCRETE ASSOCIATION, Research and Development – Buildings, Annual Report, pp. 9–14, (Slough, 1979).

Problems

14.1 State the influence of water/cement ratio and age on permeability of concrete.

14.2 Discuss the influence of silica content in aggregate on the alkali–silica reaction.

14.3 Explain the role of PFA in minimizing the alkali–silica reaction.

14.4 What are the consequences of sulphates in aggregate?

14.5 What is meant by ion concentration of chloride?

14.6 Why is the depth of cover to steel specified?

14.7 Compare the permeability of steam-cured and moist-cured concretes.

14.8 Does the aggregate grading affect the permeability of concrete?

14.9 What are the conditions necessary for the alkali–silica reaction to take place?

14.10 Why is it important to know whether a given cement is ordinary Portland (Type I) or Portland blast-furnace cement?

14.11 What is the mechanism of sulphate attack of concrete?

14.12 What is the action of acids on concrete?

14.13 Why can you not predict permeability of concrete from its porosity?

14.14 How do sulphates in soil and in groundwater affect concrete?

14.15 What are the effects of sulphates on reinforced concrete?

14.16 How does sewage attack concrete?

14.17 How does moorland water attack concrete?

14.18 When do capillary pores become segmented?

14.19 Which pores influence the permeability of concrete?

14.20 Which pores have little influence on the permeability of concrete?

14.21 Describe the corrosion of steel in concrete subject to carbonation.

14.22 Describe the corrosion of steel in concrete containing calcium chloride.

14.23 What is meant by alkali–reactive aggregate?

14.24 What cement would you use with aggregate suspected of being alkali reactive?

14.25 How would you assess the alkali reactivity of aggregate?

14.26 Compare alkali–reactive siliceous aggregate and alkali–reactive carbonaceous aggregate.

14.27 For what purpose is the minimum cement content specified?

14.28 What is meant by durability of concrete?

14.29 Why is permeability of concrete of importance with respect to durability?

14.30 Why is use of calcium chloride in concrete undesirable?

14.31 Why is permeability of concrete not a simple function of its porosity?

14.32 How is the strength of a sulphate solution expressed?

14.33 Which cements minimize sulphate attack, and why?

14.34 What is the difference between the action of a sulphate solution and of sea water on concrete?

14.35 State the measures required to prevent adverse effects of sea water on reinforced concrete.

14.36 How would you prevent acid attack?

15

Resistance to freezing and thawing

In Chapter 9, we discussed the particular problems associated with concreting in cold weather and the necessity of adequate protection of fresh concrete so that, when mature, it is strong and durable. In this chapter, we are concerned with the vulnerability of concrete, made at normal temperatures, to repeated cycles of freezing and thawing. This is a particular aspect of durability but it is so important that a separate chapter is devoted to it. The problem is linked to the presence of water in concrete but cannot be explained simply by the expansion of water on freezing.

Action of frost

While pure water in the open freezes at 0 °C (32 °F), in concrete the 'water' is really a solution of various salts so that its freezing point is lower. Moreover, the temperature at which water freezes is lower the smaller the size of the pores full of water. In concrete, pores range from very large to very small (see page 101) so that there is no single freezing point. Specifically, the gel pores are too small to permit the formation of ice, and the greater part of freezing takes place in the capillary pores. We can also note that larger voids, arising from incomplete compaction, are usually air-filled and are not appreciably subjected to the initial action of frost.

When water freezes there is an increase in volume of approximately 9 per cent. As the temperature of concrete drops, freezing occurs gradually so that the still unfrozen water in the capillary pores is subjected to hydraulic pressure by the expanding volume of ice. Such pressure, if not relieved, can result in internal tensile stresses of sufficient magnitude to cause local failure of the concrete. This would occur, for example, in porous, saturated concrete containing no empty voids into which the liquid water can move. On subsequent thawing, the expansion caused by ice is maintained so that there is now new space for additional water which may be subsequently imbibed. On re-freezing further expansion occurs. Thus repeated cycles of freezing and thawing have a cumulative effect, and it is the *repeated* freezing and thawing, rather than a single occurrence of frost, that causes damage.

There are two other processes which are thought to contribute to the increase of hydraulic pressure of the unfrozen water in the capillaries. Firstly, since there is a thermodynamic imbalance between the gel water and the ice, diffusion of gel water into capillaries leads to a growth in the ice body and thus to an increase of hydraulic pressure. Secondly, the hydraulic pressure is increased by the pressure of *osmosis* brought about by local increases in solute concentration due to the removal of frozen (pure) water from the original solution.

On the other hand, the presence of adjacent air voids and empty capillaries allows a relief of hydraulic pressure (caused by the formation of ice) by the flow of water into these spaces; this is the basis of deliberate air entrainment, which will be discussed later. The extent of relief depends on the rate of freezing, the permeability of the cement paste and on the length of path which the water has to travel. The net effect of the relief is a contraction of the concrete (see Fig. 15.1). This contraction is greater than the thermal contraction alone because of the relief of the additional pressure induced by the diffusion of the gel water and by osmosis.

The extent of damage caused by repeated cycles of freezing and thawing varies from *surface scaling* to complete disintegration as layers of ice are formed, starting at the exposed surface of the concrete and progressing through its depth. Road kerbs which remain wet for long periods are more vulnerable to frost than any other concrete. Highway slabs are also vulnerable, particularly when salts[1] are used for de-icing because, as they become absorbed by the top surface of the slab, the resulting high osmotic pressures force the water towards the coldest zone where freezing takes place. Damage can be prevented by ensuring that air-entrained concrete is not overvibrated so as to form laitance, and by using a rich mix with a low water/cement ratio; the concrete should be moist-cured for a sufficient period, followed by a period of drying before exposure (see page 290). ASTM C 672 – 76 prescribes a test for visually assessing the *scaling resistance* of concrete.

The main factors in determining the resistance of concrete to freezing and thawing are the degree of saturation and the pore structure of the cement paste; other factors are the strength, elasticity and creep of concrete.

Below some critical value of saturation (80 to 90 per cent), concrete is highly resistant to frost, while dry concrete is, of course, totally unaffected. We should note that even in a water-cured specimen, not all residual space is water-filled, and indeed this is the reason why such a specimen does not fail on *first* freezing. In practice, a large proportion of

---

[1] The de-icing salts normally used are sodium and calcium chlorides, and less frequently urea. Ammonium salts, even in small concentrations, are very harmful and should never be used.

286

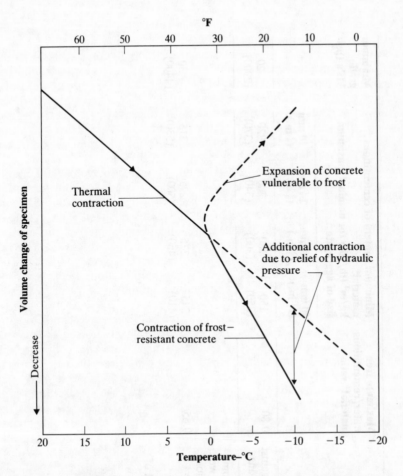

Fig. 15.1: Change in volume of frost-resistant and vulnerable concretes on
cooling
(Based on: T. C. POWERS, Resistance to weathering – freezing
and thawing, *ASTM Sp. Tech. Publicn. No. 169*, pp. 182–7 (1956).)

concrete dries partially at least at some time in its life, and on re-wetting
such concrete will not re-absorb as much water as it has lost previously
(see page 239). Indeed, this is the reason why it is prudent (if possible) to
dry out the concrete before exposure to winter conditions.

Frost-resistant concrete

In order to prevent the damage of concrete by repeated cycles of freezing
and thawing, air can be deliberately entrained within the cement paste by
the use of an air-entraining agent; this method is discussed on page 291.
Air entrainment is effective, however, only when applied to mixes with

Table 15.1: **Requirements of BS 8110: Part 1:1985 to ensure durability under specified conditions of exposure of plain concrete**

Environment	Exposure condition	Maximum free water/cementitious material* ratio	Minimum content of cementitious material* kg/m³ (lb/yd³) for nominal maximum size of aggregate				Minimum grade† MPa (psi)
			40 mm (1½ in.)	20 mm (¾ in.)	14 mm (⅝ in.)	10 mm (⅜ in.)	
Mild	Concrete surfaces protected against weather or aggressive conditions.	0.80	150 (250)	180 (300)	200 (340)	220 (370)	20 (2900)
Moderate	Concrete surfaces sheltered from severe rain or freezing whilst wet. Concrete subject to condensation. Concrete surfaces continuously under water. Concrete in contact with non-aggressive soil (see class 1 of Table 14.1).	0.65	245 (410)	275 (460)	295 (500)	315 (530)	30 (4400)

Severe	Concrete surfaces exposed to severe rain, alternating wetting and drying or occasional freezing or severe condensation.	0.60	270 (450)	300 (510)	320 (540)	340 (570)	35 (5100)
Very severe	Concrete surfaces exposed to sea water spray, de-icing salts (directly or indirectly), corrosive fumes or severe freezing conditions whilst wet.	0.55	295 (500)	325 (550)	345 (580)	365 (610)	35‡ (5100)
Extreme	Concrete surfaces exposed to abrasive action, e.g. sea water carrying solids or flowing water with pH ≤4.5 or machinery or vehicles.	0.50	320 (540)	350 (590)	370 (620)	390 (660)	45 (6500)

* Inclusive of slag or PFA.
† Grade is characteristic cube strength (see page 330).
‡ Applicable only to air-entrained concrete.

low water/cement ratios so that the cement paste has only a small volume of capillaries which are segmented or discontinuous. To achieve this latter feature, concrete should be well compacted, and substantial hydration (which requires adequate curing) must have taken place before exposure to frost.

For less severe conditions of freezing, good quality concrete without air entrainment may be sufficient. Table 15.1 gives the recommended maximum values of the water/cement ratio, minimum content of cementitious material, and minimum strength of plain concrete (see also page 280, and Tables 14.5 and 14.6) for various exposure conditions; the values refer to concrete cured for the periods specified in Table 10.1 prior to exposure and are not applicable when other destructive agencies accompany the action of frost (see Chapter 14). The use of aggregate with a large maximum size or a large proportion of flat particles is inadvisable as pockets of water may collect on the underside of the coarse aggregate.

The adequacy of resistance of a given concrete to frost attack can be determined by *freezing and thawing tests*. Two methods are prescribed by ASTM C 666 – 84. In both of these, rapid freezing is applied, but in one freezing and thawing take place in water, while in the other freezing takes place in air and thawing in water. These conditions are meant to duplicate possible practical conditions of exposure. BS 5075: Part 2: 1982 also prescribes freezing in water. Frost damage is assessed after a number of cycles of freezing and thawing by measuring the loss in mass of the specimen, the increase in its length, decrease in strength or decrease in the dynamic modulus of elasticity, the latter being the most common. With the ASTM methods, freezing and thawing are continued for 300 cycles or until the dynamic modulus is reduced to 60 per cent of its original value, whichever occurs first. The *durability factor, D_f*, is then given by

$$D_f = \frac{n}{3} \left[\frac{E_{dn}}{E_{do}} \right]$$

where n = number of cycles at the end of test,
E_{dn} = dynamic modulus at the end of test, and
E_{do} = dynamic modulus at the start of test.

The value of D_f is of interest primarily in a comparison of different concretes, preferably when only one variable (e.g. the aggregate) is changed. Generally, a value smaller than 40 means that the concrete is probably unsatisfactory, values between 40 and 60 are regarded as doubtful, while values over 60 indicate that the concrete is probably satisfactory.

The test conditions of ASTM C 666 – 84 are more severe than those occurring in practice since the prescribed heating and cooling cycle is between 4.4 and −17.8 °C (40 and 0 °F) at a rate of cooling of up to 14 °C per hour (26 °F per hour). In most parts of the world, a rate of 3 °C per hour (5 °F per hour) is rarely exceeded. However, in another test method, ASTM C 671 – 77 prescribes this slower rate of freezing with one cycle every two weeks. The test continues until the specimen has undergone either the desired number of cycles or the number of cycles after which the critical dilation has occurred. The critical dilation is defined as a dilation which is at least twice that in the preceding cycle.

Air-entraining agents

In the remainder of this chapter we are concerned with protecting concrete from damage due to alternating freezing and thawing by intentionally entraining air bubbles in the concrete by means of a suitable admixture. This air should be clearly distinguished from accidentally entrapped air, which is in the form of larger bubbles left behind during the compaction of fresh concrete.

When mixed with water, air-entraining admixtures produce *discrete* bubble cavities which become incorporated in the cement paste. The essential constituent of the air-entraining admixture is a surface-active agent which lowers the surface tension of water to facilitate the formation of the bubbles, and subsequently ensures that they are stabilized. The surface-active agents concentrate at the air/water interfaces and have hydrophobic (water-repelling) and hydrophilic (water-attracting) properties which are responsible for the dispersion and stabilization of the air bubbles. The bubbles are separate from the capillary pore system in the cement paste and they never become filled with the products of hydration of cement as gel can form only in water. The main types of air-entraining agents are:

(a) animal and vegetable fats and oils and their fatty acids;
(b) natural wood resins, which react with lime in the cement to form a soluble resinate. The resin may be pre-neutralized with NaOH so that a water-soluble soap of a resin acid is obtained; and
(c) wetting agents such as alkali salts of sulphated and sulphonated organic compounds.

Numerous proprietary brands of air-entraining admixtures are available commercially but the performance of the unknown ones should be checked by trial mixes in terms of the requirements of ASTM C 260 – 77 or BS 5075:Part 2:1982. The essential requirement of an air-entraining admixture is that it rapidly produces a system of finely divided and stable foam, the individual bubbles of which resist coalescence; also, the foam must have no harmful chemical effect on the cement.

Air-entraining agents are available as additives as well as admixtures, the former being interground with cement in fixed proportions, as in cements Type IA and IIA of ASTM standards. The additive approach allows less flexibility in altering the air content of different concrete mixes than when admixtures are used. On the other hand, with an admixture, careful control of the batching operation is required to ensure that the quantity of entrained air is within specified limits; otherwise, the advantages of air-entrained concrete may be lost. The dosage required is between 0.005 and 0.05 per cent by mass of cement, and it is necessary to pre-mix such small quantities with some of the batching water in order to facilitate a uniform dispersion of the air-entraining agent.

For protection of concrete, the required minimum volume of voids is 9 per cent of the volume of *mortar,* and it is of course essential that the air be distributed throughout the cement paste. The actual controlling factor is the spacing of the bubbles, i.e. the cement paste thickness between adjacent air voids which should be less than 0.25 mm (0.01 in.) for full protection against frost damage (Fig. 15.2). The spacing can be looked upon as representing twice the distance which the water has to travel in order to relieve pressure.

The adequacy of air entrainment in a given concrete can be estimated by a *spacing factor,* as prescribed by ASTM C 457 – 82a. This factor is an index of the maximum distance of any point in the cement paste from the periphery of a nearby air void, the factor being calculated on the

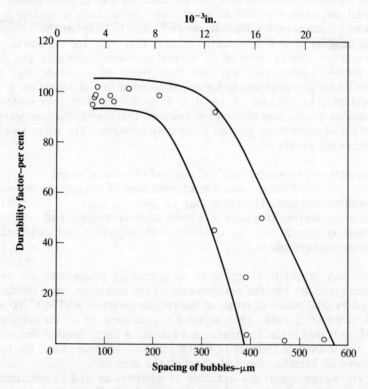

Fig. 15.2: Relation between durability and spacing of bubbles of entrained air (From: U.S. BUREAU OF RECLAMATION, The air-void systems of Highway Research Board co-operative concretes, *Concrete Laboratory Report No.* C–824 (Denver, Colorado, April 1956).)

assumption that all air voids are equal-sized spheres arranged in a simple cubic lattice. The calculation requires a knowledge of: the air content of hardened concrete, as determined, say, by a linear traverse microscope; the average number of air void sections per unit length or the average chord intercept of the voids; and the cement paste content of the concrete by volume. A maximum spacing factor of 0.2 mm (0.008 in.) is required for satisfactory frost protection.

The air bubbles should be as small as possible since the total volume of voids (porosity) affects the strength of concrete (see Chapter 6). Their size depends to a large degree on the air-entraining agent used. In fact, the voids are not all of one size (0.05 to 1.25 mm (0.002 to 0.05 in.)), and it is convenient to express their size in terms of specific surface, i.e. surface area per unit volume.

It must be remembered that accidental air is present in any concrete, whether air-entrained or not, and as the two kinds of voids cannot be readily distinguished, the specific surface represents an *average* value for all voids in a given paste. For air-entrained concrete of satisfactory quality, the specific surface of voids is usually between 16 and 24 mm^{-1} (400 and 600 in^{-1}). By contrast, the specific surface of accidental air is less than 12 mm^{-1} (300 in^{-1}).

Factors influencing air entrainment

Although entrained air is present only in the cement paste, it is usual to specify and measure the air content as a percentage of the volume of the *concrete*. Typical values of air content required for a spacing of 0.25 mm (0.01 in.) are given in Table 15.2, which indicates that richer mixes require a greater volume of entrained air than leaner mixes. Recommended air contents of concretes having different maximum aggregate sizes are given in Table 15.3.

Generally, the larger the quantity of air-entraining agent the more air is entrained but there is a dosage limit beyond which there is no further increase in the volume of voids. For a given amount of air-entraining agent, other influencing factors are as follows:

(a) a more workable mix holds more air than a drier mix;
(b) an increase in the fineness of cement decreases the effectiveness of air entraining;
(c) alkali content in cement greater than 0.8 per cent increases the amount of entrained air;
(d) an increase in carbon content of fly ash (PFA) decreases the amount of entrained air; the use of water-reducing admixtures (see Chapter 8) leads to an increase in the amount of entrained air even if the water-reducing admixture has no air-entraining properties *per se*. (The influence of superplasticizers is less clear so that tests should always be made.);
(e) an excess of very fine sand particles reduces the amount of entrained

Table 15.2: Air content required for a void spacing of 0.25 mm (0.01 in.)

Approximate cement content of concrete		Water/cement ratio	Air requirement as a percentage of volume of concrete for specific surface of voids, mm^{-1} (in.$^{-1}$), of:				
kg/m³	lb/yd³		14 (350)	18 (450)	20 (500)	24 (600)	31 (800)
445	750		8.5	6.4	5.0	3.4	1.8
390	660	0.35	7.5	5.6	4.4	3.0	1.6
330	560		6.4	4.8	3.8	2.5	1.3
445	750		10.2	7.6	6.0	4.0	2.1
390	660	0.49	8.9	6.7	5.3	3.5	1.9
330	560		7.6	5.7	4.5	3.0	1.6
280	470		6.4	4.8	3.8	2.5	1.3
445	750		12.4	9.4	7.4	5.0	2.6
390	660		10.9	8.2	6.4	4.3	2.3
330	560	0.66	9.3	7.0	5.5	3.7	1.9
280	470		7.8	5.8	4.6	3.1	1.6
225	380		6.2	4.7	3.7	2.5	1.3

From: T. C. POWERS, Void spacing as a basis for producing air-entrained concrete, J. Amer. Conc. Inst., 50, pp. 741-60 (May 1954), and Discussion, pp. 760-1-760-15 (Dec. 1954).

Table 15.3: Recommended air content of concretes containing aggregates of different maximum size, according to ACI 201.2R–77 (reaffirmed 1982) and BS 8110: Part 1: 1985

Maximum size of aggregate		Recommended total air content of concrete (per cent) for level of exposure:		
		ACI		BS*
mm	in. (approx.)	Moderate†	Severe‡	Wet conditions and de-icing salts
10	$\frac{3}{8}$	6.0	7.5	7.0
12.5	$\frac{1}{2}$	5.5	7.0	–
14	$\frac{1}{2}$	–	–	6.0
20	$\frac{3}{4}$	5.0	6.0	5.0
25	1	4.5	6.0	–
40	$1\frac{1}{2}$	4.5	5.5	4.0
50	2	4.0	5.0	–
70	3	3.5	4.5	–
150	6	3.0	4.0	–

* Applicable for concrete of grade (characteristic cube strength) (see page 330) less than 50 (7300 psi).
† Cold climate where concrete will be occasionally exposed to moisture prior to freezing, and where no de-icing salts are used, e.g. exterior walls, beams, slabs not in contact with soil.
‡ Outdoor exposure in cold climate where concrete will be in almost continuous contact with moisture prior to freezing or where de-icing salts are used, e.g. bridge decks, pavements, sidewalks, and water tanks.

air, but the material in the 300 to 600 μm range (No. 50 to 30 ASTM sieves) increases it;
(f) hard mixing water reduces the entrained-air content;
(g) mixing time should be an optimum because too short a time causes a non-uniform dispersion of the bubbles, while over-mixing gradually expels some air;
(h) very fast rotation of the mixer increases the amount of entrained air;
(i) higher temperature leads to a greater loss of air, and steam-curing of concrete may lead to incipient cracking because of the expansion of the air bubbles;
(j) transportation and prolonged vibration reduce the amount of entrained air (hence, the air content should be determined on concrete as *placed*).

295

Measurement of air content

There are three methods of measuring the *total* air content of fresh concrete: *gravimetric* (ASTM C 138–81); *volumetric* (ASTM C 173–78); and *pressure* (ASTM C 231–82 and BS 1881: Part 106: 1983). Since the entrained air cannot be distinguished in these tests from the large bubbles of accidental air, it is important that the concrete be fully compacted.

The most dependable and accurate method is the pressure method, which is based on the relation between the volume of air and the applied pressure (at a constant temperature) given by Boyle's law. The mix proportions or the material properties need not be known and the percentage of air is obtained direct. However, at high altitudes, the pressure meter must be re-calibrated, and the method is not suitable for use with porous aggregates.

A typical air meter is shown in Fig. 15.3. The procedure consists essentially of observing the decrease in volume of a sample of compacted concrete when subjected to a known pressure as applied by a small pump. When the pressure gauge shows the required value, the fall in the

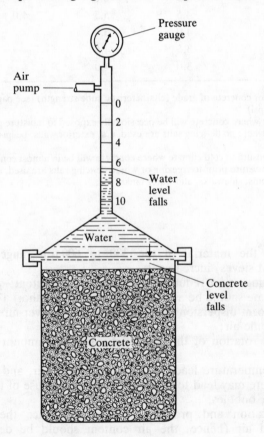

Fig. 15.3: Pressure-type air meter

level of the water in the calibrated tube above the concrete gives the decrease in volume of air in the concrete, i.e. the percentage air content.

As mentioned previously, the air content of *hardened* concrete is measured on polished sections of concrete by means of a microscope (ASTM C 457–82a). Alternatively, a high-pressure air meter can be used.

Other effects of air entrainment

As stated on page 286, the beneficial effect of air entrainment on concrete subjected to freezing and thawing cycles is to create space for the movement of water under hydraulic pressure. There are, however,

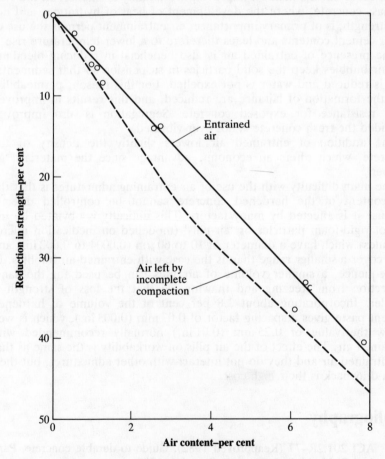

Fig. 15.4: Effect of entrained and accidental air on the strength of concrete (From: P. J. F. WRIGHT, Entrained air in concrete, *Proc. Inst. C. E.*, Part 1, 2, No. 3, pp. 337–58 (London, May 1953); TRRL, Crown copyright.)

some further effects on the properties of concrete, some beneficial, others not. One of the most important is the influence of voids on the strength of concrete at all ages. Figure 15.4 shows that when entrained air is added to a mix without any other change in mix proportions, the decrease in strength is proportional to the air content up to a level of 8 per cent. However, air-entrainment has a beneficial effect on the workability of concrete, probably because the spherical air bubbles act as fine aggregate of very low surface friction and high compressibility. Thus in order to keep the workability constant, the addition of entrained air can be accompanied by a reduction in the water/cement ratio, and this consequently compensates for the loss of strength. This compensating effect depends on the richness of the mix: the nett loss of strength of a richer mix is higher than that of a leaner mix because, in the former, the effect of air entrainment on improving workability is smaller. In the case of mass concrete, where the development of heat of hydration, and not the strength, is of primary importance, air entrainment permits the use of lower cement contents and leads therefore to a lower temperature rise.

The presence of entrained air is also beneficial in reducing bleeding: the air bubbles keep the solid particles in suspension so that sedimentation is reduced and water is not expelled. For this reason, permeability and the formation of laitance are reduced, and this results in improved frost resistance for exposed concrete. Segregation is also improved provided the fresh concrete is not over-vibrated.

The addition of entrained air lowers slightly the density of the concrete, which offers an economic advantage since the materials 'go further'.

The main difficulty with the use of air-entraining admixtures is that the air content of the hardened concrete cannot be controlled directly because it is affected by many factors. This difficulty is obviated by the use of rigid-foam particles or 'air-pills' (modelled on medication micro-capsules) which have a diameter of 10 to 60 μm (0.0004 to 0.002 in.) and thus cover a smaller range than is the case with entrained-air bubbles. In consequence, a smaller volume of air pills can be used for the same protection from freezing and thawing, so that the loss of strength is smaller. Incorporating about 2.8 per cent of the volume of hardened cement paste gives a spacing factor of 0.07 mm (0.003 in.), which is well below the value of 0.25 mm (0.01 in.) normally recommended with entrained air. The effect of the air pills on workability is the same as that of entrained air and they do not interact with other admixtures, but their main drawback is their high cost.

Bibliography

15.1 ACI 201.2R–77 (Reapproved 1982), Guide to durable concrete, Part 1, *ACI Manual of Concrete Practice*, 1990.
15.2 V. M. MALHOTRA (Editor), DEVELOPMENTS IN THE USE OF SUPERPLASTICIZERS, *Amer. Concr. Inst. Sp. Publicn.* No. 68, pp. 561 (1981).

15.3 SUPERPLASTICIZERS IN CONCRETE, *Amer. Concr. Inst. Sp. Publicn.* No. 62, pp. 427 (1979).

Problems

15.1 What is meant by air detraining?
15.2 Discuss the factors affecting the air content in air-entrained concrete.
15.3 How would you determine the air content of hardened concrete?
15.4 Explain how air entrainment improves the resistance of concrete to freezing and thawing.
15.5 What is meant by the spacing factor in cement paste?
15.6 What are the types of air in concrete determined by an air meter?
15.7 What is the effect of pumping on the air content of concrete?
15.8 What are the methods of determining the air content of concrete?
15.9 Which concrete will suffer more damage due to freezing and thawing: (a) dry or wet; (b) well-cured or poorly cured; (c) young or old? Give your reasons.
15.10 What is meant by the durability factor?
15.11 What causes more damage: one cycle of heavy frost or many cycles of light frost? Give your reasons.
15.12 What is the difference between entrapped air and entrained air?
15.13 How does the entrained-air content necessary for durability vary with the maximum aggregate size?
15.14 Explain why there is a difference between the air content measured at the mixer and after placing.
15.15 What factors affect the air content of concrete made with a given quantity of an air-entraining admixture?
15.16 What is the effect of temperature on the air content of concrete made with a given quantity of an air-entraining admixture?
15.17 Why should the air content of concrete be measured at the location of placing?
15.18 Describe the mechanism of frost attack on hardened concrete.
15.19 How do diffusion and osmosis contribute to frost attack?

16

Testing

It's obvious that it is not enough to know how to select a concrete mix so that it can be expected to have certain properties and to specify such a mix, but it is also necessary to ensure that this is indeed the case.

The basic method of verifying that concrete complies with the specification (see Chapter 17) is to test its strength using cubes or cylinders made from samples of fresh concrete. Ideally, it would be preferable to devise *compliance* tests for the mix proportions of *fresh* concrete even before it has been placed but, unfortunately, such tests are rather complex and not suitable for site work. Consequently, the strength of *hardened* concrete has to be determined, by which time a considerable amount of suspect concrete may have been placed. To offset this disadvantage, *accelerated* strength tests are sometimes used as a basis for compliance.

It must be noted that non-compliance by a single test specimen, or even by a group, does not necessarily mean that the concrete from which the test specimens have been made is inferior to that specified; the engineer's reaction should be to investigate the concrete further. This may take the form of non-destructive tests on the concrete in the structure (see BS 1881: Part 201: 1986) or of taking test cores for assessing the strength. All these topics will now be discussed.

Precision of testing

In the next chapter, we shall refer to the variability of the properties of concrete. These can be determined only by testing, and testing itself introduces error. It is important to realize this and to understand what is meant by *precision* of testing concrete. Precision is a general term used for the closeness of agreement between replicate test results, and two terms are relevant: *repeatability* and *reproducibility*.

BS 5497: Part 1: 1979 defines repeatability as the value below which the absolute difference between two single test results, obtained with the same method on identical test material under the 'same' conditions (i.e. same operator, same apparatus, same laboratory, and a short interval of time), may be expected to lie within a specified probability (usually 95 per cent). On the other hand, reproducibility is defined as the value below which the absolute difference between two single test results, obtained with the same method on identical test material under 'different' conditions (i.e. different operators, different apparatus, different laboratories and/or at different times), may be expected to lie within a specified probability (usually 95 per cent).

Values of repeatability and reproducibility are applied in a variety of ways, e.g.

(a) to verify that the experimental technique of a laboratory is up to requirement;
(b) to compare the results of tests performed on a sample from a batch of material with the specification;
(c) to compare test results obtained by a supplier and by a consumer on the same batch of material.

According to BS 5497: Part 1: 1979, the repeatability r and reproducibility R are given by:

$$r = 1.96 \, (2\sigma_r^2)^{\frac{1}{2}} = 2.8\sigma_r$$

$$R = 1.96(2 \, [\sigma_L^2 + \sigma_r^2])^{\frac{1}{2}}$$

$$= 1.96 \, (2\sigma_R^2)^{\frac{1}{2}}$$

or

$$R = 2.8\sigma_R$$

where σ_r^2 = repeatability variance,
σ_L^2 = between-laboratory variance (including between-operator and between-equipment variances), and
σ_R^2 = reproducibility variance.

In the above expressions, the coefficient of 1.96 is for a normal distribution (see page 237) with a sufficient number of test results. The coefficient $2^{\frac{1}{2}}$ is derived from the fact that r and R refer to the differences between two single test results.

Recently, standards in the UK have introduced precision compliance requirements for testing concrete. For the compressive strength, BS 1881: Part 116: 1983 calls for a repeatability of 10 per cent of the mean of pairs of 150 mm (6 in.) cubes made from the same sample of concrete, cured under the same conditions and tested at the age of 28 days. For the method of sampling fresh concrete on site, BS 1881: Part 101: 1983 controls the precision by the sampling error and testing error of compressive strength results; both values should be less than 3 per cent for a satisfactory sampling procedure. Advice is also given in BS 812: Part 101: 1984 on the use of repeatability values to screen data and to monitor performance within the laboratory. In the same standard, information is provided on the use of reproducibility values for comparison of two or more laboratories in setting specification limits.

Analysis of fresh concrete

The determination of the composition of the concrete at an early age could be of considerable benefit because, if the actual proportions correspond to those specified, there is little need for testing the strength of hardened concrete. The two properties of greatest interest are the water/cement ratio and the cement content because these are mainly responsible for ensuring that concrete is both adequately strong and durable.

In the UK, the recommendations for the assessment of the composition of fresh concrete (see Bibliography) suggest five different methods for assessing the cement content. The *buoyancy method* requires that a concrete test sample is weighed in air and in water, and then washed over a nest of sieves to separate the cement and the fines in the aggregate; fines are defined as those passing a 150 μm (No. 100 ASTM) test sieve. The washed aggregate is weighed in water and the proportion of cement is determined from the difference between the apparent mass (weight) of the sample in water and the apparent mass (weight) of aggregate in water. Calibration tests are required to determine the relative densities of the aggregates and the fraction of the aggregate which passes the 150 μm (No. 100 ASTM) sieve so that corrections can be made for silt and fine sand in the 'cement fraction'.

In the *chemical method,* a sample of concrete is weighed and washed over a nest of sieves to separate material finer than 300 μm (No. 50 ASTM) test sieve; there must be no calcareous material in the fines. A sub-sample of the suspension of cement and fines is treated with nitric acid and the concentration of calcium is determined using a flame photometer; here, calibration tests are required. The water content of the concrete is determined by estimating the dilution of a standard solution of sodium chloride. A siphon container is used to assess the coarse aggregate content, the fine aggregate being found by the difference.

The *constant volume (RAM[1]) method* requires a sample which is weighed and transferred to an elutriation column where the upward flow of water separates the material smaller than 600 μm (No. 30 ASTM) sieve. A part of this slurry is vibrated on a 150 μm (No. 100 ASTM) sieve, then flocculated and transferred to a constant volume vessel. This is weighed and, using a calibration chart, the cement content is determined. A correction for aggregate particles smaller than 150 μm

---

[1] Abbreviation for Rapid Analysis Machine.

(No. 100 ASTM) has to be made, and the calibration has to be performed for each set of materials used.

The *physical separation method* requires the concrete sample to be weighed and washed through a vibrating nest of sieves to separate the material passing a 212 μm (No. 70 ASTM) sieve. The washings are sub-sampled automatically and the solids are flocculated, collected and dried. The cement is separated from the fine sand by centrifuging a small sample in bromoform, which is a liquid with a relative density between that of cement and of a typical aggregate. Alternatively, the quantity of fine sand in the sample of cement and fine sand can be estimated from calibration tests.

In the *pressure filter method,* the sample of concrete is weighed, agitated with water and then washed over a nest of sieves to separate the cement and the fines passing a 150 μm (No. 100 ASTM) sieve. The fine material is then filtered under pressure and the separated quantities weighed. Calibration is required to determine the amount of fine sand passing the 150 μm (No. 100 ASTM) sieve. Corrections are also necessary for cement solubility and for the fraction of cement retained on the sieve. The aggregate content is determined after drying and weighing the material retained on the sieves.

The water content of the fresh concrete can be found as in the chemical method or, alternatively, a rapid drying method can be used; during heating, the sample must be continuously stirred to prevent the formation of lumps. The water content is determined by the difference in mass before and after drying but an allowance is required for absorbed water. The determination of water content is complicated also by the changes which take place as the cement hydrates.

We have described five different methods of analysis of fresh concrete but, because of the difficulties with their accuracy, *compliance* testing for cement content and water/cement ratio of fresh concrete has so far not been used. However, BS 5328: 1981 includes the analysis of fresh concrete for the purpose of determining the mix proportions with the proviso that the method of test should have an accuracy of ±10 per cent of the true value with a confidence interval of 95 per cent.[2] It should also be noted that some other properties of fresh concrete are determined in order to establish compliance: density (unit weight), workability, air content, and temperature (see Chapter 17).

Strength tests

For obvious practical reasons, the strength of concrete is determined using small specimens. As we have seen in Chapters 6 and 11, the

[2] J. B. Kennedy and A. M. Neville, *Basic Statistical Methods for Engineers and Scientists,* 3rd Edition, (Harper & Row, 1985).

strength of a given concrete specimen is influenced by several secondary factors such as the rate of loading, moisture condition, specimen size, and curing conditions. Furthermore, the type of testing machine influences the test result recorded. Consequently, we need to standardize procedures in the manufacture of test specimens and in their testing in order to assess accurately the quality of concrete.

Compressive strength

This is determined using 150×300 mm (6×12 in.) cylinders in the US and 150 mm (6 in.) cubes in the UK, although standards permit the use of smaller specimens depending on the maximum size of aggregate.

According to ASTM C 470–81, the *test cylinder* is cast either in a reusable mould, preferably with a clamped base, or in non-reusable mould. The former type of mould is made from steel, cast iron, brass and various plastics, whilst a non-reusable mould can be made from sheet metal, plastic, waterproof paper products or other materials which satisfy the physical requirements of watertightness, absorptivity and elongation. A thin layer of mineral oil has to be applied to the inside surfaces of most types of moulds in order to prevent bond between the concrete and the mould. Concrete is then placed in the mould in layers. Compaction of high-slump concrete is achieved in three layers, each layer being compacted by 25 strokes of a 16 mm ($\frac{5}{8}$ in.) diameter steel rod with a rounded end. For low-slump concrete, compaction is in two layers using internal or external vibration – details of these procedures are prescribed by ASTM C 192–81.

The top surface of a cylinder, finished by a trowel, is not plane and smooth enough for testing, and so requires further preparation. ASTM C 617–84 requires the end surfaces to be plane within 0.05 mm (0.002 in.), a tolerance which applies also to the platens of the testing machine. There are two methods of obtaining a plane and smooth surface: grinding and capping. The former method is satisfactory but expensive. For capping, three materials can be used: a stiff Portland cement paste on freshly-cast concrete, and either a mixture of sulphur and a granular material (e.g. milled fired clay) or a high-strength gypsum plaster on hardened concrete. The cap should be thin, preferably 1.5 to 3 mm ($\frac{1}{16}$ to $\frac{1}{8}$ in.) thick, and have a strength similar to that of the concrete being tested. Probably the best capping material is the sulphur-clay mixture which is suitable for concrete strengths up to 100 MPa (16 000 psi). However, the use of a fume cupboard is necessary as toxic fumes are produced.

In addition to being plane, the end surfaces of the test cylinder should be normal to its axis, and this guarantees also that the end planes are parallel to one another. However, a small tolerance is permitted, usually an inclination of the axis of the specimen to the axis of the testing machine of 6 mm in 300 mm ($\frac{1}{4}$ in. in 12 in.). No apparent loss of strength occurs as a result of such a deviation. Likewise, a small lack of parallelism between the end surfaces of a specimen does not affect its strength, provided the testing machine is equipped with a seating which can align freely, as prescribed by ASTM C 39–83b.

The curing conditions for the *standard test cylinders* are specified by ASTM C 192–81. When cast in the laboratory, the moulded specimens are stored for not less than 20, and not more than 48 hours, at a temperature of $23 \pm 1.7\,°C$ ($73 \pm 3\,°F$) so that moisture loss is prevented. Subsequently, the de-moulded cylinders are stored at the same temperature and under moist conditions or in saturated lime water until the prescribed age at testing. Because they are subjected to standard conditions, these cylinders give the *potential* strength of concrete. In addition, *service cylinders* (ASTM C 31–84) may be used to determine the *actual* quality of the concrete in the structure by being subjected to the same conditions as the structure. This procedure is of interest when we want to decide when formwork may be struck, or when further (superimposed) construction may continue, or when the structure may be put into service.

The compressive strength of the cylinders is determined according to ASTM C 39–83b at a constant loading rate of 0.15 to 0.34 MPa/s (20 to 50 psi/s) for hydraulically operated machines, or at a standard deformation rate of 1.3 mm/min. (0.05 in./min.) for mechanically operated machines. The maximum recorded load divided by the area of cross section of the specimen gives the compressive strength, which is reported to the nearest 10 psi (0.05 MPa).

In the UK, the *test cube* is cast in steel or cast-iron moulds of prescribed dimensions and planeness with narrow tolerances, with the mould and its base clamped together. BS 1881: Part 108: 1983 prescribes filling the mould in layers of approximately 50 mm. Compaction of each layer is achieved by not less than 35 strokes (for 150 mm cubes), or 25 strokes (for 100 mm cubes), of a 25 mm (1 in.) square steel punner; alternatively, vibration may be used. Further treatment of the test cubes is prescribed by BS 1881: Part 111: 1983. After the top surface has been finished by a trowel, the cube should be stored at a temperature of $20 \pm 5\,°C$ ($68 \pm 9\,°F$) when the cubes are to be tested at, or more than, 7 days or $20 \pm 2\,°C$ ($68 \pm 3.6\,°F$) when the test age is less than 7 days; the preferred relative humidity is not less than 90 per cent, but storage under damp material covered with an impervious cover is permitted. The cube is de-moulded just before testing at 24 hours. For greater ages at test, demoulding takes place between 16 to 28 hours after adding water to the mix, and the specimens are stored in a curing tank at $20 \pm 2\,°C$ ($68 \pm 3.6\,°F$) until the prescribed age. The most common age at test is 28 days, but additional tests can be made at 3 and 7 days, and less commonly, at 1, 2, and 14 days, 13 and 26 weeks and 1 year.

The foregoing curing procedure applies to *standard test cubes* but, as in the case of cylinders, *service cubes* may also be used to determine the *actual* quality of the concrete in the structure by curing the cubes under the same conditions as apply to the concrete in the structure.

BS 1881: Part 116: 1983 specifies that the cube is placed with the *cast* faces in contact with the platens of the testing machine, i.e. the position of the cubes as tested is at right angles to the position as cast. The load is applied at a constant rate of stress within the range of 0.2 to 0.4 MPa/s (29 to 58 psi/s), and the crushing strength is reported to the nearest 0.5 MPa (50 psi).

On page 100 we considered the failure of concrete subjected to (pure) uniaxial compression. This would be the ideal mode of testing, but the compression test imposes a more complex system of stress, mainly because of lateral forces developed between the end surfaces of the concrete specimen and the adjacent steel platens of the testing machine. These forces are induced by the restraint of the concrete, which attempts to expand laterally (Poisson effect), by the several-times stiffer steel, which has a much smaller lateral expansion. The degree of *platen restraint* on the concrete section depends on the friction developed at the concrete-platen interfaces, and on the distance from the end surfaces of the concrete. Consequently, in addition to the imposed uniaxial compression, there is a lateral shearing stress, the effect of which is to increase the apparent compressive strength of concrete.

The influence of platen restraint can be seen from the typical failure modes of test cubes, shown in Fig. 16.1. The effect of shear is always present, although it decreases towards the centre of the cube, so that the sides of the cube have near-vertical cracks, or completely disintegrate so as to leave a relatively undamaged central core (Fig. 16.1(a)). This happens when testing in a rigid testing machine, but a less rigid machine can store more energy so that an explosive failure is possible (Fig. 16.1(b)); here one face touching the platen cracks and disintegrates so as to leave a pyramid or a cone. Types of failure other than those of Fig. 16.1 are regarded as unsatisfactory and indicate a probable fault in the testing machine.

When the ratio of height to width of the specimen increases, the influence of shear becomes smaller so that the central part of the specimen may fail by lateral splitting. This is the situation in a standard cylinder test where the height/diameter ratio is 2. Figure 16.2 shows the possible modes of failure, of which the more usual one is by splitting and shear (Fig. 16.2(c)).

Sometimes cylinders of different height/diameter ratios are encountered, for example, with *test cores* (see page 311) cut from *in situ* concrete: the diameter depends on the core-cutting tool while the height of the core depends on the thickness of the slab or member. If the core is too long it can be trimmed to a height/diameter ratio of 2 but with too short a core it is necessary to *estimate* the strength which would have been obtained using a height/diameter ratio of 2; this is done by applying *correction factors*. Strictly speaking, the correction factors depend on the level of strength of the concrete but overall values are given by ASTM C 42–84a. Figure 16.3 shows the general pattern of the influence of the height/diameter ratio on the apparent compressive strength of a cylinder.

Since the influence of platen restraint on the mode of failure is greater in a cube than in a standard cylinder, the cube strength is *approximately*

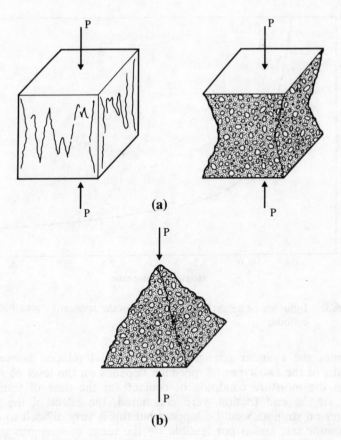

Fig. 16.1: Typical satisfactory failure modes of test cubes according to BS 1881: Part 116: 1983: (a) non-explosive, and (b) explosive

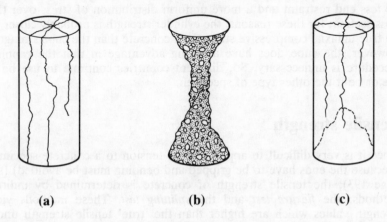

Fig. 16.2 Typical failure modes of standard test cylinders; (a) splitting, (b) shear (cone), and (c) splitting and shear (cone)

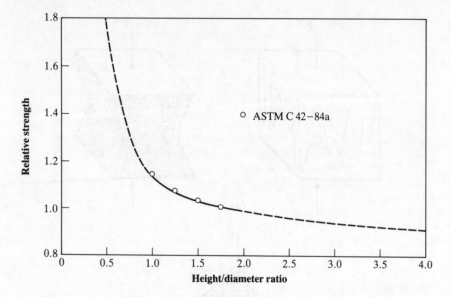

Fig. 16.3: Influence of height/diameter ratio on the apparent strength of a cylinder

1.25 times the cylinder strength, but the actual relation between the strengths of the two types of specimen depends on the level of strength and on the moisture condition of concrete at the time of testing. Of course, if the end friction were eliminated, the effect of the height/diameter on strength would disappear but this is very difficult to achieve in a routine test and is not feasible for the range of strengths normally encountered.

It is reasonable to ask whether a cube or a cylinder is a better test specimen. Compared with the cube test, the advantages of the cylinder are less end restraint and a more uniform distribution of stress over the cross section; for these reasons, the cylinder strength is probably closer to the true uniaxial compressive strength of concrete than the cube strength. However, the cube does have a strong advantage in that the capping procedure is unnecessary. So, different countries continue to use be it one or be it the other type of specimen.

Tensile strength

Since it is very difficult to apply uniaxial tension to a concrete specimen (because the ends have to be gripped and bending must be avoided) (see page 193), the tensile strength of concrete is determined by indirect methods: the *flexure test* and the *splitting test*. These methods yield strength values which are higher than the 'true' tensile strength under uniaxial loading for the reasons stated on page 194.

In the flexure test, the theoretical maximum tensile stress reached in the bottom fibre of a test beam is known as the *modulus of rupture,* which is relevant to the design of highway and airfield pavements. The test is prescribed as a compliance test by BS 5328: 1981, but in the US is thought to be unsuitable for compliance purposes because of its relative complexity. The value of the modulus of rupture depends on the dimensions of the beam and, above all, on the arrangement of loading. Nowadays, symmetrical two-point loading (at third points of the span) is used both in the UK and the US. This produces a constant bending moment between the load points so that one third of the span is subjected to the maximum stress, and therefore it is there that cracking is likely to take place.

Figure 16.4 shows the arrangement of the flexure test, as prescribed by BS 1881: Part 118: 1983. The preferred size of beam is $150 \times 150 \times 750$ mm ($6 \times 6 \times 30$ in.) but, when the maximum size of aggregate is less than 25 mm (1 in.), $100 \times 100 \times 500$ mm ($4 \times 4 \times 20$ in.) beams may be used. The making and curing of standard test beams is covered by BS 1881: Part 3: 1970; the use of sawn specimens, obtained from *in situ* concrete, is permitted by BS 1881: Part 118: 1983. The beams are tested on their side in relation to the as-cast position, in a moist condition, at a rate of increase in stress in the bottom fibre of between 0.02 and 0.10 MPa/s (2.9 and 14.5 psi/s), the lower rate being for low strength concrete and the higher rate for high strength concrete.

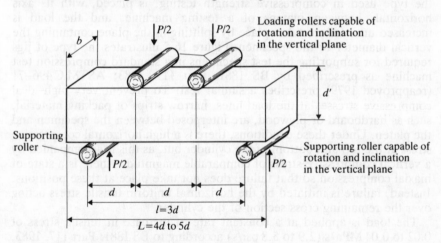

Fig. 16.4: Arrangement for the modulus of rupture test
(From: BS 1881: Part 118: 1983.)

ASTM C 78–84 prescribes a similar flexure test except that the size of the beam is $152 \times 152 \times 508$ mm ($6 \times 6 \times 20$ in.) and the loading rate is between 0.0143 and 0.020 MPa/s (2.1 and 2.9 psi/s).

If fracture occurs within the middle one-third of the beam, the modulus of rupture (f_{bt}) is calculated, to the nearest 0.1 MPa (15 psi), on the basis of ordinary elastic theory, viz.

$$f_{bl} = \frac{Pl}{bd^3}$$ (16.1)

where P = maximum total load,
l = span,
d = depth of the beam, and
b = width of the beam.

If fracture takes place outside the middle one-third, then, according to BS 1881: Part 118: 1983, the test result should be discarded. On the other hand, ASTM C 78–84 allows for failure outside the load points, say, at an average distance a from the nearest support, by the equation

$$f_{bl} = \frac{3Pa}{bd^3}.$$ (16.2)

If, however, failure occurs at a section such that $(l/3 - a) > 0.05l$, then the result should be discarded.

In the *splitting test,* a concrete cylinder (or, less commonly, cube) of the type used in compressive strength testing, is placed, with its axis horizontal, between platens of a testing machine, and the load is increased until failure takes place by splitting in the plane containing the vertical diameter of the specimen. Figure 16.5 illustrates the type of jigs required for supporting the test specimens in a standard compression test machine as prescribed by BS 1881: Part 117: 1983; ASTM C 496–71 (reapproved 1979) prescribes a similar test. To prevent very high local compressive stresses at the load lines, narrow strips of packing material, such as hardboard or plywood, are interposed between the specimen and the platen. Under these conditions, there is a high horizontal compressive stress at the top and bottom of the cylinder but, as this is accompanied by a vertical compressive stress of comparable magnitude, there is a state of biaxial compression so that failure does not take place at these positions. Instead, failure is initiated by the horizontal uniform tensile stress acting over the remaining cross section of the cylinder.

The load is applied at a constant rate of increase in tensile stress of 0.02 to 0.04 MPa/s (2.9 to 5.8 psi/s) according to BS 1881: Part 117: 1983, and 0.011 to 0.023 MPa/s (1.7 to 3.3 psi/s) according to ASTM C 496–71 (reapproved 1979). The tensile splitting strength (f_{st}) is then calculated, to the nearest 0.05 MPa (5 psi), from

$$f_{st} = \frac{2P}{\pi Ld}$$ (16.3)

where P = maximum load,
L = length of the specimen, and
d = diameter or width of the specimen.

310

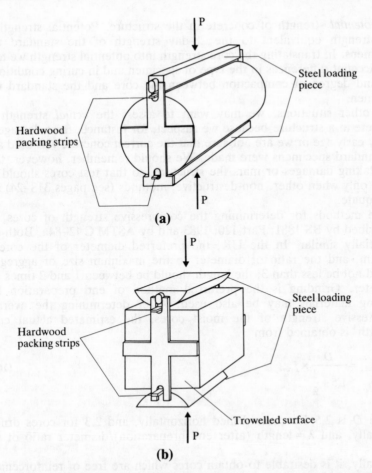

Fig. 16.5: Jigs for supporting test specimens for the determination of splitting strength according to BS 1881: Part 117: 1983: (a) cylinder and (b) cube or prism

Test cores

As we mentioned in the introduction to this chapter, the main purpose of determining the strength of concrete standard *specimens* is to ensure that the potential strength of the concrete in the *actual structure* is satisfactory. Now, if the strength of the standard compression test specimens is below the specified value (see page 330) then either the concrete in the actual structure is unsatisfactory, or else the specimens are not truly representative of the concrete in the structure. The latter possibility should not be ignored in disputes of the acceptance, or otherwise, of a doubtful part of the structure: the test specimens may have been incorrectly prepared, handled or cured, or the testing machine could be at fault. The argument is often resolved by testing cores of hardened concrete taken from the suspect part of the structure in order to estimate the *potential* strength of concrete in the structure. Potential strength is the strength equivalent to the 28-day strength of the standard test specimens. In translating the core strength into potential strength we take into account differences in the type of specimen and in curing conditions, age and degree of compaction between the core and the standard test specimen.

In other situations, we may want to assess the *actual* strength of concrete in a structure because we suspect, for instance, frost damage at a very early age or we are not sure that the correct concrete was used and no standard specimens were made. We should remember, however, that core taking damages or mars the structure, so that test cores should be taken only when other, non-destructive, methods (see pages 318–24) are inadequate.

The methods for determining the compressive strength of cores are prescribed by BS 1881: Part 120: 1983 and by ASTM C 42–84a. Both are essentially similar. In the UK, the preferred diameter of the core is 150 mm, and the ratio of diameter to the maximum size of aggregate should not be less than 3; the length should be between 1 and 2 times the diameter. Grinding is the preferred method of end preparation but capping materials may be also used. After determining the average compressive strength of the moist cores, the estimated actual cube strength[3] is obtained from

$$f_{cube} = \frac{D}{1.5 + \dfrac{1}{\lambda}} \times f_{core} \tag{16.4}$$

where D is 2.5 for cores drilled horizontally, and 2.3 for cores drilled vertically, and λ = length (after end preparation)/diameter ratio of the core.

Ideally, it is desirable to obtain cores which are free of reinforcement but, if steel is present, then we have to apply to Eq. (16.4) a correction factor for the quantity and location of the reinforcement in the core.

The procedure for estimating the potential strength is given in the Concrete Society Technical Report No. 11 (see Bibliography); concretes containing non-Portland cement, pozzolans and lightweight aggregates are excluded. When the composition, compaction and curing history of the suspect concrete are considered to be 'normal' and there is no reinforcement present in the core, the estimated potential strength of a standard cube at 28 days is

$$f_{cube} = \frac{D'}{1.5 + \dfrac{1}{\lambda}} \times f_{core} \tag{16.5}$$

---

[3] This cannot be equated to the standard 28-day cube strength.

where D' is 3.25 for cores drilled horizontally, and 3.0 for cores drilled vertically.

The term 'normal' means that the core is representative of the concrete within the structure (not within 20 per cent of the height from the top surface of the structure), the volume of voids[4] in the core (estimated by visual assessment or by density measurements) does not exceed that of a well-made cube of the same concrete, and the curing conditions are typical of those in the UK. If any one of these three is considered abnormal, then correction factors have to be applied to Eq. (16.5), together with a factor for the presence of any reinforcement in the core.

ACI 318–83 considers that concrete in the part of the structure represented by test cores is adequate if the average strength of three cores is equal to at least 85 per cent of the specified strength and if no single core has a strength lower than 75 per cent of the specified value. It should be noted that, according to the ACI, the cores are tested in a dry state, which leads to a higher strength than when tested in a moist condition (as prescribed by ASTM and BS standards) so that the ACI requirements are fairly liberal.

We have considered, so far, the use of test cores for strength determination, but they are also taken for a variety of other purposes, as listed in Table 16.1. Tests, for example, to determine the *composition of hardened concrete* are used mainly in resolving disputes and not as a means of controlling the quality of concrete. ASTM C 1084–87 and BS 1881: Part 124: 1988 describe chemical tests for determining the cement content, while the same UK standard gives a method for the determination of the original water/cement ratio.

Accelerated curing

A major disadvantage of the standard compression test is the length of time needed before the results are known, i.e. 28 days or even 7 days, by which time a considerable quantity of additional concrete may have been placed in the structure. Consequently, it is then rather late for remedial action if the concrete is too weak; if it is too strong then the mix was probably uneconomical.

Clearly, it would be advantageous to be able to predict the 28-day strength within a few hours of casting. Unfortunately, the 1- to 3-day strength of a given mix cured under normal conditions is not reliable in this respect because it is very sensitive to small variations in temperature during the first few hours of casting and to variation in the fineness of cement. To predict the 28-day strength it is, therefore, necessary for the concrete to have achieved, within a few hours of casting, a greater

---

[4] Cores containing honeycombed concrete should not be used.

Table 16.1: **Tests, other than compressive strength, which may be made on cores to provide information to assist interpretation of strength data and for other purposes as given by the Concrete Society**

	Coarse aggregate	Nominal maximum size Grading – continuous or discontinuous Particle shape Mineralogy, Group Classification Relative proportions, distribution in concrete
	Fine aggregate	Nominal maximum size Grading – fine or coarse Type – natural, crushed or mixture Particle shape Relative proportion, distribution Mineralogy
Direct visual examination of core before trimming and capping (by naked eye or possibly hand lens)	Cement	Colour of matrix of concrete
	Concrete	Compaction, segregation, porosity, honeycombing General composition, apparent coarse aggregate to mortar proportions Depth of carbonation Evidence of bleeding Evidence of plastic settlement, loss of bond Presence of entrained air Applied finishes, depth and other visible features Abrasion resistance Crack depth, width, other features Concrete depth, thickness Inclusions, particularly impurities Cold joints
	Reinforcement	Type (round, square, twisted, deformed) Size, number, depth/cover
Non-destructive	Core drilling faults	Bowing Ridges

314

Indirect visual examination of core before trimming and capping (by microscopic or petrographic techniques)	Mineralogy Air and sand contents, voids size and spacing Microcracking Surface texture of coarse aggregates Fine aggregate particle shape, maximum size, grading Degradation
Routine physical tests of cores before capping	Density Water absorption Ultrasonic pulse velocity
Special physical tests of companion cores	Indirect tensile strength Abrasion resistance (surface only) Frost resistance Movement characteristics
Routine chemical tests of cores after crushing for strength	Aggregate/cement ratio Type of cement Aggregate grading (recovered) Sulphates Chlorides Contaminants Admixtures
Routine chemical tests of companion core (not to be used for compressive strength)	Water/cement ratio
Special tests on core after crushing for strength	Sulphate attack Cement and other minerals and mineral phases, and molecular groupings such as NaCl, $CaCl_2$, SO_3, C_3A etc. Contaminants Chloride attack High alumina conversion Aggregate reactivity

CONCRETE SOCIETY, Concrete core testing for strength, *Technical Report No.* 11, pp. 44 (London, 1976).

proportion of its 28-day strength. This can be done by tests based on accelerated curing methods.

ASTM C 684–81 prescribes three methods of accelerated curing of test cylinders. In the *warm water method,* covered cylinders are immersed in water at 35 °C (95 °F) and, after capping, tested at the age of 24 hours. The *boiling water method* requires pre-curing in a moist environment at 21 °C (70 °F) for 23 hours, before curing in boiling water for $3\frac{1}{2}$ hours; after cooling for 1 hour, the cylinder is capped and tested at the age of $28\frac{1}{2}$ hours. The third method, known as the *autogenous method,* uses curing by insulation for 48 hours before capping and testing at the age of approximately 49 hours.

BS 1881: Part 112: 1983 also describes three methods, all of which involve curing covered test cubes in water heated to 35 °C (95 °F), 55 °C (131 °F) and 82 °C (180 °F), respectively. The 35 °C (95 °F) method requires the cubes to be stored for 24 hours at the required temperature except for a period not exceeding 15 min immediately after immersion of the specimens. The 55 °C (131 °F) method requires the specimens to stand undisturbed at 20 °C (68 °F) for at least 1 hour before immersing for a period of approximately 20 hours at the required temperature; the cubes are tested after cooling in water at 20 °C (68 °F) for between 1 and 2 hours. The 82 °C (180 °F) method requires the cubes to stand undisturbed for at least 1 hour before placing in an empty curing tank. The tank is filled with water at ambient temperature, which is raised to 82 °C (180 °F) in a period of 2 hours, and maintained at that temperature for a further 14 hours. The water is then discharged quickly and the cubes are tested within 1 hour whilst they are still hot.

The accelerated strengths achieved by any of the foregoing methods are all different and are lower than the 28-day strength of standard specimens. However, for a *given* mix, the accelerated strength determined by any one method can be correlated with the 7- or 28-day strength of standard specimens (see Fig. 16.6). The relation between the two strengths has to be established prior to placing the concrete in the structure so that the accelerated test can be used as a rapid quality control test for detecting variations in the mix proportions (see page 337).

On the other hand, in Canada, there has been established a relation between accelerated strength, R_a, and the 28-day cylinder strength, R_{28}, which is independent of the cement type, mix proportions and type of admixture, viz.

in psi: $R_{28} = \dfrac{26\,160\,R_a}{R_a + 11\,620}$

(16.6a)

in MPa: $R_{28} = \dfrac{180\,R_a}{R_a + 80}$

(16.6b)

The procedure requires delaying the accelerated curing until a fixed set

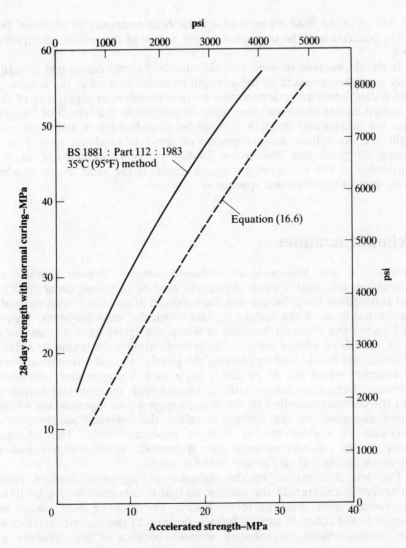

Fig. 16.6: Typical relations between strength determined by accelerated curing and 28-day strength of concrete with normal curing

has occurred, as measured by the Proctor needle penetration[5] under 3500 psi (24 MPa). After a delay of 20 min., the moulded standard cylinder is placed in boiling water for 16 hours, then is demoulded and left to cool for 30 min. The strength is determined (after capping the cylinder) 1 hour after removal from the boiling water. The disadvantages

[5] See ASTM C 403–80.

of this so-called *fixed-set method* are the time necessary to ascertain the set of concrete and the somewhat erratic nature of the needle penetration test.

It should be remembered that the standard compression test is really only a *relative* measure of the strength of concrete used in the structure, and it can therefore be argued that there is no inherent superiority of the standard 28-day test over other tests. In fact, there is a school of thought that the accelerated strength test can be considered as a test in its own right, i.e. as a basis for acceptance of concrete, and not merely as a means of predicting the 7- or 28-day strength, particularly as the variability of the accelerated strength results is the same as, or smaller than, that of standard test specimens.

Schmidt hammer

This test is also known as the *rebound hammer, impact hammer* or *sclerometer* test, and is a *non-destructive method* of testing concrete. The test is based on the principle that the rebound of an elastic mass depends on the hardness of the surface against which the mass impinges. Figure 16.7 shows the rebound hammer in which the spring-loaded mass has a fixed amount of energy imparted to it by extending the spring to a fixed position; this is achieved by pressing the plunger against a smooth surface of concrete which has to be firmly supported. Upon release, the mass rebounds from the plunger (still in contact with the concrete surface), and the distance travelled by the mass, expressed as a percentage of the initial extension of the spring, is called the *rebound number*; it is indicated by a rider moving along a graduated scale. The rebound number is an arbitrary measure since it depends on the energy stored in the given spring and on the size of the mass.

The test is sensitive to the presence of aggregate and of voids immediately underneath the plunger so that it is necessary to take 10 to 12 readings over the area to be tested. The plunger must always be normal to the concrete surface but the position of the hammer relative to the vertical affects the rebound number because of the influence of

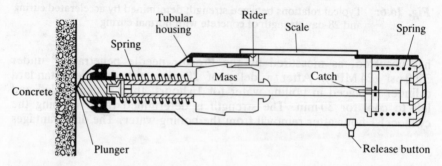

Fig. 16.7: Rebound hammer

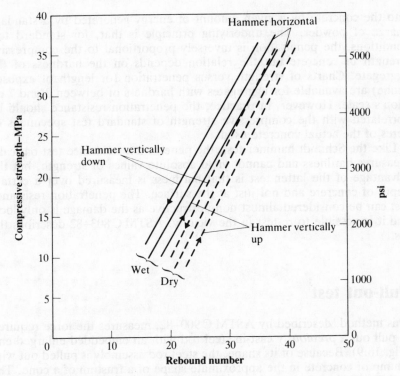

Fig. 16.8: Typical relations between compressive strength and rebound number with the hammer horizontal and vertical on a dry and a wet surface of concrete

gravity on the moving mass. Thus, for a given concrete, the rebound number of a floor is smaller than that of a soffit (see Fig. 16.8), while inclined and vertical surfaces yield intermediate values; the actual variation is best determined experimentally.

There is no unique relation between hardness and strength of concrete but experimental relationships can be determined for a given concrete; the relationship is dependent upon factors affecting the concrete surface, such as degree of saturation (see Fig. 16.8) and carbonation. In consequence, the Schmidt hammer test is useful as a measure of uniformity and relative quality of concrete in a structure or in the manufacture of a number of similar precast members but not as an acceptance test. ASTM C 805–79 and BS 1881: Part 202: 1986 describe the test.

Penetration resistance

This test, known commercially as the *Windsor probe* test, estimates the strength of concrete from the depth of penetration by a metal rod driven into the concrete by a given amount of energy generated by a standard charge of powder. The underlying principle is that, for standard test conditions, the penetration is inversely proportional to the compressive strength of concrete but the relation depends on the hardness of the aggregate. Charts of strength versus penetration (or length of exposed probe) are available for aggregates with hardness of between 3 and 7 on Moh's scale. However, in practice, the penetration resistance should be correlated with the compressive strength of standard test specimens or cores of the actual concrete used.

Like the Schmidt hammer test, the penetration resistance test basically measures hardness and cannot yield absolute values of strength, but the advantage of the latter test is that hardness is measured over a certain depth of concrete and not just at the surface. The penetration resistance test can be considered almost non-destructive as the damage is only local and it is possible to re-test in the vicinity. ASTM C 803–82 describes the test.

Pull-out test

This method, described by ASTM C 900–82, measures the force required to pull out a *previously* cast-in steel rod with an embedded enlarged end (Fig. 16.9). Because of its shape, the steel rod assembly is pulled out with a lump of concrete in the approximate shape of a frustum of a cone. The *pull-out strength* is calculated as the ratio of the force to the idealized area of the frustum, the strength being close to that of the shearing strength of concrete. However, the pull-out force or strength correlates

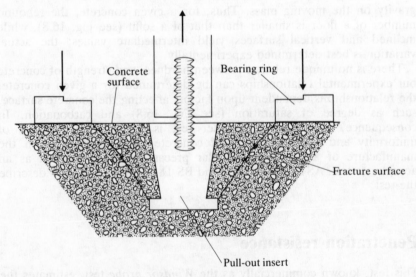

Fig. 16.9: Diagrammatic representation of the pull-out test

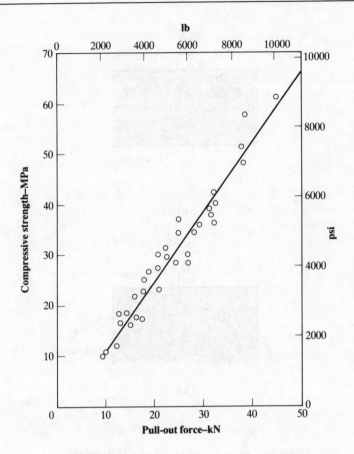

Fig. 16.10: Relation between compressive strength of cores and pull-out force
for actual structures
(From: U. BELLANDER, Strength in concrete structures, *CBI
Report* 1:78, pp. 15 (Swedish Cement and Concrete Research Inst.
1978).

well with the compressive strength of cores or standard cylinders for a
wide range of curing conditions and ages; Fig. 16.10 shows typical results.

Ultrasonic pulse velocity test

The principle of this test is that the velocity of sound in a solid material,
V, is a function of the square root of the ratio of its modulus of elasticity,
E, to its density, ρ, viz.

$$V = f \left[\frac{gE}{\rho} \right]^{\frac{1}{2}} \qquad (16.7)$$

321

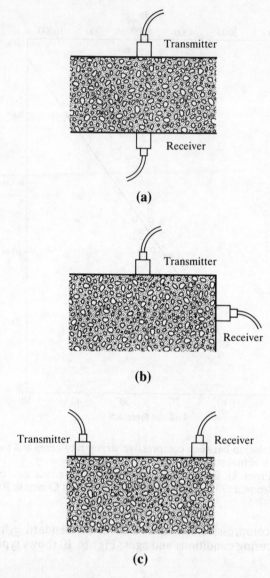

Fig. 16.11: Methods of propagating and receiving ultrasonic pulses: (a) direct transmission, (b) semi-direct transmission, and (c) indirect or surface transmission

where *g* is the acceleration due to gravity. This relation can be used for the determination of the modulus of elasticity of concrete if Poisson's ratio is known (see page 215) and hence as a means of checking the quality of concrete.

The apparatus generates a pulse of vibrations at an ultrasonic frequency which are transmitted by an electro-acoustic transducer held in

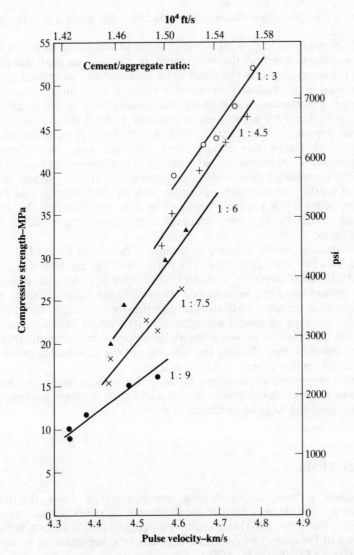

Fig. 16.12: Relation between compressive strength and ultrasonic pulse velocity of concrete cubes for concretes of different mix proportions (From: R. JONES and E. N. GATFIELD, Testing concrete by an ultrasonic pulse technique, *DSIR Road Research Tech. Paper No.* 34 (London, HMSO, 1955.)

contact with the surface of the concrete under test. After passing through the concrete, the vibrations are received and converted to an electrical signal by a second electro-acoustic transducer, the signal being fed through an amplifier to a cathode-ray oscilloscope. The time taken by the pulse to travel through the concrete is measured by an electrical timing-unit with an accuracy of ±0.1 microsecond and, knowing the

323

length of path travelled through the concrete, the pulse velocity can be calculated.

It is necessary to have a high-energy pulse of vibrations to give a sharp onset waveform because the boundaries of the various material phases within the concrete cause the pulse to be reflected and weakened; in fact, longitudinal (compression), transverse (shear), and surface waves are produced. For maximum sensitivity, the leading edge of the longitudinal waves is detected by a receiving transducer located on the face of the concrete opposite to the emitting transducer; this is direct transmission. Figure 16.11 shows this arrangement, together with two alternative arrangements of transducers: semi-direct transmission, and indirect or surface transmission; these utilize the presence of transverse and surface waves. Clearly, the alternative positions can be used when access to two opposite sides of a concrete member is not possible but the energy received and, hence, the accuracy are lower than with direct transmission.

The ultrasonic pulse velocity technique is described by ASTM C 597–83 and BS 1881: Part 203: 1986. The main use of the method is in quality control of similar concrete: both lack of compaction and a change in the water/cement ratio can be detected. However, the pulse velocity cannot be used as a general indicator of compressive strength because, for example, the type of coarse aggregate and its content in concrete greatly influence the relation between the pulse velocity and strength (see Fig. 16.12). Other factors affecting the relation are the moisture content, age, presence of reinforcement and temperature.

Further important applications of the pulse velocity technique are the detection of crack development in structures such as dams, and checking deterioration due to frost or chemical action.

Other tests

Specialized techniques for testing concrete range from the use of electromagnetic devices for the measurement of *cover* to reinforcement (BS 1881: Part 204: 1988) to *gamma radiography* for the determination of variation in the quality of concrete, e.g. lack of compaction or location of voids (BS 1881: Part 205: 1986).

Bibliography

16.1   ACI COMMITTEE 214.1R–81, Use of accelerated strength testing, Part 2: Construction Practices and Inspection Pavements, *ACI Manual of Concrete Practice*, 1990.

16.2   P. M. BARBER, Analysis of fresh concrete, *Concrete*, Vol. 17, No. 6, pp. 12–13 (June 1983), (Discussion of 'Recommendations for the assessment of the composition of fresh concrete').

16.3   CONCRETE SOCIETY, Concrete core testing for strength,
       Technical Report No. 11, pp. 44 (London, 1976).
16.4   V. M. MALHOTRA, Accelerated strength testing of concrete
       specimens, *Progress in Construction Science and Technology No.*
       2, pp. 59–82 (Medical and Technical Publishing Co. Ltd.,
       Manchester, 1973).
16.5   P. SMITH and B. CHOJNACKI, Accelerated strength testing of
       concrete cylinders, *Proc. ASTM,* 63, pp. 1079–1101 (1963).

Problems

16.1   Comment on the use of the Schmidt hammer on surfaces with
       different inclinations.
16.2   Some precast concrete members were subjected to freezing at a
       very early age, others were not. How would you investigate
       whether frost damage had occurred?
16.3   What is meant by repeatability and reproducibility?
16.4   Discuss the possible reasons for a difference between the strength
       of test cylinders and cores from the same concrete.
16.5   Why is there a difference between the modulus of rupture and the
       splitting tensile strength of a given concrete?
16.6   What are the advantages and disadvantages of the pull-out test?
16.7   What are the advantages and disadvantages of the Windsor probe?
16.8   How would you investigate a suspected existence of voids in a
       concrete slab?
16.9   What test would you use to determine the age for early striking of
       soffit formwork?
16.10 How do you convert the strength of a concrete core to the
       estimated strength of a test cube?
16.11 What is the stress distribution in a specimen subjected to indirect
       tension just prior to failure?
16.12 What is the influence of cracks on the ultrasonic pulse velocity of
       concrete?
16.13 Why are compression test cylinders capped?
16.14 How does the splitting tensile strength relate to the modulus of
       rupture?
16.15 Explain the difference in strength of large and small cylinders
       made from the same mix.
16.16 Why is precision important in testing?
16.17 How is the ultrasonic pulse velocity of concrete determined?
16.18 What is the purpose of determining the ultrasonic pulse velocity of
       concrete?
16.19 What is the influence of the moisture content of concrete on its
       ultrasonic pulse velocity?
16.20 How do you determine the splitting strength of concrete?
16.21 Why is the direct tensile strength of concrete not normally
       determined?

325

16.22 How is the potential strength of concrete determined?

16.23 How is the actual strength of concrete determined?

16.24 What is the difference between the actual strength of concrete specimens and the *in situ* strength?

16.25 What are the non-destructive tests for the determination of the strength of concrete?

16.26 Explain how deficiencies in a compression testing machine can affect the test result.

16.27 Explain how incorrect curing of compression test specimens can affect the test result.

16.28 Explain how incorrect making of compression test specimens can affect the test result.

16.29 Why is the strength of a test specimen subjected to accelerated curing different from the strength of a similar standard test specimen?

16.30 Why does a standard test specimen not give adequate information about the strength of concrete in the structure?

16.31 Why are standard compression test specimens not tested at the age of 1 or 2 days?

16.32 Discuss the advantages and disadvantages of cube- and cylinder-shaped test specimens?

16.33 Discuss the various types of tests for the tensile strength of concrete.

16.34 What is meant by non-destructive methods of testing concrete?

16.35 Describe the pull-out test.

16.36 Why are standard test cubes cured in a standard manner?

16.37 What type of cube failure is unsatisfactory?

16.38 Discuss the difference in the strength obtained from a cube test and from a standard cylinder test.

16.39 What are the uses of the ultrasonic pulse velocity test?

16.40 Suggest a non-destructive test to investigate a suspected presence of voids in a large concrete mass. Give your reasons.

16.41 Suggest a non-destructive test to compare the quality of precast floor units. Give your reasons.

16.42 Why does the core strength differ from the standard cube strength?

16.43 Briefly describe two methods of analysing fresh concrete.

16.44 Describe three standard methods of accelerated curing.

17

Compliance with specifications

The design of concrete structures is based on the assumption of certain minimum (occasionally, maximum) properties of concrete, such as strength, but the actual strength of the concrete produced, whether on site or in the laboratory, is a variable quantity. The sources of variability are many: variations in mix ingredients, changes in concrete making and placing, and also, with respect to test results, the variations in the sampling procedure and the very testing. It is important to minimize this variability by quality control measures and by adopting the standard testing procedures described in Chapter 16. Moreover, knowledge of the variability is required so that we can interpret strength values properly or, in other words, detect statistically significant changes in strength, as opposed to random fluctuations.

The knowledge of variability forms the basis of devising a satisfactory compliance scheme for the strength of designed mixes. In other cases, properties such as mix proportions, density, air content and workability have to comply with specifications so as to satisfy both strength and durability requirements.

Variability of strength

Since strength is a variable quantity, when designing a concrete mix, we must aim at a mean strength higher than the minimum required from the structural standpoint so that we can expect every part of the structure to be made of concrete of adequate strength.

Let us suppose that we have a large sample of similar test specimens which represent all of the concrete in a structure. The results of testing will show a scatter or a distribution of strengths about the mean strength. This can be represented by a *histogram* in which the number of specimens falling within an interval of strength (frequency) is plotted against the interval of strength. Figure 17.1 shows such a histogram in which the distribution of strength is approximated by the dashed curve, which is called the *frequency distribution curve*. For the strength of concrete, this curve can be assumed to have a characteristic form called the *normal* or

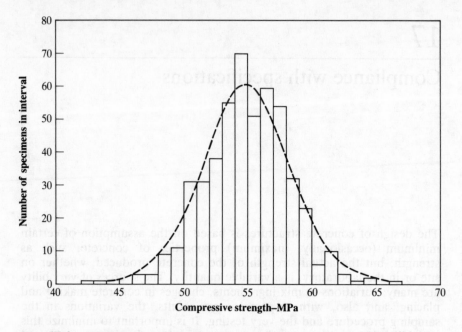

Fig. 17.1: A histogram of strength values

Gaussian distribution. This curve is described in terms of the mean strength f_m and the *standard deviation s,* the latter being a measure of the scatter or dispersion of strength about the mean, defined as

$$s = \left[\frac{\sum_{1}^{n} (f_i - f_m)^2}{n-1} \right]^{\frac{1}{2}} \qquad (17.1)$$

or

$$s = \left[\frac{n \sum_{1}^{n} f_i^2 - \left(\sum_{1}^{n} f_i \right)^2}{n(n-1)} \right]^{\frac{1}{2}} \qquad (17.1a)$$

where f_i = strength of test specimen i,

$$f_m = \frac{\sum_{1}^{n} f_i}{n}$$

and n = number of test specimens.

The theoretical normal distribution is represented graphically in Fig. 17.2. It can be seen that the curve is symmetrical about the mean value and extends to plus and minus infinity. In practice, these very low and very high values of strength do not occur in concrete but these extremes

328

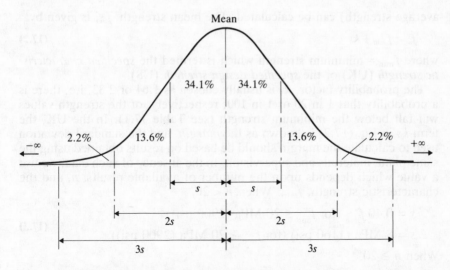

Fig. 17.2: Normal distribution curve; percentage of specimens in intervals of one standard deviation shown

can be ignored because most of the area under the curve (99.6 per cent) lies within $\pm 3s$ and can be taken to represent all the strength values of concrete. In other words, we can say that the *probability* of a value of strength falling within $\pm 3s$ from the mean value is 99.6 per cent. Likewise, the probability of a value falling between any given limits about the mean value $(f_m \pm ks)$ can be stated. Table 17.1 lists values of probability for various values of k (*probability factor*) together with the probability of encountering a strength below $(f_m - ks)$.

The methods of mix design are discussed in Chapter 19 but it is appropriate at this stage to outline the first step in designing a mix, viz. the use of standard deviation so that the *mean strength* (or required

Table 17.1: Probability of strength values in the range $f_m \pm ks$ and below $f_m - ks$ for normal distribution

Probability factor k	Probability of strength in the range $f_m \pm ks$, per cent	Probability of strength below $f_m - ks$ (risk), per cent
1.00	68.2	15.9 (1 in 6)
1.64	90.0	5.0 (1 in 20)
1.96	95.0	2.5 (1 in 40)
2.33	98.0	1.0 (1 in 100)
3.00	99.7	0.15 (1 in 700)

average strength) can be calculated. The mean strength, f_m, is given by:

$$f_m = f_{min} + ks \qquad (17.2)$$

where f_{min} = minimum strength which is termed the *specified characteristic strength* (UK) or the *specified design strength* (US).

The probability factor, k, is usually chosen as 1.64 or 2.33, i.e. there is a probability that 1 in 20 or 1 in 100, respectively, of the strength values will fall below the minimum strength (see Table 17.1). In the UK, the term ks in Eq. (17.2) is known as the *margin*, and the standard deviation used to calculate the margin should be based on results obtained using the same plant, materials and supervision. In the absence of such data, we use a value which depends upon the number of available results, n, and the characteristic strength, f_{min}. When $n < 20$,

$$s = 0.40 f_{min} \text{ (for } f_{min} \leq 20 \text{ MPa (2900 psi))}$$
$$s = 8 \text{ MPa (1160 psi) (for } f_{min} \geq 20 \text{ MPa (2900 psi))} \qquad (17.3)$$

When $n \geq 20$,

$$s = 0.20 f_{min} \text{ (for } f_{min} \leq 20 \text{ MPa (2900 psi))}$$
$$s = 4 \text{ MPa (580 psi) (for } f_{min} \geq 20 \text{ MPa (2900 psi))} \qquad (17.4)$$

Standard deviations estimated from Eqs (17.3) and (17.4) should be used only until adequate production data have become available.

In the British method of mix design for air-entrained concrete, it is assumed that a loss of 5.5 per cent in compressive strength results for each 1 per cent by volume of air entrained in the mix (see page 297). This reduction in strength is taken into account by aiming for a higher mean strength, viz

$$f_m = \frac{f_{min} + ks}{1 - 0.055a} \qquad (17.5)$$

Where a is the percentage of air entrained.

The approach of the ACI Building Code Requirements for Reinforced Concrete (ACI 318–83) is based on several criteria. When at least 30 consecutive test results in one series are available for similar materials and conditions, and for which their specified design strength is within 7 MPa (1000 psi) of that now required, the standard deviation is calculated from Eq. (17.1). If two test series are used to obtain at least 30 test results, the standard deviation used shall be the *statistical* average, $\bar{s}$, of the values calculated from each record, as follows:

$$\bar{s} = \left[\frac{(n_1 - 1)s_1^2 + (n_2 - 1)s_2^2}{n_1 + n_2 - 2} \right]^{\frac{1}{2}} \qquad (17.6)$$

where s_1, s_2 are the standard deviations calculated from the two test series, and n_1, n_2 are the numbers of tests in each test series. Note that $\bar{s}$ is not the arithmetic mean of s_1 and s_2.

If the number of test results is between 15 and 29, the calculated

| Table 17.2: | Modification factor for standard deviation given by ACI 318–83 |

Number of tests	Factor for standard deviation
15	1.16
20	1.08
25	1.03
30 or more	1.00

standard deviation is increased by the factors given in Table 17.2. When a suitable record of test results is not available, the required average strength must exceed the specified design strength by an amount which depends on the specified design strength (see Table 17.3), but as data become available during construction the standard deviation may be calculated in accordance with the appropriate number of test results.

Once the standard deviation has been determined, the required average strength is obtained from the larger of the following equations:

$$f_{cr}' = f_c' + 1.34s \qquad (17.7)$$

and

in psi units: $f_{cr}' = f_c' + 2.33s - 500$ \hfill (17.8a)

in MPa: $\qquad f_{cr}' = f_c' + 2.33s - 3.5$ \hfill (17.8b)

where $\qquad f_{cr}' = $ required average compressive strength, and

$f_c' = $ specified compressive strength.

| Table 17.3: | Required increase in strength for specified compressive strength when no tests records are available, according to ACI 318–83 |

Specified compressive strength		Required increase in strength	
MPa	psi	MPa	psi
less than 21	less than 3000	7	1000
21 to 35	3000 to 5000	8.5	1200
35 or more	5000 or more	10.0	1400

Acceptance and compliance

Let us now return to the main topic of this chapter: acceptance and compliance with the specified strength. According to BS 5328: 1981, compliance with the characteristic strength is based on groups of consecutive test results, as well as on single results. Each *result* is the average of two cubes, made in the specified manner from concrete which is sampled at a prescribed rate, and normally tested at 28 days. Compliance is assumed if *both* of the following requirements are satisfied:

(a)   The mean strength, determined from the first two, three or four consecutive test results or from any group of four consecutive results, complies with the limits of Table 17.4.

(b)   No individual test result falls short of the specified characteristic strength by more than the value given in Table 17.4.

Table 17.4:   **Compliance requirements for compressive strength according to BS 5328: Part 4: 1990**

Specified characteristic strength (grade)		Group of results	Minimum value by which the mean strength of the group of test results should exceed the grade strength		Maximum value by which any individual test result falls short of the grade strength	
MPa	psi		MPa	psi	MPa	psi
7.5	1100	first 2	0	0	2	290
to	to	first 3	1	150	2	290
15	2180	4*	2	290	2	290
20	2900	first 2	1	150	3	440
and	and	first 3	2	290	3	440
above	above	4*	3	440	3	440

* any consecutive four

If only one result (average of two cubes) fails to meet the second requirement, we can assume that that result represents only the particular batch of concrete from which the cubes were made, provided that the average strength of the group satisfies the first requirement. If the average strength of any group of four consecutive results fails to meet the first requirement, then *all* the concrete in all the batches represented by the test cubes is deemed not to comply with the strength requirements. In such a case, the mix proportions of subsequent batches of concrete should be modified to increase the strength.

When flexural strength or indirect tensile strength is specified, the compliance criteria of BS 5328: 1981 are:

(a) The average strength determined from any group of four consecutive test results should exceed the specified characteristic strength by 0.3 MPa (50 psi); and

(b) The strength determined from any test result does not fall short of the specified characteristic strength by more than 0.3 MPa (50 psi).

Since it is not practicable to inspect all the concrete produced, compliance with the specification is determined using sampling schemes based on random selection; minimum sampling rates are recommended in codes of practice. However, as the sampling and testing themselves are also variable procedures, the overall variability of the test results includes the variability of the sampling and testing as well as the variability in the production of concrete. Hence, there may exist a situation when the true average quality of the concrete is within the specification but test results may indicate non-compliance; this situation is the *producer's risk*. On the other hand, the test results may comply but the true average quality is outside the specification; this situation is the *consumer's risk*. A good compliance scheme ensures that these risks are at an acceptable level and they are properly distributed between the producer and the consumer.

To evaluate a compliance scheme we can use the *operating–characteristic (O–C) curve* associated with the compliance criterion (see Fig. 17.3). An ideal testing plan would be one in which the concrete of a quality equal to that specified (e.g. when not more than 5 per cent of the total is defective) is always accepted so that the producer's risk is zero, and in which concrete of a quality worse than specified (e.g. when more than 5 per cent of the total is defective) is always rejected so that the consumer's risk is also zero. This ideal situation is shown in Fig. 17.3 but can be achieved only with an infinite population of results. When the compliance criteria are fixed, the O–C curve depends on the standard deviation but, as an example, the O–C curve for the average-of-any-four-test-results clause of the BS 5328: 1981 for a standard deviation of 4 MPa (600 psi) is also shown in Fig. 17.3. Here, there is a producer's risk of 13 per cent (or 1 in 8) of non-compliance if he is making concrete exactly of the quality specified, i.e. 5 per cent defective to the characteristic strength. Also there is a consumer's risk of 50 per cent (or an even chance) of detecting concrete which is 13 per cent defective. These values illustrate the practical meaning of a testing scheme based on an apparent assumption of tolerance of 5 per cent of defective concrete, used in British Standards.

The evaluation and acceptance of concrete laid down by the ACI Building Code Requirements for Reinforced Concrete (ACI 318.R-83) is based on a strength test as the average of *two* cylinders made from the same sample of concrete and usually tested at 28 days. The strength of a given concrete is considered satisfactory if *both* of the following require-

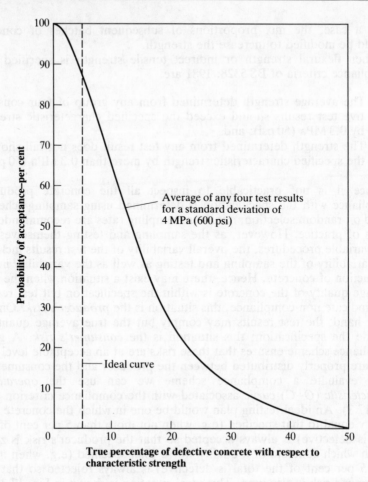

Average of any four test results
for a standard deviation of
4 MPa (600 psi)

Ideal curve

Probability of acceptance–per cent (y-axis)

True percentage of defective concrete with respect to characteristic strength (x-axis)

Fig. 17.3: Example of an operating-characteristic curve for the testing plan of BS 5328: 1981

ments are met:

(a) The average of all sets of three consecutive tests is at least equal to the specified design strength.
(b) No individual strength test falls below the specified design strength by more than 3.5 MPa (500 psi).

It should be stressed that non-compliance does not automatically mean rejection of concrete; it merely serves as a warning to the engineer that further investigation is warranted. The factors to be considered are of two kinds. First, the validity of the test results has to be studied: were the specimens sampled and tested according to the prescribed procedures or was the testing machine at fault? (See page 306.) Second, is the non-complying strength likely to cause structural failure or serviceability defects or to impair durability?

If, after these considerations, further action is necessary this should be in the form of non-destructive tests, followed by testing cores taken from the structure and finally by load tests on the structure. If the concrete is deemed to be unsatisfactory then the structure would have to be strengthened or, in the extreme, demolished.

Compliance requirements for other properties

In the preceding section, we discussed the compliance requirements for the strength of *designed mixes,* that is mixes whose performance is specified by the designer but the actual mix proportions are determined by the concrete producer. To satisfy strength and durability requirements, limiting values of water/cement ratio, cement content, air content and density are often stipulated. Workability and temperature of fresh concrete may also be specified. Consequently, compliance is also necessary for all these properties.

On the other hand, BS 5328: 1990 requires no compliance testing for the strength of *prescribed mixes* (that is those with specified mix proportions) but compliance is required for the mix proportions and workability, with an expectation that the concrete is likely to have a sufficient strength. Prescribed mixes are used for special purposes where strength is usually of secondary importance, e.g. to obtain a special finish, whilst standard mixes are generally used on small jobs when the 28-day strength does not exceed 30 MPa (4350 psi) and the water/cement ratio is within the limits of Tables 14.1, 14.3 and 15.1.

The compliance requirements of BS 5328: 1990 for the various properties are as follows:

(a) Minimum or maximum cement content

Compliance may be assessed by observation of the batching or from autographic records. In either case, the cement content should not be less than 95 per cent of the specified minimum or more than 105 per cent of the specified maximum.

Where compliance is assessed by the analysis of fresh concrete, the limits of the cement content should be agreed. Alternatively, the equivalent grade, given in BS 5328: Part 1: 1990, can be used as the controlling factor in applying Table 17.4.

(b) Maximum free water/cement ratio

Compliance may be assessed from workability tests, provided that there is a proven relation between workability and free water/cement ratio, in which case the former should comply with BS 5328: Part 4: 1990.

When compliance is assessed by the analysis of fresh concrete, the limits of the free water/cement ratio should be agreed. Alternatively, the equivalent grade can be used as in (a) above. Compliance may also be assessed from records: the mean free water/cement ratio should not exceed the specified maximum, and no individual value should exceed the specified value by more than 5 per cent.

(c) Workability

For designed and prescribed mixes, the workability requirements are those of Table 17.5.

335

Table 17.5: Compliance requirements for workability of fresh concrete for designed and prescribed mixes according to BS 5328: 1990

Workability test	Specified value	Compliance requirement
Slump (standard sample)	–	±25 mm (1 in.) or ±$\frac{1}{3}$ of specified value, whichever is the greater
Slump (truck discharge sample)	10 mm ($\frac{1}{2}$ in.)	+35 mm (1$\frac{1}{2}$ in.), −10 mm ($\frac{1}{2}$ in.)
	25 mm (1 in.)	+35 mm (1$\frac{1}{2}$ in.), −25 mm (1 in.)
	50 mm (2 in.)	±35 mm (1$\frac{1}{2}$ in.)
	75 mm (3 in.) and above	±$\frac{1}{3}$ of specified value plus 10 mm ($\frac{1}{2}$ in.)
Vebe	–	±3 sec or ±$\frac{1}{5}$ of specified value whichever is the greater
Compacting factor	0.90 and above	±0.03
	between 0.80 and 0.90	±0.04
	below 0.80	±0.05
Flow	–	±50 mm (2 in.)

(d) Air content[1]

The percentage air content, as determined from individual samples taken at the point of placing the concrete and representing the given batch of concrete, should be within ±2 of the specified value. Also, the average of four consecutive measurements from separate batches should be within ±1.5 of the specified value.

(e) Temperature

At the time of delivery, the temperature of fresh concrete should not fall below the specified minimum or exceed the specified maximum.

(f) Density (unit weight)

At the time of delivery, the density of concrete should not be less than 97.5 per cent of the specified minimum or more than 102.5 per cent of the specified maximum. For hardened concrete, the corresponding limits are 95 and 105 per cent.

BS 5328: Part 4: 1990 also specifies alkali content (see page 273) and chloride content (see page 278).

---

[1] The pressure method of determining the air content is not suitable for some lightweight aggregate concretes (see page 296).

336

Quality control charts

The desirability of controlling the quality of concrete as closely as possible arises not only from the need to comply with the specification but also for economic reasons for the concrete producer. For example, poor quality control will result in a higher standard deviation, and therefore a higher mean strength will have to be achieved by using more cement in order to meet the specified design strength or characteristic strength.

The purpose of *quality control* is to measure and control the variation of the mix ingredients and to measure and control the variation of those operations which affect the strength or the uniformity of concrete: batching, mixing, placing, curing, and testing. Quality control is distinguished from *quality assurance,* which is defined as the systematic action necessary to provide adequate confidence that a product will perform satisfactorily in service.

Quality control charts are widely used by the suppliers of ready-mixed

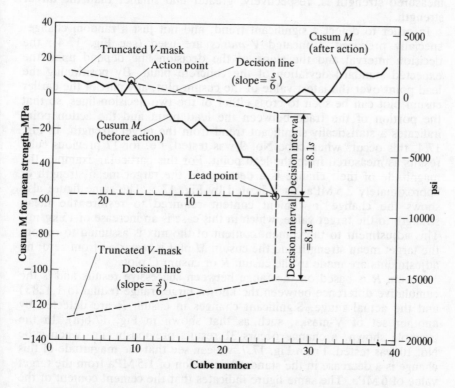

Fig. 17.4: The British Ready Mixed Concrete Association cusum chart for mean strength, illustrating the use of a truncated *V*-mask to determine a significantly lower mean strength than the target mean strength; *s* = target standard deviation, which is assumed to be 6 MPa (870 psi) for the example given

concrete and by the engineer on site to assess continually the strength of concrete. There are many methods in existence but, in this section, we shall consider only one.

In the UK, the cumulative sum or *cusum method* is often used to control the estimated 28-day cube strength as predicted from the 24-hour accelerated strength, using a previously established relation. The procedure is to take duplicate cubes at about daily intervals, one of which is stored for 28 days whilst the other is tested after 24 hours of accelerated curing. The cusum analysis is then carried out to monitor, concurrently, the mean strength (cusum M), the variability or range (cusum R) and the prediction method (cusum C).

First, let us consider the mean strength. As each successive result is obtained, the differences between the known target value and the actual value are recorded on a cumulative basis. An example is shown in Fig. 17.4, in which the cusum M is plotted on a standard chart which has an appropriate scale for detecting significant trends of the actual strength in relation to the target strength; positive and negative slopes mean that the measured strength is, respectively, greater and smaller than the target strength.

In order to detect a significant trend, and not just a random change, specially prepared, truncated *V-masks* are used (see Fig. 17.4); the decision interval and the slope of the decision line depend upon the expected standard deviation of the concrete plant. By positioning the lead point over the latest value of the cusum plot, any part of the earlier cusum plot can be seen to cross either of the two decision lines, so that the portion of the trace between the lead point and the action point indicates a statistically significant trend from the target strength. In Fig. 17.4, this occurs when cube No. 9 was tested, i.e. for 17 previous cube results as measured from the lead point. For this particular example, the magnitude of their change is a decrease in the target mean strength of approximately 2.5 MPa, as indicated by Fig. 17.5. The same figure also shows the change in cement content required to restore the mean strength to the target value, which in this case is an increase of 15 kg/m^3. This adjustment to the cement content of the mix is assumed to restore the target mean strength and the cusum M plot is restarted from zero; no adjustments are made to the cusum R or cusum C plots.

Cusum R is based on the range between successive results and is the cumulative difference between the known target range (equal to 1.128 s) and the actual range. Significant changes in cusum R are detected by another set of V-masks, such as that shown in Fig. 17.6(a). In the example shown, a change occurs 31 results previously, i.e. when cube No. 10 was tested. From Fig. 17.7, we can see that the magnitude of this change is a decrease in the standard deviation of 1.5 MPa from the target value of 6 MPa. The same figure indicates that the cement content of the mix should be decreased by 20 kg/m^3. This decrease in cement content is assumed to reduce the target mean strength by 3 MPa (see footnote to Fig. 17.7) so that, if control is based on the cusum R, then no adjustment is necessary to either the cusum M or cusum C charts. However, the cusum R plot is restarted from zero and, because of the decrease in

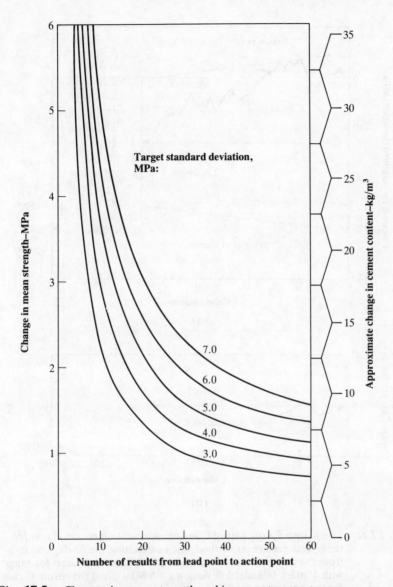

Fig. 17.5: Changes in mean strength and in cement content necessary to restore the mean strength to the target strength, as used by the British Ready Mixed Concrete Association. This figure assumes that a change in cement content of 6 kg/m³ results in a change in the mean strength of 1 MPa
(To convert MPa to psi multiply by 145; to convert kg/m³ to lb/yd³ multiply by 1.68.)
(From: BRMCA, *The Authorization Scheme for Ready Mixed Concrete*, 5th Edn, pp. 42, (March 1982).)

339

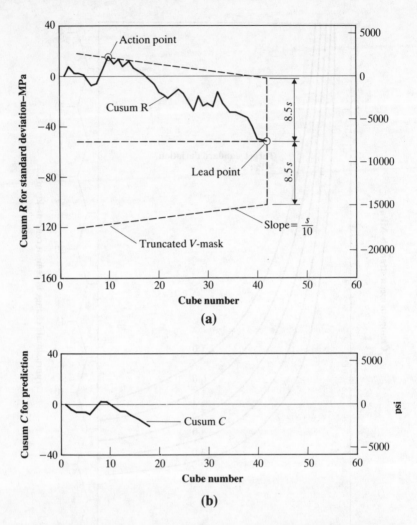

Fig. 17.6: The British Ready Mixed Concrete Association cusum charts for range and prediction method when estimating the 28-day strength from the accelerated 24 hour strength: (a) cusum R chart for range with V mask (standard deviation $s = 6$ MPa), and (b) cusum C chart for the prediction method

standard deviation, a new V-mask has to be used for the future analysis of cusum R and cusum M.

In carrying out the above procedures, an experienced operator will check both the cusum M and cusum R at the same time and make a judgement on the adjustment of cement content on the basis of both charts.

The third chart, cusum C, is necessary because the prediction method is sensitive to changes in materials and in curing conditions. The

340

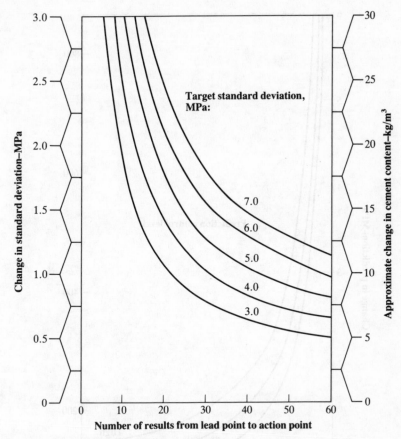

Fig. 17.7: Changes in standard deviation and in cement content necessary to maintain the target range, as used by the British Ready Mixed Concrete Association. This figure assumes that a change in cement content of 6 kg/m³ results in a change in the mean strength of 1 MPa (To convert MPa to psi multiply by 145; to convert kg/m³ to lb/yd³ multiply by 1.68.)
(From: BRMCA, *The Authorization Scheme for Ready Mixed Concrete*, 5th Edn, pp. 42, (March 1982).)

cumulative sum of the difference between the actual and predicted 28-day strengths is plotted as shown in Fig. 17.6(b), any significant variation in the prediction method being detected with a V-mask having three decision intervals and decision lines (fine, normal and high) which depend on the sensitivity of standard deviation of the manufacturing plant. When a change is detected, the chart is adjusted by an amount indicated in Fig. 17.8. At the same time, all the estimated 28-day strength values have to be recalculated back to the action point where a change in the prediction method occurred, or to the last point of a change in cement content, whichever is the more recent. The new corrected values of estimated 28-day strength are then re-plotted on the cusum *M* chart,

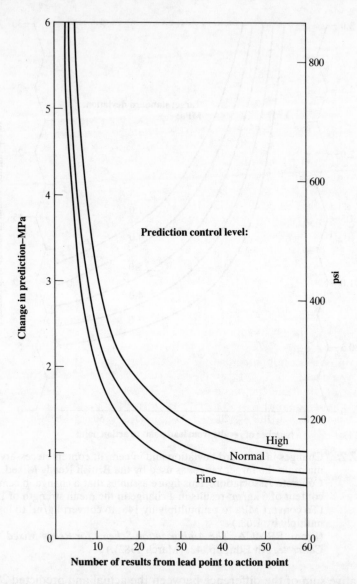

Fig. 17.8: Change in the prediction method for estimating the 28-day strength from the 24-hour accelerated strength, as used by the British Ready Mixed Concrete Association

checked with the appropriate V-mask and corrected, if necessary, by a change in cement content.

The preceding brief outline shows how interpretation of a continuing set of data can be economically beneficial, and is therefore superior to a simple use of individual test results.

Bibliography

17.1 ACI COMMITTEE 214–77 (reaffirmed 1983), Recommended practice for evaluation of strength test results of concrete, Part 2: Construction Practices and Inspection Pavements, *ACI Manual of Concrete Practice,* 1990.

17.2 ACI COMMITTEE 318–89, Building code requirements for reinforced concrete, Part 3: *ACI Manual of Concrete Practice,* 1990.

17.3 BRITISH READY MIXED CONCRETE ASSOCIATION, The authorization scheme for ready mixed concrete, Fifth Edition, pp. 42, (March 1982).

17.4 BRITISH STANDARDS INSTITUTION, BS 8110: Parts 1 and 2: 1985, Structural use of concrete (BSI 1985). Also Handbook to British Standard BS 8110: 1985, Palladian Publications, 1987.

Problems

17.1 What is meant by an operating characteristic curve?

17.2 How can operating characteristic curves be used to compare different compliance testing schemes?

17.3 What is the term used to measure the scatter or dispersion of strength about the mean?

17.4 What is meant by producer's risk in testing?

17.5 What is meant by consumer's risk in testing?

17.6 Describe the cusum technique for strength testing.

17.7 Explain the probability factor.

17.8 Why should you aim for a higher mean strength than the specified strength in design?

17.9 What are the compliance requirements of compressive strength in the UK and in the US?

17.10 What is a designed mix?

17.11 What is quality control of concrete?

17.12 What is a prescribed mix?

17.13 What is quality assurance of concrete?

17.14 Can quality assurance replace site supervision?

17.15 What properties of concrete other than strength are required to comply with specifications in the UK?

17.16 What is meant by characteristic strength of concrete?

17.17 What is the difference between the design strength and the mean strength of concrete?

17.18 Explain the difference between prescription and performance specifications for concrete.

17.19 What steps would you take if a single strength test falls short of the specified value?

17.20 What is meant by the margin? If the specified characteristic strength is 15 MPa, calculate the mean strength for 80 cube results, assuming a probability factor of 2.33.

Answer: 22.0 MPa

17.21 A series of 20 test cylinder results had a standard deviation of 400 psi. A second series of 15 test results had a standard deviation of 600 psi. Calculate the average standard deviation for the two series and the required mean compressive strength for a specified compressive strength of 3000 psi.

Answer: 495 psi; 3663 psi

18

Lightweight concrete

This chapter deals with insulating concrete and with structural concrete whose density is appreciably lower than the usual range of concretes made with normal weight aggregates. Those features which distinguish lightweight concrete from normal weight concrete are specifically considered; the types of lightweight aggregate are also described.

Classification of lightweight concretes

It is convenient to classify the various types of lightweight concrete by their method of production. These are:

(a) By using porous lightweight aggregate of low apparent specific gravity, i.e. lower than 2.6. This type of concrete is known as lightweight aggregate concrete.

(b) By introducing large voids within the concrete or mortar mass; these voids should be clearly distinguished from the extremely fine voids produced by air entrainment. This type of concrete is variously known as *aerated, cellular, foamed* or *gas* concrete.

(c) By omitting the fine aggregate from the mix so that a large number of interstitial voids is present; normal weight coarse aggregate is generally used. This concrete is known as *no-fines* concrete.

In essence, the decrease in density of the concrete in each method is obtained by the presence of voids, either in the aggregate or in the mortar or in the interstices between the coarse aggregate particles. It is clear that the presence of these voids reduces the strength of lightweight concrete compared with ordinary, normal weight concrete, but in many applications high strength is not essential and in others there are compensations (see page 346).

Because it contains air-filled voids, lightweight concrete provides good thermal insulation and has a satisfactory durability but is not highly resistant to abrasion. In general, lightweight concrete is more expensive than ordinary concrete, and mixing, handling and placing require more care and attention than ordinary concrete. However, for many purposes the advantages of lightweight concrete outweigh its disadvantages, and there is a continuing world-wide trend towards more lightweight concrete in applications such as prestressed concrete, high-rise buildings and even shell roofs.

Lightweight concrete can also be classified according to the purpose for which it is to be used: we distinguish between structural lightweight concrete (ASTM C 330–82a), concrete used in masonry units (ASTM C 331–81), and insulating concrete (ASTM C 332–83). This classification of structural lightweight concrete is based on a minimum strength: according to ASTM C 330–82a, the 28-day cylinder compressive strength should not be less than 17 MPa (2500 psi). The density (unit weight) of such concrete (determined in the dry state) should not exceed 1840 kg/m³ (115 lb/ft³), and is usually between 1400 and 1800 kg/m³ (85 and 110 lb/ft³). On the other hand, masonry concrete generally has a density between 500 and 800 kg/m³ (30 and 50 lb/ft³) and a strength between 7 and 14 MPa (1000 and 2000 psi). The essential feature of insulating concrete is its coefficient of *thermal conductivity* which should be below about 0.3 J/m²s °C/m (0.2 Btu/ft²h °F/ft), whilst density is generally lower than 800 kg/m³ (50 lb/ft³), and strength is between 0.7 and 7 MPa (100 and 1000 psi).

In concrete construction, self-weight usually represents a very large proportion of the total load on the structure, and there are clearly considerable advantages in reducing the density of concrete. The chief of these are a reduction in dead load and therefore in the total load on the various members and the corresponding reduction in the size of foundations. Furthermore, with lighter concrete, the formwork need withstand a lower pressure than would be the case with ordinary concrete, and also the total mass of materials to be handled is reduced with a consequent increase in productivity. Thus, the case for the use of structural lightweight concrete rests primarily on economic considerations.

Types of lightweight aggregate

The first distinction can be made between aggregates occurring in nature and those manufactured. The main *natural* lightweight aggregates are diatomite, pumice, scoria, volcanic cinders, and tuff; except for diatomite, all of these are of volcanic origin. Pumice is more widely employed than any of the others but, because they are found only in some areas, natural lightweight aggregates are not extensively used.

Pumice is a light-coloured, froth-like volcanic glass with a bulk density in the region of 500 to 900 kg/m³ (30 to 55 lb/ft³). Those varieties of pumice which are not too weak structurally make a satisfactory concrete with a density of 700 to 1400 kg/m³ (45 to 90 lb/ft³) and with good insulating characteristics, but high absorption and high shrinkage.

Scoria, which is a vesicular glassy rock, rather like industrial cinders, makes a concrete of similar properties.

Artificial aggregates are known by a variety of trade names, but are best classified on the basis of the raw material used and the method of manufacture.

In the first type are included the aggregates produced by the application of heat in order to expand clay, shale, slate, diatomaceous shale, perlite, obsidian and vermiculite. The second type is obtained by special cooling processes through which an expansion of blast-furnace slag is obtained. Industrial cinders form the third and last type.

Expanded clay, shale, and slate are obtained by heating suitable raw materials in a rotary kiln to incipient fusion (temperature of 1000 to 1200 °C (about 1800 to 2200 °F)) when expansion of the material takes place due to the generation of gases which become entrapped in a viscous pyrpoplastic mass. This porous structure is retained on cooling so that the apparent specific gravity of the expanded material is lower than before heating. Often, the raw material is reduced to the desired size before heating, but crushing after expansion may also be applied. Expansion can also be achieved by the use of a sinter strand. Here, the moistened material is carried by a travelling grate under burners so that heating gradually penetrates the full depth of the bed of the material. Its viscosity is such that the expanded gases are entrapped. As with the rotary kiln, either the cooled mass is crushed or initially pelletized material is used.

The use of pelletized material produces particles with a smooth shell or 'coating' (50 to 100 μm (0.002 to 0.004 in.) thick) over the cellular interior. These nearly spherical particles with a semi-impervious glaze have a lower water absorption than uncoated particles whose absorption ranges from about 12 to 30 per cent. Coated particles are easier to handle and to mix, and produce concrete of higher workability but are dearer than the uncoated aggregate.

Expanded shale and clay aggregates made by the sinter strand process have a density of 650 to 900 kg/m^3 (85 to 110 lb/ft^3), and 300 to 650 kg/m^3 (20 to 40 lb/ft^3) when made in a rotary kiln. They produce concrete with a density usually within the range of 1400 to 1800 kg/m^3 (85 to 110 lb/ft^3), although values as low as 800 kg/m^3 (50 lb/ft^3) have been obtained. Concrete made with expanded shale or clay aggregates generally has a higher strength than when any other lightweight aggregate is used.

Perlite is a glassy volcanic rock found in America, Ulster, Italy and elsewhere. When heated rapidly to the point of incipient fusion (900 to 1100 °C) it expands owing to the evolution of steam and forms a cellular material with a bulk density as low as 30 to 240 kg/m^3 (2 to 15 lb/ft^3). Concrete made with perlite has a very low strength, a very high shrinkage (because of a low modulus of elasticity – see page 242) and is used primarily for insulation purposes. An advantage of such concrete is that it is fast drying and can be finished rapidly.

Vermiculite is a material with a platey structure, somewhat similar to that of mica, and is found in America and Africa. When heated to a temperature of 650 to 1000 °C (about 1200 to 1800 °F), vermiculite expands to several, or even as many as 30, times its original volume by exfoliation of its thin plates. As a result, the bulk density of exfoliated vermiculite is only 60 to 130 kg/m^3 (4 to 8 lb/ft^3) and concrete made with it is of very low strength and exhibits high shrinkage but is an excellent heat insulator.

Expanded blast-furnace slag is produced in two ways. In one, a limited amount of water in the form of a spray comes into contact with the molten slag as it is being discharged from the furnace (in the production of pig-iron). Steam is generated and it bloats the still plastic slag, so that the slag hardens in a porous form, rather similar to pumice. This is the water-jet process. In the machine process, the molten slag is rapidly agitated with a controlled amount of water. Steam is entrapped and there is also some formation of gases due to chemical reactions of some of the slag constituents with water vapour.

Expanded, or foamed, slag has been used for many years and is produced with a bulk density varying between 300 and 1100 kg/m³ (20 and 70 lb/ft³), depending on the details of the cooling process and, to a certain degree, on the particle size and grading. Expanded slag for aggregate is covered by BS 3797: 1990 while BS 1047: 1983 specifies the non-expanded air-cooled slag aggregate. (Concrete made with expanded slag has a density of 950 to 1750 kg/m³ (60 to 110 lb/ft³).

Pelletized slag is also available. This is made by expanding the molten slag under water and breaking it into particles which are propelled through the air to form smooth pellets.

Clinker aggregate, known in the US as *cinders,* is made from well-burnt residue of industrial high-temperature furnaces, fused or sintered into lumps. It is important that the clinker be free from harmful varieties of unburnt coal, which may undergo expansion in the concrete, thus causing unsoundness. BS 1165: 1985 lays down the limits of loss on ignition and of soluble sulphate content in clinker aggregate to be used in plain concrete for general purposes and in *in situ* interior concrete not normally exposed to damp conditions. The standard does not recommend the use of clinker aggregate in reinforced concrete or in concrete required to have a specially high durability.

Iron or pyrites in the clinker may result in staining of surfaces, and should therefore be removed. Unsoundness due to hard-burnt lime can be avoided by allowing the clinker to stand wet for a period of several weeks: the lime will become slaked and will not expand in the concrete.

Breeze is the name given to a material similar to clinker but more lightly sintered and less well burnt. There is no clear-cut demarcation between breeze and clinker.

When cinders are used as both fine and coarse aggregate, concrete with a density of about 1100 to 1400 kg/m³ (70 to 85 lb/ft³) is obtained, but often natural sand is used in order to improve the workability of the mix: the density of the resulting concrete is then 1750 to 1850 kg/m³ (110 to 115 lb/ft³).

It should be noted that, in contrast to normal weight aggregate, the finer particles of lightweight aggregate generally have a higher apparent specific gravity than the coarser ones. This is caused by the crushing process: fracture occurs through the larger pores so that the smaller the particle the smaller the pores in it.

One general feature of artificial aggregates is worth noting. Since the

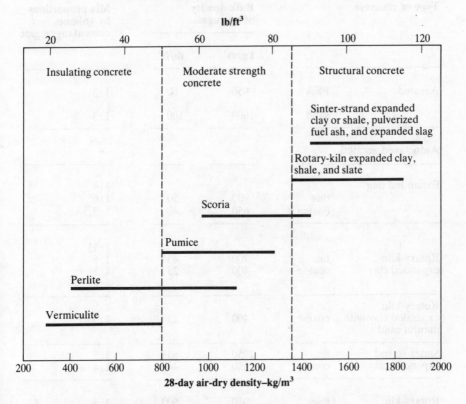

Fig. 18.1: Typical ranges of density of concretes made with various lightweight aggregates

particles are factory made under closely controlled conditions, they are less variable than many natural aggregates.

Typical ranges of density of concretes made with various lightweight aggregates, largely based on the ACI classification, are shown in Fig. 18.1. Some general requirements for lightweight aggregates are prescribed by ASTM Standards C 330–82a, C 331–81, and C 332–83 and by BS 3797: Part 2: 1976.

Properties of lightweight aggregate concrete

The various types of lightweight aggregate available allow the density of concrete to range from a little over 300 up to 1850 kg/m^3 (20 to 115 lb/ft^3), with a corresponding strength range of 0.3 and 40 MPa (50 to 6000 psi) and sometimes even higher. Strengths up to 60 MPa (9000 psi) can be obtained with very high cement contents (560 kg/m^3 (950 lb/yd^3)). Generally, with lightweight aggregate, the cement content varies from

349

Table 18.1: **Typical properties of lightweight concretes**

Type of concrete		Bulk density of aggregate		Mix proportions by volume cement : aggregate
		kg/m³	lb/ft³	
Aerated	PFA*	950	60	1:3
	sand	1600	100	1:3
Autoclaved aerated		–	–	–
Expanded slag				1:8
	fine	900	50 ⎫	1:6
	coarse	650	40 ⎭	1:3.5
Rotary-kiln				1:11
expanded clay	fine	700	45 ⎫	1:6
	coarse	400	25 ⎭	1:5
				1:4
Rotary-kiln expanded clay with natural sand	coarse	400	25	1:5
Sinter-strand	fine	1050	65 ⎫	1:5
expanded clay	coarse	650	40 ⎭	1:4
Rotary-kiln	fine	950	60 ⎫	1:6
expanded slate	coarse	700	45 ⎭	1:4.5
Sintered				1:5.9
pulverized	fine	1050	65 ⎫	1:5.3
fuel ash	coarse	800	50 ⎭	1:4.5
				1:3.1
Sintered pulverized fuel ash with natural sand	coarse	800	50	1:6.1
				1:5.5
				1:5.0
				1:3.6
Pumice		500–800	30–50	1:6
				1:4
				1:2
Exfoliated vermiculite		65–130	4–8	1:6
Perlite		95–130	6–8	1:6

* PFA = pulverized fuel ash (fly ash).

Dry density of concrete		Compressive strength		Drying shrinkage	Thermal conductivity	
kg/m³	lb/ft³	MPa	psi	10^{-6}	J/m²s °C/m	Btu/ft²h °F/ft
750	47	3	500	700	0.19	0.11
900	55	6	800		0.22	0.13
800	55	4	600	800	0.25	0.14
1700	105	7	1000	400	0.45	0.26
1850	115	21	3000	500	0.69	0.40
2100	130	41	6000	600	0.76	0.44
650–1000	42–62	3–4	400–600	–	0.17	0.10
1100	70	14	2000	550	0.31	0.18
1200	75	17	2500	600	0.38	0.22
1300	80	19	2800	700	0.40	0.23
1350–1500	85–95	17	2500	–	0.57	0.33
1500	95	24	3500	600	0.55	0.32
1600	100	31	4500	750	0.61	0.35
1700	105	28	4000	400	0.61	0.35
1750	110	35	5000	450	0.69	0.40
1490	93	20	2900	300	–	–
1500	95	25	3600	300	–	–
1540	96	30	4400	350	–	–
1570	98	40	5800	400	–	–
1670	104	20	2900	300	–	–
1700	106	25	3600	300	–	–
1750	109	30	4400	350	–	–
1790	112	40	5800	400	–	–
1200	74	14	2000	1200	–	–
1250	77	19	2800	1000	0.14	0.08
1450	90	29	4200	–	–	–
300–500	20–30	2	300	3000	0.10	0.06
–	–	–	–	2000	0.05	0.03

the same as with normal weight aggregate to 70 per cent more for the same strength of concrete.

Lightweight aggregates, even similar in appearance, may produce concretes varying widely in properties so that a careful check on the performance of each new aggregate is necessary. Classification of concrete according to the type of aggregate used is difficult as the properties of the concrete are affected also by the grading of the aggregate, the cement content, the water/cement ratio, and the degree of compaction. Typical properties are listed in Table 18.1, and thermal conductivity is shown as a function of density in Fig. 18.2. The grading requirements for coarse and fine aggregate are given in Tables 18.2 to 18.4.

The suitability of a lightweight concrete is governed by the desired properties: density, cost, strength, and thermal conductivity. The low

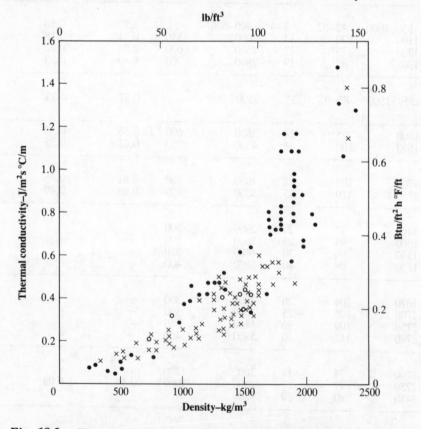

Fig. 18.2: Thermal conductivity of lightweight aggregates concretes of various types
(From: N. DAVEY, Concrete mixes for various building purposes, *Proc. of a Symposium on Mix Design and Quality Control of Concrete,* pp. 28–41 (London, Cement and Concrete Assoc., 1954).)

Table 18.2: Grading requirements for coarse lightweight aggregate given by BS 3797: Part 2:1976 (amended 1981)

| Sieve size | Percentage by mass passing BS sieves | | | | | | |
| | Nominal size of graded aggregate, mm | | | Nominal size of single-sized aggregate, mm | | | |
mm	20 to 5	14 to 5	10 to 2.36	20	14	10	6
37.5	100	–	–	100	–	–	–
20.0	95–100	100	–	85–100	100	–	–
14.0	–	95–100	100	–	90–100	100	–
10.0	30–60	50–90	85–100	5–25	20–45	85–100	100
6.3	–	–	–	–	–	–	70–100
5.0	0–10	0–15	15–50	0–5	0–10	15–35	0–35
2.36	–	–	0–15	–	–	0–5	0–10

Table 18.3: Grading requirements for fine lightweight aggregate given by BS 3797: Part 2: 1976

Sieve size	Percentage by mass passing BS sieves	
	Grading Zone L1	Grading Zone L2
10.00 mm	100	100
5.00 mm	90–100	90–100
2.36 mm	55–95	60–100
1.18 mm	35–70	40–80
600 μm	20–50	30–60
300 μm	10–30	25–40
150 μm	5–19	20–35

Note: Certain departures from a zone are permitted except for the maximum percentage passing the 150 μm sieve in the case of Grading Zone L1 and the minimum in the case of Grading Zone L2.

thermal conductivity of lightweight aggregate concrete is clearly advantageous for applications requiring very good insulation, but the same property causes a higher temperature rise under mass-curing conditions, which is relevant to the possibility of early-age thermal cracking (see Chapter 13).

Other properties which have to be considered are workability, absorption, drying shrinkage, and moisture movement. For equal workability (ease of compaction), lightweight aggregate concrete registers a lower slump and a lower compacting factor than normal weight concrete because the work done by gravity is smaller in the case of the lighter material. A consequential danger is that, if a higher workability is used, there is a greater tendency to segregation.

The porous nature of lightweight aggregates means that they have high and rapid water absorption. Thus, if the aggregate is dry at the time of mixing, it will rapidly absorb water and the workability will quickly decrease. The remedy lies in mixing the aggregate with at least one-half of the mixing water before adding the cement; knowing the water absorption (see Chapter 3), the *effective* water/cement ratio (see page 55) can be calculated. However, such a procedure will cause the density of the concrete to rise and the thermal insulation to fall. To overcome these disadvantages it is possible to waterproof the aggregate by a coating of bitumen, using a special process, but this is rarely done.

Lightweight aggregate mixes tend to be *harsh,* but harshness can be reduced by air entrainment: water requirement is reduced and so is the tendency to bleeding and segregation. The usual *total* air contents by volume are: 4 to 8 per cent for 20 mm ($\frac{3}{4}$ in.) maximum size of aggregate, and 5 to 9 per cent for 10 mm ($\frac{3}{8}$ in.) maximum size of aggregate. Air contents in excess of these values lower the compressive strength by about 1 MPa (150 psi) for each additional percentage point of air.

The use of lightweight fines, as well as of lightweight coarse aggregate,

Table 18.4: Grading requirements for lightweight aggregate for structural concrete given by ASTM C 330-82a

ASTM sieve size	Percentage by weight passing ASTM sieves						
	Nominal size of graded coarse aggregate				Fine aggregate	Nominal size of combined fine and coarse aggregate	
	1 in. to No. 4 (25 mm to 4.75 mm)	$\frac{3}{4}$ in. to No. 4 (19 mm to 4.75 mm)	$\frac{1}{2}$ in. to No. 4 (12.5 mm to 4.75 mm)	$\frac{3}{8}$ in. to No. 8 (9.5 mm to 2.36 mm)	No. 4 down (4.75 mm down)	$\frac{1}{2}$ in. (12.5 mm)	$\frac{3}{8}$ in. (9.5 mm)
1 in. (25.0 mm)	95–100	–	–	–	–	–	–
$\frac{3}{4}$ in. (19.0 mm)	–	100	–	–	–	–	–
$\frac{1}{2}$ in. (12.5 mm)	25–60	90–100	100	–	–	100	–
$\frac{3}{8}$ in. (9.5 mm)	–	–	90–100	100	–	95–100	100
4 (4.75 mm)	0–10	10–50	40–80	80–100	100	50–80	90–100
8 (2.36 mm)	–	0–15	0–20	5–40	85–100	–	65–90
16 (1.18 mm)	–	–	0–10	0–20	40–80	–	35–65
50 (300 µm)	–	–	–	0–10	10–35	5–20	10–25
100 (150 µm)	–	–	–	–	5–25	2–15	5–15

aggravates the problem of low workability. It may, therefore, be preferable to use normal weight fines with lightweight coarse aggregate. Such concrete is referred to as *semi-lightweight* (or sand-lightweight) concrete, and, of course, its density and thermal conductivity are higher than when all-lightweight aggregate is used. Typically, for the same workability, semi-lightweight concrete will require 12 to 14 per cent less mixing water than lightweight aggregate concrete. The modulus of elasticity of semi-lightweight concrete is higher and its shrinkage is lower than when all-lightweight aggregate is used.

We should note that a *partial* replacement of fine aggregate by normal weight fines is also possible. In any case, replacement should be on an equal *volume* basis.

When lightweight aggregate is used in reinforced concrete, special care should be taken to protect the reinforcement from corrosion because the depth of carbonation (see page 241), i.e. the depth within which corrosion can occur under suitable conditions, can be up to twice that with normal weight aggregate. The behaviour of different aggregates varies considerably but generally, with lightweight aggregate, an additional cover of 10 mm to reinforcement is desirable. Alternatively, the use of a rendered finish or coating of reinforcement with rich mortar has been found useful. In the case of clinker aggregate, there is an additional danger of corrosion due to the presence of sulphur in the clinker, and coating of the steel is necessary, although the use of this aggregate is not common.

All concretes made with lightweight aggregate exhibit a higher *moisture movement* (see page 239) than is the case with normal weight concrete.

Table 18.5 shows that the coefficient of thermal expansion of lightweight aggregate concrete is generally lower than in the case of normal weight concrete. This can create some problems when lightweight aggregate and ordinary concretes are used side by side but, on the other hand, lightweight aggregate concrete members have a lower tendency to warp or buckle due to differential temperature gradients.

Table 18.5: **Coefficient of thermal expansion of lightweight aggregate concrete**

Type of aggregate used	Linear coefficient of thermal expansion (determined over a range of −22 °C to 52 °C (−7 °F to 125 °F))	
	10^{-6} per °C	10^{-6} per °F
Pumice	9.4 to 10.8	5.2 to 6.0
Perlite	7.6 to 11.0	4.2 to 6.1
Vermiculite	8.3 to 14.2	4.6 to 7.9
Cinders	about 3.8	about 2.1
Expanded shale	6.5 to 8.1	3.6 to 4.5
Expanded slag	7.0 to 11.2	3.9 to 6.2

Some other properties of lightweight aggregate concretes as compared with normal weight concrete may be of interest:

(a) For the same strength, the modulus of elasticity is lower by 25 to 50 per cent; hence, deflections are greater.

(b) Resistance to freezing and thawing is greater because of the greater porosity of the lightweight aggregate, *provided* the aggregate is not saturated before mixing.

(c) Fire resistance is greater because lightweight aggregates have a lesser tendency to spall; the concrete also suffers a lower loss of strength with a rise in temperature.

(d) Lightweight concrete is easier to cut or to have fitments attached.

(e) For the same compressive strength, the shear strength is lower by 15 to 25 per cent and the bond strength is lower by 20 to 50 per cent. These differences have to be taken into account in the design of reinforced concrete beams.

(f) The tensile strain capacity (see page 169) is about 50 per cent greater than in normal weight concrete. Hence, the ability to withstand restraint to movement, e.g. due to internal temperature gradients, is greater for lightweight concrete.

(g) For the same strength, creep of lightweight aggregate concrete is about the same as that of normal weight concrete.

Aerated concrete

As mentioned earlier, one means of obtaining lightweight concrete is to introduce gas bubbles into the plastic mix of mortar (cement and sand) in order to produce a material with a cellular structure, containing voids between 0.1 and 1 mm (0.004 to 0.04 in.) in size. The 'skin' of the voids or cells must be able to withstand mixing and compaction. The resulting concrete is known as aerated or *cellular* concrete, although strictly speaking, the term concrete is inappropriate as usually no coarse aggregate is present.

There are two basic methods of producing aeration, an appropriate name being given to each end product.

(a) *Gas* concrete is obtained by a chemical reaction generating a gas in the fresh mortar, so that when it sets it contains a large number of gas bubbles. The mortar must be of the correct consistence so that the gas can expand the mortar but does not escape. Thus, the speed of gas evolution, consistence of mortar and its setting time must be matched. Finely divided aluminium powder is most commonly used, its proportion being of the order of 0.2 per cent of the mass of cement. The reaction of the active powder with calcium hydroxide or the alkalis liberates hydrogen bubbles. Powdered zinc or aluminium alloy can also be used. Sometimes hydrogen peroxide is employed to entrain oxygen bubbles.

(b) *Foamed* concrete is produced by adding a foaming agent (usually some form of hydrolyzed protein or resin soap) to the mix. The agent introduces and stabilizes air bubbles during mixing at high speed. In some processes, a stable pre-formed foam is added to the mortar during mixing in an ordinary mixer.

Aerated concrete may be made without sand, but only for non-structural purposes such as for heat insulation when a density range of 200 to 300 kg/m^3 (12 to 20 lb/ft^3) can be obtained. More usual mixes (with sand) have densities between 500 and 1100 kg/m^3 (30 and 70 lb/ft^3) when a mixture of cement and very fine sand is used.

As in other lightweight concretes, strength varies in proportion to density, and so does the thermal conductivity. A concrete with a density of 500 kg/m^3 (30 lb/ft^3) would have a strength in the region of 3 to 4 MPa (450 to 600 psi) and a thermal conductivity of about 0.1 J/m^{2}s °C/m (0.06 Btu/ft^{2}h °F/ft). For a concrete with a density of 1400 kg/m^3 (90 lb/ft^3), the corresponding values would be approximately 12 to 14 MPa (1800 to 2000 psi) and 0.4 J/m^{2}s °C/m (0.23 Btu/ft^{2}h °F/ft). By comparison, the conductivity of ordinary concrete is about 10 times larger. It should be noted that thermal conductivity increases linearly with the moisture content: when this is 20 per cent, the conductivity is typically almost double that when the moisture content is zero.

The modulus of elasticity of aerated concrete is usually between 1.7 and 3.5 GPa (0.25 and 0.5 × 10^6 psi). Creep, expressed on the basis of stress/strength ratio, is sensibly the same as for ordinary concrete; however, on the basis of equal stress, the specific creep of aerated concrete is higher (see Chapter 12). Compared with lightweight aggregate concrete of the same strength, aerated concrete has higher thermal movement, higher shrinkage and higher moisture movement, but these may be reduced by autoclaving (high-pressure steam curing), which also improves the compressive strength.

Aerated concrete is principally used for heat insulation purposes because of its low thermal conductivity, and for fireproofing as it has a better fire resistance than ordinary concrete. Structurally, aerated concrete is used mostly in the form of autoclaved blocks or precast members but it can also be used for floor construction instead of a hollow tile floor. For light, insulating courses, flowing aerated concrete can be obtained using a superplasticizer (see page 156).

Other advantages of aerated concrete are that it can be sawn, it holds nails, and it is reasonably durable for, although its water absorption is high, the rate of water penetration through aerated concrete is low as the larger pores will not fill by suction. For this reason, aerated concrete has a comparatively good resistance to frost, and, if rendered, can be used in wall construction.

Unprotected reinforcement in aerated concrete would be vulnerable to corrosion even when the external attack is not very severe. The reinforcement should therefore be treated by dipping in a suitable anti-corrosive liquid: bituminous solutions and epoxy resins have been found to be successful without any adverse effect on bond.

No-fines concrete

This concrete is obtained by omitting fine aggregate from the mix so that there is an agglomeration of nominally one-size coarse aggregate particles, each surrounded by a coating of cement paste up to about 1.3 mm (0.05 in.) thick. There exist, therefore, large pores within the body of the concrete, which are responsible for its low strength but, as with aerated concrete, their large size means that no capillary movement of water can take place and, consequently, the rate of water penetration is low.

For a given type of aggregate, the density of no-fines concrete depends primarily on the grading of the aggregate. With one-size aggregate, the density is about 10 per cent lower than when a well-graded aggregate of the same specific gravity is used. The usual aggregate size is 10 to 20 mm ($\frac{3}{8}$ to $\frac{3}{4}$ in.); 5 per cent oversize and 10 per cent undersize are allowed but no material should be smaller than 5 mm ($\frac{3}{16}$ in.). A no-fines concrete with a density as low as 640 kg/m^3 (40 lb/ft^3) can be obtained using lightweight aggregate. On the other hand, with normal weight aggregate, the density varies between 1600 and 2000 kg/m^3 (100 and 125 lb/ft^3) (see Table 18.6). Sharp-edged coarse aggregate should be avoided as, otherwise, local crushing can occur under load.

Table 18.6: Typical data for 10 to 20 mm ($\frac{3}{8}$ to $\frac{3}{4}$ in.) no-fines concrete

Aggregate/cement ratio by volume	Water/cement ratio by mass	Density		28-day compressive strength	
		kg/m^3	lb/ft^3	MPa	psi
6	0.38	2020	126	14	2100
7	0.40	1970	123	12	1700
8	0.41	1940	121	10	1450
10	0.45	1870	117	7	1000

Compared with ordinary concrete, no-fines concrete compacts very little and, in fact, vibration should be applied for very short periods only as, otherwise, the cement paste would run off. Rodding is not recommended as it can lead to high local density, but care is required to prevent *arching* across the form. There are no workability tests for no-fines concrete – a visual check to ensure even coating of all particles is adequate. Since no-fines concrete does not segregate, it can be dropped from a considerable height and placed in high lifts.

The compressive strength of no-fines concrete varies generally between 1.4 and 14 MPa (200 and 2000 psi), depending mainly on its density, which is governed by the cement content (see Table 18.6). Practical mixes vary rather widely, with a lean limit for the cement/aggregate ratio by *volume* of between 1:10 and 1:20; the corresponding cement contents are approximately 130 kg/m^3 (220 lb/yd^3) and 70 kg/m^3 (120 lb/yd^3). Whereas for normal concrete made with well-graded aggregate, the water/cement ratio is the controlling factor in strength (see Chapter 6), this is not the case for no-fines concrete, in which there is a narrow optimum value of the water/cement ratio for any given aggregate. A water/cement ratio higher than the optimum would make the cement paste drain away from the aggregate particles while with too low a water/cement ratio the paste would not be sufficiently adhesive, and a uniform composition of concrete would not be achieved. Typically, the optimum water/cement ratio is between 0.38 and 0.52, depending on the cement content necessary for a sufficient coating of the aggregate.

The actual strength of the concrete has to be determined by tests but the increase in strength with age is similar to that of ordinary concrete. However, the modulus of rupture (see page 309) of no-fines concrete is about 30 per cent of the compressive strength, a proportion which is higher than for normal weight concrete.

Because no-fines concrete exhibits very little cohesion, formwork must remain in place until sufficient strength has been developed. Moist curing is important, especially in a dry climate or under windy conditions, because of the small thickness of cement paste involved.

Shrinkage of no-fines concrete is considerably lower than that of normal concrete because contraction is restrained by the large volume of aggregate relative to the paste. However, the initial rate of shrinkage is high because a large surface area of the cement paste is exposed to the air. Typical values of shrinkage after 1 month of drying lie between 120×10^{-6} and 200×10^{-6}.

The thermal movement of no-fines concrete is about 70 per cent of that of normal weight concrete, the actual value of the coefficient of thermal expansion depending, of course, on the type of aggregate used. As with aerated concrete, an advantage of no-fines concrete is its low thermal conductivity: approximately $0.22 \text{ J/m}^2\text{s }^\circ\text{C/m}$ ($0.13 \text{ Btu/ft}^2\text{h }^\circ\text{F/ft}$) with lightweight aggregate and $0.80 \text{ J/m}^2\text{s }^\circ\text{C/m}$ ($0.46 \text{ Btu/ft}^2\text{h }^\circ\text{F/ft}$) with normal weight aggregate. However, a high moisture content of the concrete appreciably increases the thermal conductivity.

Because of the absence of capillaries, no-fines concrete is highly resistant to frost, provided that the pores are not saturated, in which case freezing would cause a rapid disintegration. High absorption of water (up to 12 per cent by mass) however, makes no-fines concrete unsuitable for use in foundations. Although under less severe conditions the absorption is less, it is still necessary to render external walls on both sides, a practice which reduces the permeability to air as well as the sound-absorbing properties of no-fines concrete. Where the acoustic properties are considered to be of paramount importance, one side of a wall should *not* be rendered.

Although the strength of no-fines concrete is considerably lower than that of normal weight concrete, this strength, coupled with the lower self-weight, is sufficient for use in buildings, even of many storeys, and in many other applications. No-fines concrete is not normally used in reinforced concrete, but if this is required the reinforcement has to be coated with a thin layer (about 3 mm ($\frac{1}{8}$ in.)) of cement paste to improve the bond characteristics and to prevent corrosion. The easiest way to coat the reinforcement is by shotcreting (see page 141).

Since sand is absent and the cement content of no-fines concrete is low, its cost is comparatively low: in lean mixes, the cement content can be as little as 70 to 130 kg of cement per cubic metre (120 to 220 lb/yd³) of concrete.

Bibliography

18.1   ACI COMMITTEE 213.R–87, Guide for structural lightweight aggregate concrete, Part 1: Materials and General Properties of Concrete, *ACI Manual of Concrete Practice,* 1990.

18.2   ACI COMMITTEE 304.5R–82, Batching, mixing, and job control of lightweight concrete, Part 2: Construction Practices and Inspection Pavements, *ACI Manual of Concrete Practice,* 1990.

18.3   ACI COMMITTEE 523.1R–86 (Reapproved 1987), Guide for cast-in-place low density concrete;
ACI COMMITTEE 523.2R–68 (Reapproved 1987), Guide for low density precast concrete floor, roof, and wall units;
ACI COMMITTEE 523.3R–75 (Reapproved 1987), Guide for cellular concretes above 50 lb/ft³ and for aggregate concretes above 50 lb/ft³ with compressive strengths less than 2500 psi, Part 5: Masonry, Precast Concrete, and Special Processes, *ACI Manual of Concrete Practice,* 1990.

18.4   A. LOW, A bridge designer's view of lightweight concrete, *Concrete,* pp. 7–9 (London, Jan. 1985).

Problems

18.1   Discuss the use of no-fines concrete.

18.2   Discuss the properties of insulating concrete.

18.3   What is the difference between lightweight concrete and lightweight aggregate concrete?

18.4   What is meant by structural lightweight concrete? How is it classified?

18.5   What happens when aluminium powder is put into a concrete mix?

18.6   What is meant by artificial lightweight aggregate?

18.7   Describe some methods of manufacturing lightweight aggregate.

18.8   What are the main categories of lightweight concrete?

18.9   In what respect is the stress–strain behaviour of lightweight concrete different from normal concrete?

18.10 What are the main differences between lightweight and normal weight aggregates?

18.11 For what purposes are lightweight aggregates used?

18.12 What is a non-structural lightweight aggregate?

18.13 What is semi-lightweight concrete?

18.14 What are the advantages of semi-lightweight concrete compared with lightweight concrete?

18.15 What are the advantages of semi-lightweight concrete compared with normal weight concrete?

18.16 Compare the moduli of elasticity of concretes made with lightweight and with normal weight aggregates.

18.17 Compare the shrinkage of concretes made with lightweight and with normal weight aggregates.

18.18 Compare the creep of concretes made with lightweight and with normal weight aggregates.

18.19 How does lightweight aggregate affect the corrosion of reinforcement?

18.20 Give alternative terms for aerated concrete.

18.21 Discuss the effect of density and moisture content on thermal conductivity of concrete.

18.22 Comment on the shear strength of concrete made with lightweight aggregate.

18.23 Compare the variability of normal weight aggregate and artificial lightweight aggregate.

19

Mix design

How do we decide what concrete we need in any particular case? The required properties of *hardened* concrete are specified by the designer of the structure and the properties of *fresh* concrete are governed by the type of construction and by the techniques of placing and transporting. These two sets of requirements make it possible to determine the composition of the mix, taking also account of the degree of control exercised on site. Mix design can, therefore, be defined as the process of selecting suitable ingredients of concrete and determining their relative quantities with the purpose of producing an economical concrete which has certain minimum properties, notably workability, strength and durability.

In the previous chapters, we have discussed in detail the various factors which influence the properties of concrete. In this chapter, we shall briefly summarize those important factors which are taken into account in the process of mix design. Nowadays, we use *designed mixes,* rather than prescribed mixes (see page 335), for which specifications lay down limiting values for a range of properties that must be satisfied. These properties are usually: the maximum water/cement ratio, minimum cement content, minimum strength, minimum workability, and maximum size of aggregate, and air content within specified limits.

It should be explained that design in the strict sense of the word is not possible: the materials used are variable in a number of respects and many of their properties cannot be assessed truly quantitatively, so that we are really making no more than an intelligent guess at the optimum combinations of the ingredients on the basis of the relationships established in the earlier chapters. It is not surprising, therefore, that in order to obtain a satisfactory mix we must check the estimated proportions of the mix by making trial mixes and, if necessary, make appropriate adjustments to the proportions until a satisfactory mix has been obtained.

In the subsequent sections, we shall concentrate on the American (ACI 211.1-81) method of mix design for normal weight aggregate concrete; ACI 211.1-81 also covers heavy weight and mass concretes. In addition, we shall discuss the British method for normal weight concrete, developed for the Department of the Environment in 1975 and revised in

1988. Examples of the use of both methods are given. The final section deals with the design of lightweight aggregate concrete, mainly as prescribed by ACI 211.2–81.

Factors to be considered

The economic and technical factors and the procedures of estimating the mix quantities will now be outlined. Virtually always, the strength of concrete has to be considered. The actual cost of concrete is related to the materials required to produce a certain mean strength but, as we have seen in Chapter 17, it is the minimum strength which is specified by the structural designer. Normally, the strength for structural purposes is required at 28 days, but other considerations may dictate the strength at other ages, e.g. formwork striking times. The expected or known variability then determines the mean strength. By adopting quality control techniques, the variability of strength can be minimized so that the lowest mean strength is needed for a given minimum specified strength. However, the cost of implementing and operating a more elaborate quality control scheme has to be weighed against the possible savings in cement resulting from a lower mean strength.

Water/cement ratio

The water/cement ratio required to produce a given mean compressive strength is best determined from previously established relations for mixes made from similar ingredients or by carrying out tests using trial mixes made with the actual ingredients to be used in the construction, including admixtures. However, Tables 19.1 and 19.2, and Fig. 19.1 and Table 19.3 may be used to estimate the approximate water/cement ratio for the cements listed for each set of values; for other cements, the water/cement ratio must be established from trial mixes.

In the case of the British method of mix design, knowing the type of coarse aggregate, the type of cement and the age when the mean strength is required, the mean strength for a free (or effective) water/cement ratio of 0.5 is obtained from Table 19.3. This value is entered in Fig. 19.1, and a strength *versus* water/cement ratio curve is plotted by interpolation between the curves adjacent to the entered value. The free water/cement ratio for the desired mean strength can now be read from the graph.

It is important that the water/cement ratio selected on the basis of strength is satisfactory also for the durability requirements. Moreover, this water/cement ratio for durability should be established prior to the commencement of the structural design because, if it is lower than necessary from structural considerations, advantage of the use of a higher strength of concrete can be taken in the design calculations.

When pozzolan or slag is used in concrete, the water/cementitious

Table 19.1: Relation between water/cement ratio and average compressive strength of concrete, according to ACI 211.1–81

Average compressive strength at 28 days*		Effective water/cement ratio (by mass)	
MPa	psi	Non-air-entrained concrete	Air-entrained concrete
45	–	0.38	–
–	6000	0.41	–
40	–	0.43	–
35	5000	0.48	0.40
30	–	0.55	0.46
–	4000	0.57	0.48
25	–	0.62	0.53
–	3000	0.68	0.59
20	–	0.70	0.61
15	–	0.80	0.71
–	2000	0.82	0.74

* Measured on standard cylinders. The values given are for a maximum size of aggregate of 20 to 25 mm ($\frac{3}{4}$ to 1 in.), for concrete containing not more than the percentage of air shown in Table 19.5, and for ordinary Portland (Type I) cement.

material ratio by mass has to be considered. With pozzolans, the ACI 211.1–81 approach is to treat the water/cementitious material ratio as equivalent to the water/cement ratio of a Portland cement mix either by having the same mass of cementitious material or by having the same volume of cementitious material as the cement in the Portland-cement-only mix. With the mass method, the water/cementitious material ratio is equal to the water/cement ratio of the Portland-cement-only mix but, because pozzolans have a lower specific gravity than Portland cement, the volume of the cementitious material is greater than that of the Portland cement mix. With the volume method, the mass of the cementitious material is smaller than that of the cement in the Portland-cement-only mix so that the water/cementitious material ratio is greater than in the Portland-cement-only mix.

Whichever approach is used, a partial replacement of cement by pozzolan generally reduces the strength at early ages (see Fig. 2.6). For this reason, the ACI 211.1–81 mix design is used mainly for mass concrete, in which the reduction of the heat of hydration is of paramount importance (see page 169) and the early strength is of lesser significance.

The use of PFA (fly ash) – an artificial pozzolan – in concrete at large, and not for a specific purpose, has recently become more common. In consequence, the mix has to be designed so as to meet the strength requirement at 28 days as well as, of course, the durability requirements. In the UK, one approach is to recognize two factors resulting from the

Table 19.2: Relation between water/cement ratio and specified compressive strength of concrete according to ACI 318–83

Specified compressive strength at 28 days*		Absolute water/cement ratio (by mass)	
MPa	psi	Non-air-entrained concrete	Air-entrained concrete
–	4500	0.38	–
30	–	0.40	–
–	4000	0.44	0.35
25	–	0.50	0.39
–	3500	0.51	0.40
–	3000	0.58	0.46
20	–	0.60	0.49
17	–	0.66	0.54
–	2500	0.67	0.54

* Measured on standard cylinders. Applicable for cements: ordinary Portland (Types I, IA), modified Portland (Types II, IIA) cement, rapid-hardening Portland (Types III, IIIA), sulphate-resisting Portland (Type V); also Portland blast-furnace (Types IS, IS-A) and Portland-pozzolan (Types IP, P, I(PM), IP-A) including moderate sulphate-resisting cements (MS).

NOTES: The use of admixtures, other than air-entraining, or of low density aggregate is not permitted. The values of absolute water/cement ratio are conservative and include any water absorbed by the aggregates. Hence, with most materials, these water/cement ratios will provide average strengths which are greater than the specified strength.

Table 19.3: Approximate compressive strengths of concrete mixes made with a free water/cement ratio of 0.5 according to the 1986 British method

Type of cement	Type of coarse aggregate	Compressive strength* (MPa (psi)) at an age of (days):			
		3	7	28	91
Ordinary Portland (Type I)	Uncrushed	22 (3200)	30 (4400)	42 (6100)	49 (7100)
Sulphate-resisting Portland (Type V)	Crushed	27 (3900)	36 (5200)	49 (7100)	56 (8100)
Rapid-hardening Portland (Type III)	Uncrushed	29 (4200)	37 (5400)	48 (7000)	54 (7800)
	Crushed	34 (4900)	43 (6200)	55 (8000)	61 (8900)

* Measured on cubes.
Building Research Establishment, Crown copyright

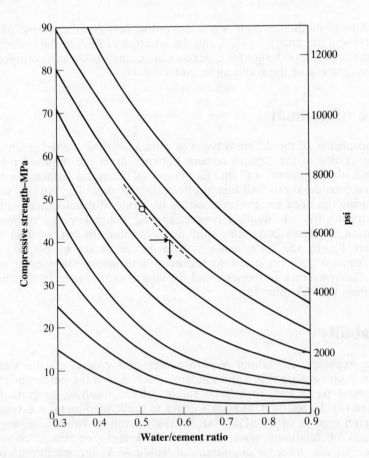

Fig. 19.1: Relation between compressive strength and free water/cement ratio
for use in the British mix design method.
(Building Research Establishment, Crown copyright)
An example is shown where sulphate-resisting Portland (Type V)
cement is used to make concrete. According to Table 19.3, for a
free water/cement ratio of 0.5, the 28-day strength is 49 MPa
(7100 psi). This point is plotted and the dashed curve is interpolated
between the adjacent curves. From this curve, the water/cement
ratio for a strength of 40 MPa (5800 psi) is estimated to be 0.57.

use of PFA: improved workability and therefore a reduction in the water
requirement, and a reduction in early strength. For the same workability
as a Portland-cement-only mix, the reduction in water requirement of a
PFA mix increases with an increase in the level of replacement.
However, to offset the reduction in early strength, the mass of the
cementitious material (compared with the mass of cement in a Portland-
cement-only mix) has to be increased. In consequence, the water/
cementitious material ratio is lower than the water/cement ratio of a
Portland-cement-only mix. This approach can be described as a modified

replacement method, which is a compromise between the simple partial replacement of cement by PFA and the addition of PFA to the cement. A new method of mix design for concrete containing PFA was developed by the Department of the Environment in 1988.

Type of cement

The properties of the different types of cement were discussed in Chapter 2. The choice of the type of cement depends upon the required rate of strength development, on the likelihood of chemical attack, and on thermal considerations. All this has been discussed earlier, but it is worth reiterating the need for a cement with a high rate of development of heat of hydration for cold-weather concreting and with a low rate of heat of hydration for mass concreting and for concreting in hot weather (see Chapter 9). In the latter case, it may be necessary to use a lower water/cement ratio in order to ensure a satisfactory strength at early ages. The resistance to freezing and thawing is not a factor in the choice of cement (see Chapter 15).

Durability

Severe exposure conditions require a stringent control of the water/cement ratio because this is the fundamental factor in the permeability of the cement paste and, to a large extent, of the resulting concrete (see Chapter 14). In addition, adequate cover to embedded metal is essential. The requirements of BS 8110: Part 1: 1985, given in Table 14.5, are on the basis of minimum cover for reinforced and prestressed concrete exposed to the different conditions of Table 15.1; the minimum cover should always be greater than the size of the reinforcing bar and than the maximum nominal size of aggregate. The quality of the concrete in the cover is specified by the maximum free water/cement ratio, the minimum cement content and the minimum grade of concrete (characteristic strength). For plain concrete (see Table 15.1), the requirements of durability for different exposures are given in terms of the free water/cement ratio, the minimum cement content and the strength. The reason for specifying minimum cement content and strength is that the water/cement ratio is not easy to measure and control for compliance purposes. However, the water/cement ratio can be assessed indirectly through the workability of the mix, the cement content and strength. We should recall that, for a given workability, increasing the maximum size of aggregate lowers the water requirement of the mix, so that, if the water/cement ratio is fixed from durability requirements, then the cement content can be reduced by the use of larger-size aggregate (see page 66).

The requirements of ACI 318–83 for reinforced concrete are given in Tables 14.6 and 14.7; here, the maximum water/cement ratio is specified for normal weight concrete while the minimum strength is specified for lightweight aggregate concrete because of the uncertainty of knowing the free water/cement ratio. It must be remembered that air entrainment is essential under conditions of freezing and thawing or exposure to de-icing salts (see Table 15.3), although entrained air will not protect concrete containing coarse aggregate that undergoes disruptive volume changes when frozen in a saturated condition.

Table 14.2 lists the appropriate types of cement and the maximum water/cement ratios (or minimum strengths in the case of lightweight aggregate concrete) for various conditions of exposure to sulphates according to ACI 318–83. BS 8110: Part 1: 1985 stipulates the minimum cement content for various maximum sizes of aggregate, as well as the free water/cement ratio for durable concrete in the presence of sulphates (see Table 14.1).

It should be stressed that, in addition to the correct selection of the type of cement, of water/cement ratio and of air entrainment, also essential for durable concrete are adequate compaction and sufficient moist curing.

Workability and water content

So far we have considered the requirements for the concrete to be satisfactory in the hardened state but, as said before, its properties when being handled and placed are equally important. One essential at this stage is a satisfactory workability.

Table 19.4: **Recommended values of slump for various types of construction as given by ACI 211.1–81**

Type of construction	Range of slump*	
	mm	in.
Reinforced foundation walls and footings	20–80	1–3
Plain footings, caissons and substructure walls	20–80	1–3
Beams and reinforced walls	20–100	1–4
Building columns	20–100	1–4
Pavements and slabs	20–80	1–3
Mass concrete	20–80	1–2

* The upper limit of slump may be increased by 20 mm (1 in.) for compaction by hand.

Table 19.5: Approximate requirements for mixing water and air content for different workabilities and nominal maximum sizes of aggregates according to ACI 211.1–81

Workability or air content	Water content, kg/m³ (lb/yd³) of concrete for indicated maximum aggregate size							
	10 mm ($\frac{3}{8}$ in.)	12.5 mm ($\frac{1}{2}$ in.)	20 mm ($\frac{3}{4}$ in.)	25 mm (1 in.)	40 mm ($1\frac{1}{2}$ in.)	50 mm (2 in.)	70 mm (3 in.)	150 mm (6 in.)
	Non-air entrained concrete							
Slump:								
30–50 mm (1–2 in.)	205 (350)	200 (335)	185 (315)	180 (300)	160 (275)	155 (260)	145 (220)	125 (190)
80–100 mm (3–4 in.)	225 (385)	215 (365)	200 (340)	195 (325)	175 (300)	170 (285)	160 (245)	140 (210)
150–180 mm (6–7 in.)	240 (410)	230 (385)	210 (360)	205 (340)	185 (315)	180 (300)	170 (270)	–
Approximate entrapped air content, per cent	3	2.5	2	1.5	1	0.5	0.3	0.2

Air-entrained concrete

Slump:								
30–50 mm (1–2 in.)	180 (305)	175 (295)	165 (280)	160 (270)	145 (250)	140 (240)	135 (205)	120 (180)
80–100 mm (3–4 in.)	200 (340)	190 (325)	180 (305)	175 (295)	160 (275)	155 (265)	150 (225)	135 (200)
150–180 mm (6–7 in.)	215 (365)	205 (345)	190 (325)	185 (310)	170 (290)	165 (280)	160 (260)	–
Recommended average total air content, per cent:								
Mild exposure	4.5	4.0	3.5	3.0	2.5	2.0	1.5*	1.0*
Moderate exposure	6.0	5.5	5.0	4.5	4.5	4.0	3.5*	3.0*
Extreme exposure†	7.5	7.0	6.0	6.0	5.5	5.0	4.5*	4.0*

Slump values for concrete containing aggregate larger than 40 mm (1½ in.) are based on slump tests made after removal of particles larger than 40 mm (1½ in.) by wet-screening.

Water contents for nominal maximum size of aggregate of 70 mm (3 in.) and 150 mm (6 in.) are average values for reasonably well-shaped coarse aggregates, well graded from coarse to fine.

* For concrete containing large aggregate which will be wet-screened over the 40 mm (1½ in.) sieve prior to testing of air content, the percentage of air expected in the material smaller than 40 mm (1½ in.) should be as tabulated in the 40 mm (1½ in.) column. However, initial proportioning calculations should be based on the air content as a percentage of the whole mix.

† These values are based on the criterion that a 9 per cent air content is needed in the mortar phase of the concrete.

The workability that is considered desirable depends on two factors. The first of these is the size of the section to be concreted and the amount and spacing of reinforcement; the second is the method of compaction to be used.

It is clear that when the section is narrow and complicated, or when there are numerous corners or inaccessible parts, the concrete must have a high workability so that full compaction can be achieved with a reasonable amount of effort. The same applies when embedded steel sections or fixtures are present, or when the amount and spacing of reinforcement make placing and compaction difficult. Since these features of the structure are determined during its design, the designer of the mix is presented with fixed requirements and has little choice. On the other hand, when no such limitations are present, workability may be chosen within fairly wide limits, but the means of compaction must be decided upon accordingly; it is important that the prescribed method of compaction is actually used during the entire progress of construction. A guide to workability for different types of construction is given in Tables 5.1 and 19.4.

The cost of labour is largely influenced by the workability of the mix: a workability which is inadequate for the available means of compaction results in high labour costs if the concrete is to be compacted sufficiently. For a given cement content, the optimum workability is controlled by the fine/coarse aggregate ratio and the maximum aggregate size. Workability can be improved by adding more cement and water or by using admixtures (see Chapter 8) but cost has to be considered.

Having chosen the workability, we can estimate the water content of the mix (mass of water per unit volume of concrete). ACI 211.1–81 gives the water content for various maximum sizes of aggregate and workabilities, with and without air entrainment (see Table 19.5). The values

Table 19.6: **Approximate free water content required to give various levels of workability according to the 1988 British method**

Aggregate		Water content kg/m³, (lb/yd³) for:				
Max. size mm (in.)	Type	Slump mm (in.)	0–10 (0–½)	10–30 (½–1)	30–60 (1–2½)	60–180 (2½–7)
		Vebe s	>12	6–12	3–6	0–3
10 (⅜)	Uncrushed		150 (255)	180 (305)	205 (345)	225 (380)
	Crushed		180 (305)	205 (345)	230 (390)	250 (420)
20 (¾)	Uncrushed		135 (230)	160 (270)	180 (305)	195 (330)
	Crushed		170 (285)	190 (320)	210 (355)	225 (380)
40 (1½)	Uncrushed		115 (195)	140 (235)	160 (270)	175 (295)
	Crushed		155 (260)	175 (295)	190 (320)	205 (345)

Building Research Establishment, Crown copyright

372

Table 19.7 : **Reductions in the free water contents of Table 19.6 when using PFA**

Percentage of PFA in cementitious material	Reduction in water content, kg/m³ (lb/yd³) for:			
Slump (mm) (in.)	0–10 (0–½)	10–30 (½–1)	30–60 (1–2½)	60–100 (2½–7)
Vebe s	> 12	6–12	3–6	0–3
10	5 (10)	5 (10)	5 (10)	10 (20)
20	10 (20)	10 (20)	10 (20)	15 (25)
30	15 (25)	15 (25)	20 (35)	20 (35)
40	20 (35)	20 (35)	25 (40)	25 (40)
50	25 (40)	25 (40)	30 (50)	30 (50)

Building Research Establishment, Crown copyright

apply for well-shaped angular coarse aggregates and, although the water requirement is influenced by the texture and shape of the aggregate, the values given are sufficiently accurate for a first estimate. The 1988 British method of mix design uses a similar approach to estimate the free water content but uncrushed and crushed aggregates are differentiated (see Table 19.6). In the case of air-entrained concrete, the free water content is selected for the next less-workable category of Table 19.6, e.g. the water content for a required slump of 30–60 mm ($1-2\frac{1}{2}$ in.) is selected from the 10–30 mm ($\frac{1}{2}$–1 in.) slump category.

For a given workability, the water content of a mix containing PFA depends upon level of replacement of Portland cement. Therefore, we adjust the estimated water content of a Portland-cement-only mix by the amounts of Table 19.7. It should be noted that, for a given workability, a PFA mix has a lower slump than a Portland-cement-only mix; in practice, a general rule is to allow a reduction in slump of approximately 25 mm (1 in.).

Choice of aggregate

As we stated earlier, in reinforced concrete, the maximum size of aggregate which can be used is governed by the width of the section and the spacing of the reinforcement. With this proviso, it is generally considered desirable to use as large a maximum size of aggregate as possible. However, it should be remembered that the improvement in the properties of concrete with an increase in the size of aggregate does not extend beyond about 40 mm ($1\frac{1}{2}$ in.) so that the use of even larger sizes may not be advantageous (see page 66).

Furthermore, the use of a larger maximum size means that a greater number of stockpiles has to be maintained and the batching operations become correspondingly more complicated. This may be uneconomical on small sites, but where large quantities of concrete are to be placed the extra handling cost may be offset by a reduction in the cement content of the mix.

The choice of the maximum size of aggregate may also be governed by the availability of material and by its cost. For instance, when various sizes are screened from a pit it is generally preferable not to reject the largest size, provided this is acceptable on technical grounds.

The remarks in the preceding paragraph apply equally to the considerations of aggregate grading, as it is often more economical to use the material available locally even though it may require a richer mix (but provided it will produce concrete free from segregation), rather than to bring in a better graded aggregate from farther afield.

An important feature of satisfactory aggregate is the uniformity of its grading. In the case of coarse aggregate, this is achieved comparatively easily by the use of separate stockpiles for each size fraction. However, considerable care is required in maintaining the uniformity of grading of fine aggregate, and this is especially important when the water content of the mix is controlled by the mixer operator on the basis of a constant workability: a sudden change toward finer grading requires additional water for the workability to be preserved, and this means a lower strength of the batch concerned. Also, an excess of fine aggregate may make full compaction impossible and thus lead to a drop in strength.

In general terms, we can say that, while narrow specification limits for aggregate grading may be unduly restrictive, it is essential that the grading of aggregate varies from batch to batch within prescribed limits only.

As we stated on page 64, there is no ideal combined grading of fine and coarse aggregates because of the influence of several interacting factors on workability. Instead, practical gradings are recommended; those for fine aggregate are given in Table 3.8 and for coarse aggregate in Tables 3.9 and 3.10.

For mass concrete with a maximum size of aggregate larger than 40 mm ($1\frac{1}{2}$ in.), ACI 211.1–81 recommends the combination of coarse aggregate fractions to give maximum density and minimum voids. In such a case, a parabolic grading curve for the percentage of material passing a sieve size represents the 'ideal' grading, viz.

$$\text{in SI units: } P = \frac{d^x - 3.76^x}{D^x - 3.76^x} \times 100 \tag{19.1a}$$

$$\text{in US units } P = \frac{d^x - 0.1875^x}{D^x - 0.1875^x} \times 100 \tag{19.1b}$$

where P = the cumulative percentage passing the d-size sieve,
$\quad d$ = sieve size in mm (in.),
$\quad D$ = nominal maximum size of aggregate in mm (in.), and
$\quad x$ = exponent (0.5 for rounded and 0.8 for crushed aggregate).

374

Table 19.8: 'Ideal' combined grading for coarse aggregate of nominal maximum size of 150 mm (6 in.) and 75 mm (3 in.) as given by Eq. (19.1)

Sieve size		Cumulative percentage passing for nominal maximum size of aggregate in mm (in.)			
		150 (6)		75 (3)	
mm	in.	Crushed	Rounded	Crushed	Rounded
150	6	100	100	–	–
125	5	85	89	–	–
100	4	70	78	–	–
75	3	55	64	100	100
50	2	38	49	69	75
37.5	$1\frac{1}{2}$	28	39	52	61
25	1	19	28	34	44
19	$\frac{3}{4}$	13	21	25	33
9.5	$\frac{3}{8}$	5	9	9	14

Table 19.8 gives the idealized combined grading for the 150 mm (6 in.) and 75 mm (3 in.) nominal maximum sizes of aggregate, according to Eq. (19.1).

To demonstrate the proportioning of fractions of crushed coarse aggregate so as to obtain the ideal combined grading of the first column of Table 19.8, let us consider four size fractions: 150 to 75 mm (6 to 3 in.), 75 to 37.5 mm (3 to $1\frac{1}{2}$ in.), 37.5 to 19 mm ($1\frac{1}{2}$ to $\frac{3}{4}$ in.) and 19 to 4.76 mm ($\frac{3}{4}$ to No. 4). The gradings of these fractions are given in Table 19.9.

Let a be the proportion of 150 to 75 mm (6 to 3 in.), b the proportion of 75 to 37.5 mm (3 to $1\frac{1}{2}$ in.), c the proportion of 37.5 to 19 mm ($1\frac{1}{2}$ to $\frac{3}{4}$ in.) and d the proportion of 19 to 4.76 mm ($\frac{3}{4}$ to No. 4), all in the total coarse aggregate. To satisfy the condition that 55 per cent of the combined aggregate passes the 75 mm (3 in.) sieve, we have

$$0.10a + 0.92b + 1.0c + 1.0d = 0.55(a + b + c + d).$$

The condition that 28 per cent of the combined aggregate passes the 37.5 mm ($1\frac{1}{2}$ in.) sieve requires

$$0.06b + 0.94c + 1.0d = 0.28(a + b + c + d).$$

Similarly, for the 19 mm ($\frac{3}{4}$ in.) sieve,

$$0.04c + 0.92d = 0.13(a + b + c + d).$$

Table 19.9: **Example of grading of individual crushed coarse aggregate fractions to be combined into an 'ideal' grading for mass concrete**

Sieve size		Cumulative percentage passing for fraction			
		150–75 mm (6–3 in.)	75–37.5 mm (3–1½ in.)	37.5–19 mm (1½–¾ in.)	19–4.76 mm (¾ in.–No. 4)
mm	in.	(1)	(2)	(3)	(4)
175	7	100	–	–	–
150	6	98	–	–	–
100	4	30	100	–	–
75	3	10	92	–	–
50	2	2	30	100	–
37.5	1½	0	6	94	–
25	1	0	4	36	100
19	¾	0	0	4	92
9.5	⅜	0	0	2	30
4.76	No. 4	0	0	0	2

To solve the above equations for a to d, we can assume that $a = 1$ so that b, c and d can be calculated as a fraction of 1. Such a procedure yields the proportions

$$a:b:c:d = 1:0.50:0.28:0.29$$

or, in other words, we require 48 per cent of the fraction 150 to 75 mm (6 to 3 in.), 24 per cent of 75 to 37.5 mm (3 to 1½ in.), 14 per cent of 37.5 to 19 mm (1½ to ¾ in.) and 14 per cent of 19 to 4.76 mm (¾ to No. 4 ASTM).

To check the grading of the combined aggregate, we multiply columns (1), (2), (3) and (4) of Table 19.8 by 1, 0.50, 0.28 and 0.29, respectively. The four products of each line are then summed and divided by the sum of $a + b + c + d$, viz. 2.07, to give the combined grading of the aggregate.

Figure 19.2 compares the calculated combined grading with the 'ideal' curve. It can be seen that exact agreement occurs at the specified cumulative percentages passing the sieve sizes which we chose in the derivation of the constants b to d. However, the method does not yield perfect agreement for the other sieve sizes but the deviations are not unduly large.

The method is, therefore, useful in estimating the batch quantities of the various size fractions to make up the coarse aggregate for a given grading. This applies for any maximum size of aggregate but for more than four size fractions the procedure is laborious. For mass concrete, ACI 211.1–81 uses a trial and error method for this purpose.

When the maximum size of aggregate is 40 mm (1½ in.) or smaller, an approximate guide for dividing the coarse aggregate into size fractions (as

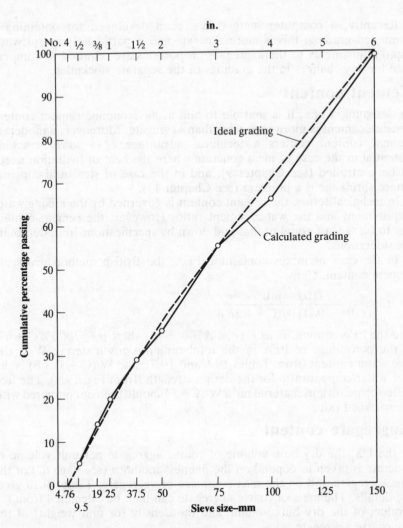

Fig. 19.2: Comparison of the calculated combined grading of four different crushed coarse aggregate fractions with the 'ideal' grading of Table 19.8 for a nominal maximum size of aggregate of 150 mm (6 in.)

a percentage) is as follows:

Total coarse aggregate	5–10 mm ($\frac{3}{16}$–$\frac{3}{8}$ in.)	10–20 mm ($\frac{3}{8}$–$\frac{3}{4}$ in.)	20–40 mm ($\frac{3}{4}$–$1\frac{1}{2}$ in.)
100	33	67	—
100	18	27	55

Recently, a computer method has been developed for obtaining a combined grading; this is more accurate and is particularly useful when rapid adjustments to the aggregate proportions are required to compensate for any changes in the gradings of the separate stockpiles.

Cement content

In designing a mix, it is sensible to aim at an economic cement content because cement is more expensive than aggregate. Moreover, a moderate cement content confers a technical advantage of a lower cracking potential in the case of mass concrete where the heat of hydration needs to be controlled (see Chapter 9), and in the case of structural concrete where shrinkage is a problem (see Chapter 13).

In technical terms, the cement content is governed by the mixing water requirement and the water/cement ratio. However, the cement content has to be at least equal to that laid down by specifications from durability considerations.

In the case of mixes containing PFA, the British method gives the cement content, C, as

$$C = \frac{(100 - p)W}{(100 - 0.7)[W/(C + 0.3F)]}$$

and the PFA content, F, as $F = pC/(100 - p)$, where $p = 100\,F/(C + F)$ is the percentage of PFA in the total cementitious material; W is the free water content (from Tables 19.6 and 19.7); and $W/(C + 0.3\,F)$ is the free water/cement ratio for the design strength (from Fig. 19.1). The free water/cementitious material ratio $W/(C+F)$ should then be compared with the specified value.

Aggregate content

In the US, the dry bulk volume of coarse aggregate per unit volume of concrete is taken to depend on the fineness modulus (see page 62) of the fine aggregate and on the maximum size of aggregate; Table 19.10 gives the details. The mass of coarse aggregate can then be calculated from the product of the dry bulk volume and the density (or unit weight) of the dry coarse aggregate.

The fine aggregate content per unit volume of concrete is then estimated using either the mass method or the volume method. In the former, the sum of the masses of cement, coarse aggregate and water is subtracted from the mass of a unit volume of concrete, which is often known from previous experience with the given materials. However, in the absence of such information, Table 19.11 can be used as a first estimate; adjustment is made after trial mixes. A more precise estimate is obtained from the following equation:

in kg/m^3: $\rho = 10\,\gamma_a(100 - A) + C\left(1 - \dfrac{\gamma_a}{\gamma}\right) - W(\gamma_a - 1)$ **(19.2a)**

in lb/yd^3: $\rho = 16.85\gamma_a(100 - A) + C\left(1 - \dfrac{\gamma_a}{\gamma}\right) - W(\gamma_a - 1)$ **(19.2b)**

where ρ = density (unit weight) of fresh concrete, kg/m^3 (lb/yd^3);

Table 19.10: **Dry bulk volume of coarse aggregate per unit volume of concrete as given by ACI 211.1–81**

Maximum size of aggregate		Dry bulk volume of rodded coarse aggregate per unit volume of concrete for fineness modulus of sand of:			
mm	in.	2.40	2.60	2.80	3.00
10	$\frac{3}{8}$	0.50	0.48	0.46	0.44
12.5	$\frac{1}{2}$	0.59	0.57	0.55	0.53
20	$\frac{3}{4}$	0.66	0.64	0.62	0.60
25	1	0.71	0.69	0.67	0.65
40	$1\frac{1}{2}$	0.75	0.73	0.71	0.69
50	2	0.78	0.76	0.74	0.72
70	3	0.82	0.80	0.78	0.76
150	6	0.87	0.85	0.83	0.81

The values given will produce a mix with a workability suitable for reinforced concrete construction. For less workable concrete, e.g. that used in road construction, the values may be increased by about 10 per cent. For more workable concrete, such as may be required for placing by pumping, the values may be reduced by up to 10 per cent.

Table 19.11: **First estimate of density (unit weight) of fresh concrete as given by ACI 211.1–81**

Maximum size of aggregate		First estimate of density (unit weight) of fresh concrete			
		Non-air-entrained		Air-entrained	
mm	in.	kg/m³	lb/yd³	kg/m³	lb/yd³
10	$\frac{3}{8}$	2285	3840	2190	3690
12.5	$\frac{1}{2}$	2315	3890	2235	3760
20	$\frac{3}{4}$	2355	3960	2280	3840
25	1	2375	4010	2315	3900
40	$1\frac{1}{2}$	2420	4070	2355	3960
50	2	2445	4120	2375	4000
70	3	2465	4160	2400	4040
150	6	2505	4230	2435	4120

γ_a = weighted average bulk specific gravity (SSD) of combined fine and coarse aggregates. Clearly, this needs to be determined from tests;

A = air content, per cent;

C = cement content, kg/m³ (lb/yd³);

γ = specific gravity of cement (generally 3.15[1]); and
W = mixing water requirement, kg/m^3 (lb/yd^3).

The volume method is an exact procedure for calculating the required amount of fine aggregate. Here, the mass of fine aggregate, A_f, is given by:

in kg/m^3: $A_f = \gamma_f \left[1000 - \left(W + \dfrac{C}{\gamma} + \dfrac{A_c}{\gamma_c} + 10A \right) \right]$ (19.3a)

in lb/ft^3: $A_f = \gamma_f \left[27 - \left(W + \dfrac{C}{\gamma} + \dfrac{A_c}{\gamma_c} + 0.27A \right) \right]$ (19.3b)

where A_c = coarse aggregate content, kg/m^3 (lb/ft^3);
  γ_f = bulk specific gravity (SSD) of fine aggregate; and
  γ_c = bulk specific gravity (SSD) of coarse aggregate.

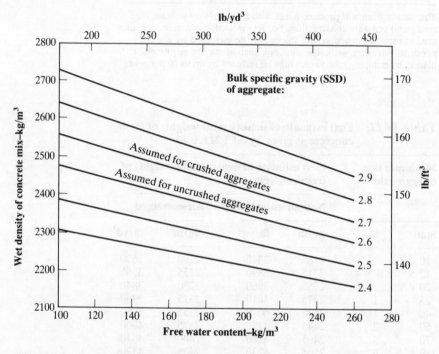

Fig. 19.3: Estimated wet density for fully compacted concrete
(Crown copyright)
(From: D. C. TEYCHENNÉ, J. C. NICHOLLS, R. E. FRANKLIN and D. W. HOBBS, *Design of Normal Concrete Mixes*, pp. 42 (Building Research Establishment, Department of the Environment, London, HMSO, 1988).)

---

[1] For Portland cements.

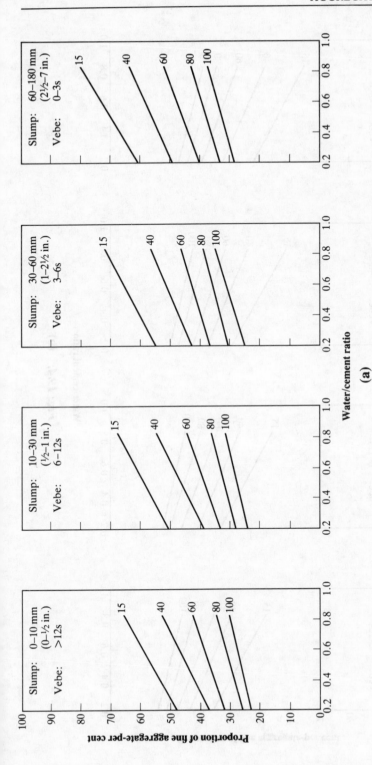

Fig. 19.4: Recommended percentage of fine aggregate in total aggregate as a function of free water/cement ratio for various values of workability and maximum size of aggregate: (a) 10 mm ($\frac{3}{8}$ in.), (b) 20 mm ($\frac{3}{4}$ in.), and (c) 40 mm ($1\frac{1}{2}$ in.). Numbers on each graph are the percentage of fines passing a 600 μm sieve. (Building Research Establishment, Crown copyright)

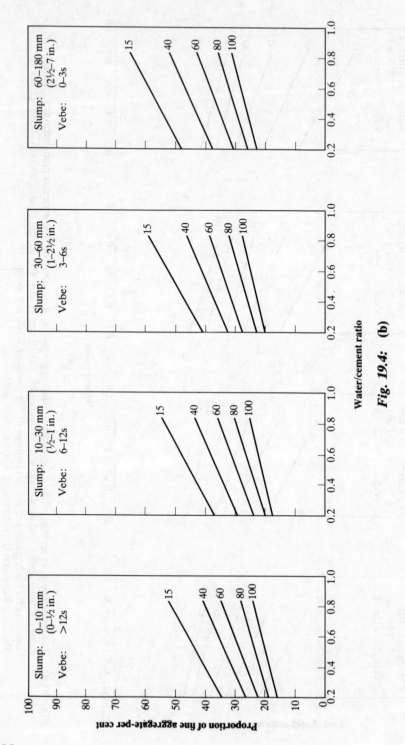

Fig. 19.4: (b)

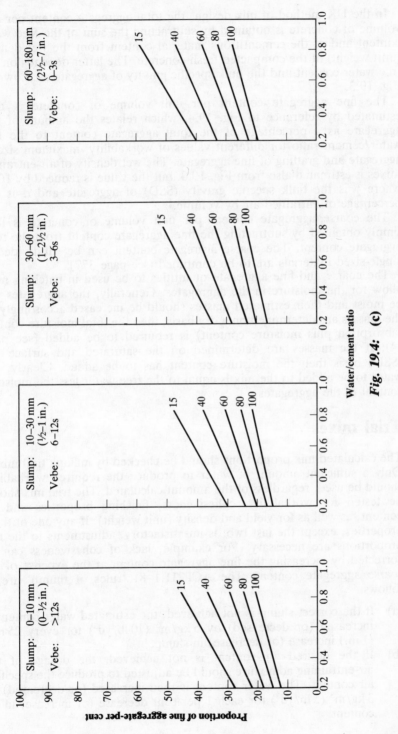

Water/cement ratio

Fig. 19.4: (c)

In the UK method of mix design, the total aggregate content per unit volume of concrete is obtained by subtracting the sum of the free water content and of the cementitious material content from the wet density (unit weight) of the compacted fresh concrete. The latter depends on the free water content and the bulk specific gravity of aggregate as shown in Fig. 19.3.

The fine aggregate content per unit volume of concrete is then estimated by reference to Fig. 19.4, which relates the amount of fine aggregate as a percentage of the total aggregate content to the free water/cement ratio for different values of workability, maximum size of aggregate and grading of fine aggregate. The wet density of air-entrained mixes is estimated also from Fig. 19.3, but the value is reduced by $10\gamma a$, where γ is the bulk specific gravity (SSD) of aggregate and a is the percentage of entrained air by volume.

The coarse aggregate content per unit volume of concrete is then simply obtained by subtracting the fine aggregate content from the total aggregate content. The coarse aggregate content can be subdivided if single-sized materials are to be combined (see page 377).

The coarse and fine aggregate quantities to be used in batching must allow for the moisture in the aggregates. Generally, the aggregates will be moist and their estimated masses should be increased accordingly. If the masses are determined on a dry basis, then the total moisture content (absorption plus moisture content) is required to be added (see page 54); if the masses are determined on the saturated and surface dry (SSD) basis then the moisture content has to be added. Clearly, the water to be added to the mix is equal to the free water less the moisture content of the aggregates.

Trial mixes

The calculated mix proportions should be checked by making trial mixes. Only a sufficient amount of water to produce the required workability should be used, regardless of the amount calculated. The trial mix should be tested for workability, cohesiveness, finishing properties and air content, as well as for yield and density (unit weight). If any one of these properties, except the last two, is unsatisfactory, adjustments to the mix proportions are necessary. For example, lack of cohesiveness can be corrected by increasing the fine aggregate content at the expense of the coarse aggregate content. The ACI 211.1–81 'rules of thumb' are as follows:

(a) If the correct slump is not achieved, the estimated water content is increased (or decreased) by $6 \, \text{kg/m}^3$ ($10 \, \text{lb/yd}^3$) for every 25 mm (1 in.) increase (or decrease) in slump.

(b) If the desired air content is not achieved, the dosage of the air-entraining admixture should be adjusted to produce the specified air content. The water content is then increased (or decreased) by $3 \, \text{kg/m}^3$ ($5 \, \text{lb/yd}^3$) for each 1 per cent decrease (or increase) in air content.

384

(c)   If the estimated density (unit weight) of fresh concrete by the mass method given on page 373 is not achieved and is of importance, mix proportions should be adjusted, allowance being made for a change in air content.

American method — Examples

Example I

Concrete is required for the internal columns of a building. The specified strength is 20 MPa (2900 psi) at 28 days and it is required that no more than 1 test result[2] in 20 will fall below the specified strength. The size of the column sections and the spacing of the reinforcement require a slump of 50 mm (2 in.) and a maximum size of aggregate of 20 mm ($\frac{3}{4}$ in.). Both the coarse and fine aggregate conform to the grading requirements of ASTM C 33–84, the fine aggregate having a fineness modulus of 2.60. Preliminary tests indicate that both aggregates have a specific gravity (SSD) of 2.65 and have negligible absorption and moisture content; the bulk density of coarse aggregate is 1600 kg/m^3 (100 lb/ft^3).

The quantities of ingredients are estimated as follows:

(a)   Since there are no special exposure conditions, ordinary Portland (Type I) cement will be used without air entrainment.

(b)   From previous experience of making concrete with a specified strength of 25 MPa (3600 psi) with similar materials to those proposed, the standard deviation of 20 standard cylinder test results (each being an average of two cylinders) is 3.5 MPa (500 psi). The required probability of inadequate strength (risk) is 1 in 20 and so the appropriate probability factor is 1.64 (see page 330 and Table 17.1). However, since only 20 test results are available, the standard deviation has to be increased by a factor of 1.08 (see Table 17.2). Now, the required average strength, f_m, can be estimated from Eq. (17.2):

$$f_m = 20 + (1.64 \times 3.5 \times 1.08) = 26 \text{ MPa (3800 psi)}.$$

From Table 19.1, the estimated water/cement ratio is 0.60. (It can be noted that, for the specified strength of 20 MPa (2900 psi), Table 19.2 also gives a water/cement ratio of 0.60.)

(c)   From Table 19.5, the mixing water content for non-air-entrained concrete with a slump of 50 mm (2 in.) and a maximum size of aggregate of 20 mm ($\frac{3}{4}$ in.) is 185 kg/m^3 (315 lb/yd^3) of concrete. Also, we note that the approximate entrapped air content is 2 per cent.

(d)   The required cement content is $\frac{185}{0.6} = 308$ kg/m^3 (519 lb/yd^3) of concrete.

(e)   From Table 19.10, for a maximum size of aggregate of 20 mm ($\frac{3}{4}$ in.) and a fine aggregate with a fineness modulus of 2.60, the dry bulk volume of coarse aggregate is 0.64 per unit volume of concrete.

---

[2] Average of two cylinders.

Hence, to produce a workable concrete, the quantity of coarse aggregate required is $0.64 \times 1600 = 1024 \, \text{kg/m}^3$ $(1725 \, \text{lb/yd}^3)$ of concrete. Since the absorption of the coarse aggregate is negligible, no adjustment is necessary to obtain the mass on the SSD basis.

(f) The fine aggregate content can now be estimated using the mass method. From Table 19.11, for the appropriate maximum size of aggregate, the first estimate of the density of non-air-entrained concrete is $2355 \, \text{kg/m}^3$ $(3960 \, \text{lb/yd}^3)$. Alternatively, Eq. (19.2) may be used to estimate the density of concrete, ρ. Assuming the specific gravity of cement to be 3.15, we have

$$\rho = 10 \times 2.65(100 - 2) + 308\left(1 - \frac{2.65}{3.15}\right) - 185(2.65 - 1)$$

or $\rho = 2292 \, \text{kg/m}^3$ $(3861 \, \text{lb/yd}^3)$.

Although this value is lower than the first estimate $(2355 \, \text{kg/m}^3$ $(3960 \, \text{lb/yd}^3))$ the difference is not significant when we remember that we are dealing with estimates which are made only for the purpose of making trial mixes. Moreover, in some other practical instances the specific gravities of coarse and fine aggregates will be different so that it would be necessary to determine the weighted average of the combined aggregate (γ_a) before Eq. (19.2) can be used.

Using the first estimate of density of concrete, we obtain the mass of fine aggregate per unit volume of concrete as

$$2355 - (185 + 308 + 1024) = 838 \, \text{kg/m}^3 \, (1412 \, \text{lb/yd}^3) \text{ of concrete.}$$

The volume method of estimating the mass of fine aggregate per unit volume of concrete is more precise. Equation (19.3) gives

$$A_f = 2.65\left[1000 - \left(185 + \frac{308}{3.15} + \frac{1024}{2.65} + 20\right)\right]$$

or

$$A_f = 824 \, \text{kg/m}^3 \, (1388 \, \text{lb/yd}^3) \text{ of concrete}$$

(g) Since all the masses obtained are on the SSD basis and the aggregates have a negligible moisture content, no further adjustments are necessary.

(h) The estimated quantities in kg per cubic metre (lb per cubic yard) of concrete are as follows:

cement:	308	(519)
fine aggregate:	824	(1388)
coarse aggregate:	1024	(1725)
added water:	185	(315)
Total:	2341	(3947).

(i) A trial mix with the above proportions was then prepared to produce 0.02 m³ (0.026 yd³) of concrete. For this, the expected quantity of water was 3.7 kg but this gave a negligible slump so that it was necessary to increase the water to 4.0 kg (8.8 lb) for a satisfactorily workable mix, which had a measured slump of 25 mm (1 in.) and a measured density (unit weight) of 2320 kg/m³ (3909 lb/yd³). Thus, the quantities used were:

cement:	6.16 kg	(13.55 lb)
fine aggregate:	16.48 kg	(36.26 lb)
coarse aggregate:	20.48 kg	(45.06 lb)
added water:	4.00 kg	(8.80 lb)
Total:	47.12 kg	(103.67 lb).

Now, the volume of the above total mass of concrete (or yield³) was $\frac{47.12}{2320} = 0.0203$ m³ (0.0266 yd³). Hence, the mixing water requirement per unit volume of concrete becomes $\frac{4.0}{0.0203} = 197$ kg/m³ (332 lb/yd³) of concrete. However, this mass of water must be increased by another 6 kg/m³ (10 lb/yd³) of concrete because the specified slump of 50 mm (2 in.) was not achieved in the trial mix (see page 335). Therefore, the quantity of mixing water becomes $197 + 6 = 203$ kg/m³ (341 lb/yd³) of concrete.

The cement content will have to be increased to maintain the same water/cement ratio; the content is $\frac{203}{0.6} = 338$ kg/m³ (569 lb/yd³) of concrete.

Since the workability of the trial mix was found to be satisfactory, the mass of coarse aggregate per unit volume of concrete will be the same as in the trial mix, i.e. $\frac{20.48}{0.0203} = 1009$ kg/m³ (1697 lb/yd³) of concrete.

To re-estimate the fine aggregate content per unit volume of concrete, we can assume the density (unit weight) of concrete to be that measured in the trial mix, viz. 2320 kg/m³ (3909 lb/yd³). Using the mass method, the fine aggregate content is $2320 - (203 + 338 + 1009) = 770$ kg/m³ (1295 lb/yd³) of concrete. Hence, the adjusted masses in kg per cubic metre (lb per cubic yard) of concrete are as follows:

cement:	338	(569)
fine aggregate:	770	(1295)
coarse aggregate:	1009	(1697)
added water:	203	(341)
Total:	2320	(3902).

---

³ See Eq. (5.1).

Example II

Concrete is required for foundations where it will be severely exposed to sulphates. The specified strength is 25 MPa (3630 psi) at 28 days with a slump of 80 to 100 mm ($3\frac{1}{4}$ to 4 in.). The available coarse aggregate has a maximum size of 40 mm ($1\frac{1}{2}$ in.), a dry-rodded bulk density (unit weight) of 1600 kg/m³ (100 lb/ft³), a bulk specific gravity (SSD) of 2.68, absorption of 0.5 per cent and total moisture content of 2 per cent. The fine aggregate has a bulk specific gravity (SSD) of 2.65, absorption of 0.7 per cent, total moisture content of 6 per cent and a fineness modulus of 2.80. The aggregates conform to the ASTM C 33–84 requirements for grading.

The quantities of ingredients are estimated as follows:

(a) For severe exposure to sulphates, Table 14.2 indicates that sulphate-resisting Portland (Type V) cement should be used with a maximum water/cement ratio of 0.45. This water/cement ratio would probably be sufficient to prevent damage due to freezing and thawing but air entrainment will be used anyway because this will provide a more workable concrete. Table 15.3 suggests that 5.5 per cent is appropriate for the maximum aggregate size of 40 mm ($1\frac{1}{2}$ in.).

(b) From Table 19.2, to produce a specified compressive strength of 25 MPa (3630 psi), the water/cement ratio is 0.39, which also satisfies the requirement (a) above and is, therefore, the value to use.

(c) From Table 19.5, the appropriate mixing water content for entrained-air concrete with a slump of 80 to 100 mm ($3\frac{1}{4}$ to 4 in.) and maximum size of aggregate of 40 mm ($1\frac{1}{2}$ in.) is 160 kg/m³ (275 lb/yd³) of concrete.

(d) Hence, the required cement content is $\dfrac{160}{0.39} = 410$ kg/m³ (691 lb/yd³) of concrete.

(e) From Table 19.10, for the appropriate fineness modulus and maximum size of aggregate, the dry bulk volume of coarse aggregate is 0.71 per unit volume of concrete. Hence, to produce a workable concrete, the quantity of dry coarse aggregate is 0.71 × 1600 = 1136 kg/m³ (1914 lb/yd³) of concrete.
We now have to allow for the absorption of the coarse aggregate in order to obtain the mass on the SSD basis, viz. 1136 × 1.05 = 1193 kg/m³ (2010 lb/yd³) of concrete.

(f) The fine aggregate content will now be obtained using the mass method with the first estimate of density (unit weight) of concrete, which is given in Table 19.11 as 2355 kg/m³ (3960 lb/yd³). Since the total content of water plus cement plus coarse aggregate is 160 + 410 + 1193 = 1763 kg/m³ (2971 lb/yd³) of concrete, the fine aggregate content is 2355 − 1763 = 592 kg/m³ (997 lb/yd³) of concrete. (Note that the volume method can also be used; this gives 555 kg/m³ (935 lb/yd³) of concrete).

(g) The quantities obtained are those on the SSD basis. Because the aggregates contain surface moisture, we have to adjust their masses.

Since the total moisture content of the coarse and fine aggregate is 2 and 6 per cent, respectively, their moisture contents are the respective differences between the total moisture content and absorption, viz. $2 - 0.5 = 1.5$ and $6 - 0.7 = 5.3$.

Hence, their wet quantities become

coarse aggregate: $1193 \times 1.015 = 1211 \text{ kg/m}^3$ (2040 lb/yd^3) of concrete

fine aggregate: $592 \times 1.053 = 623 \text{ kg/m}^3$ (1050 lb/yd^3) of concrete.

It should be noted that, although the water absorbed in the aggregates does not take part in the hydration of cement, the surface water does. Consequently, the surface water held by the aggregates has to be deducted from the estimated water to be added to the mix. Therefore, the adjusted mixing water content becomes, $160 - [(1211 - 1193) + (623 - 592)] = 111 \text{ kg/m}^3$ (187 lb/yd^3) of concrete.

(h) The estimated quantities in kg per cubic metre (lb per cubic yard) of concrete are as follows:

cement:	410	(691)
fine aggregate:	623	(1050)
coarse aggregate:	1211	(2040)
added water:	111	(187)
Total:	2355	(3968).

(i) A trial mix is now prepared.

British method — Examples

Example III

A mix is required for a reinforced concrete wall which will be exposed to the 'moderate' conditions of Table 15.1. A mean compressive strength of 30 MPa (4350 psi) is required at the age of 28 days, and the size of the section and reinforcement dictate a nominal cover of 25 mm (1 in.), using a maximum size of aggregate of 20 mm ($\frac{3}{4}$ in.).

The available coarse aggregate is uncrushed, and both the fine and coarse aggregates conform to the gradings of BS 882: 1983, the fine aggregate corresponding to the M grade, of which 50 per cent passes a 600 μm sieve. The aggregates have an absorption of 1 per cent, a total moisture content of 3 per cent, and bulk specific gravity of 2.65.

(a) From Table 19.3, the compressive strength of a mix made with ordinary Portland (Type I) cement and a water/cement ratio of 0.50 is 42 MPa (6100 psi). Hence, from Fig. 19.1, the free water/cement ratio for the required mean compressive strength of 30 MPa (4350 psi) is 0.62. However, by referring to Table 14.5, we find that, for the nominal cover of 25 mm (1 in.) and 'moderate' exposure conditions, the maximum free water/cement ratio permitted is 0.50. Therefore, this lower water/cement ratio will be used.

(b) From Table 5.1, the appropriate workability requires a slump of 75 mm (3 in.) so that the approximate free water content required for uncrushed coarse aggregate with a maximum size of 20 mm ($\frac{3}{4}$ in.) is 195 kg/m³ (328 lb/yd³) of concrete (see Table 19.6). The cement content is, therefore $\frac{195}{0.5} = 390$ kg/m³ (657 lb/yd³) of concrete.

According to Table 14.5, the minimum cement content required is 350 kg/m³ (590 lb/yd³) of concrete. Thus, our estimated cement content is satisfactory.

(c) From Fig. 19.3, for aggregate with a bulk specific gravity of 2.65, the wet density (unit weight) of concrete is 2400 kg/m³ (150 lb/ft³). Hence, the total aggregate content is $2400 - (195 + 390) = 1815$ kg/m³ (3058 lb/yd³) of concrete.

(d) For a free water/cement ratio of 0.50, a slump of 75 mm (3 in.), and maximum size of aggregate of 20 mm ($\frac{3}{4}$ in.), the fine aggregate content is 40 per cent of the total aggregate content (see Fig. 19.4(b)). Hence, the fine aggregate content is $0.4 \times 1815 = 726$ kg/m³ (1223 lb/yd³) of concrete.

(e) The coarse aggregate content is $1815 - 726 = 1089$ kg/m³ (1835 lb/yd³) of concrete.

(f) Since the aggregates contain $3 - 1 = 2$ per cent of surface moisture, we have to adjust the estimated coarse and fine aggregate contents calculated on the SSD basis. Hence, the fine aggregate content becomes $726 \times 1.02 = 741$ kg/m³ (1248 lb/yd³) of concrete, and the coarse aggregate content becomes $1089 \times 1.02 = 1111$ kg/m³ (1872 lb/yd³) of concrete. As always, the surface water on the aggregates is assumed to be available for the hydration of cement so that the mass of this water has to be deducted from the estimated mixing water content, viz. $195 - [(741 - 726) + (1111 - 1089)] = 158$ kg/m³ (266 lb/yd³) of concrete.

(g) The estimated quantities in kg per cubic metre (lb per cubic yard) are as follows:

cement:	390	(657)
fine aggregate:	741	(1248)
coarse aggregate:	1111	(1872)
added water:	158	(266)
Total:	2400	(4043).

(h) Trial mixes are now required.

Example IV

Concrete is required as in Example III except that 30 per cent of the cementitious material is specified as PFA.

We shall use the equations given in the section on Cement Content on page 378. The free water/cement ratio to be used in Fig. 19.1 is $W/(C + 0.3F)$.

(a) From Table 19.3, the compressive strength of an ordinary Portland (Type I) cement/PFA mix with a free $W/(C + 0.3F)$ ratio of 0.5 is 42 MPa (6100 psi). For a mean strength of 30 MPa (4350 psi), Fig. 19.1 indicates a free $W/(C + 0.3F)$ ratio of 0.62. (This value is not to be compared with the maximum permitted value of 0.50 in Table 14.5, but is used only for strength purposes.)

(b) As for Example III, the approximate free water content is 195 kg/m³ (328 lb/yd³) but this requires to be reduced by 20 kg/m³ (34 lb/yd³) according to Table 19.7. The equations on page 378 give the cement content as

$$C = \frac{(100 - 30)(175)}{(100 - 0.7 \times 30)(0.62)} = 250 \text{ kg/m}^3 \ (420 \text{ lb/yd}^3),$$

and the PFA content as

$$F = \frac{30 \times 250}{(100 - 30)} = 107 \text{ kg/m}^3 \ (180 \text{ lb/yd}^3).$$

Hence, the cementitious material content is $250 + 107 = 357$ kg/m³ (600 lb/yd³) and the free water/cementitious material ratio is $175/357 = 0.49$. Both the cementitious material content and water/cementitious material ratio satisfy the specified values of Table 14.5.

(c) As for Example III, the wet density is 2400 kg/m³ (150 lb/ft³). Hence, the total aggregate content is $2400 - (250 + 107 + 175) = 1868$ kg/m³ (3142 lb/yd³) of concrete.

(d) For a free water/cementitious ratio of 0.49, the fine aggregate content is 40 per cent of the total aggregate content (Fig. 19.4(b)), i.e., $0.40 \times 1868 = 747$ kg/m³ (1257 lb/yd³) of concrete.

(e) Therefore, the coarse aggregate content is $1868 - 747 = 1121$ kg/m³ (1885 lb/yd³) of concrete.

(f) Allowing for the surface moisture of the aggregates, the fine aggregate content becomes $747 \times 1.02 = 762$ kg/m³ (1282 lb/yd³), the coarse aggregate content becomes $1121 \times 1.02 = 1143$ kg/m³ (1923 lb/yd³), and the added mixing water content is $175 - [(762 - 747) + (1143 - 1121)] = 138$ kg/m³ (232 lb/yd³) of concrete.

(g) The estimated batch quantities in kg per cubic metre (lb per cubic yard) are as follows:

cement	:	250 (420)
PFA	:	107 (180)
fine aggregate	:	762 (1282)
coarse aggregate	:	1143 (1923)
added water	:	138 (232)
Total	:	2400 (4037)

(h) Trial mixes are now required.

Design of lightweight aggregate mixes

The influence of the water/cement ratio on strength applies to concrete made with lightweight aggregate in the same way as to normal aggregate concrete, and the same procedure of mix design can, therefore, be used when lightweight aggregate is employed. This is the approach in the UK and also in the US for semi-lightweight aggregate concrete (see page 356). However, it is very difficult to determine the SSD bulk specific gravity of lightweight aggregate because of its high absorption (up to 20 per cent), and also because the *rate* of absorption may vary considerably, in some cases the absorption continuing for several days. In consequence, it is difficult to calculate the free water/cement ratio at the time of mixing.

Lightweight aggregate produced artificially is usually bone-dry. If it is saturated before mixing, the strength of the resulting concrete is about 5 to 10 per cent lower than when dry aggregate is used, for the same cement content and workability, and, of course, with an appropriate allowance for the absorbed water in the calculation of the effective water/cement ratio. The explanation lies in the fact that, in the case of bone-dry aggregate, some of the mixing water is absorbed after mixing, but prior to setting, so that the effective water/cement ratio is further reduced. Furthermore, the density of concrete made with saturated aggregate is higher, and the resistance of such concrete to freezing and thawing is impaired. On the other hand, when aggregate with a high absorption is used without pre-soaking, it is difficult to obtain a sufficiently workable and yet cohesive mix. In general, aggregates with absorption of over 10 per cent should be pre-soaked, and air entrainment is recommended.

For many lightweight aggregates, the apparent specific gravity (see page 50) varies with the particle size, the finer particles being heavier than the large ones. Since proportioning is on a *mass* basis, but it is the *volumetric* proportions that govern the physical distribution of material, the percentage of finer material is greater than appears from calculations. Hence, the final volume of voids, the cement paste content and the workability of the mix are affected. We should bear this in mind. If we achieve a well-graded aggregate with a minimum volume of voids the concrete will require only a moderate amount of cement and will exhibit a comparatively small drying shrinkage and thermal movement. Grading limits are given in Tables 18.2 to 18.4.

The ACI 211.2–81 method of mix design is applicable to lightweight aggregate concrete with a compressive strength greater than 17 MPa (2500 psi) at 28 days and an air-dry density (unit weight) of not more than 1840 kg/m³ (115 lb/ft³). The method also applies to semi-lightweight aggregate concrete, provided the above requirements are met.

Trial mixes form the basis of design by either the cement content–strength method or the mass method. The former constitutes a volumetric approach and is applicable to both lightweight and semi-lightweight aggregate concrete, while the mass method is applicable to semi-lightweight aggregate concrete only. The mass method is similar in approach to the mix design of normal weight aggregate concrete, described earlier. The cement content–strength method will now be presented, and this will be followed by a worked example.

If the slump is not specified, an appropriate value for beams, reinforced concrete walls, building columns and floor slabs can be selected from Table 19.4; for trial mixes, the highest value should be used. The maximum size of aggregate should not exceed $\frac{1}{5}$ of the smallest dimension of the member, $\frac{1}{3}$ of the depth of the slab, or $\frac{3}{4}$ of the minimum spacing between reinforcing bars or bundles of bars.

Table 19.12 gives the volume of entrapped air to be expected in non-air-entrained concrete, and the recommended contents of entrained air for durability requirements.

The cement content can be roughly estimated from Table 19.13 but the aggregate producer may be able to provide a closer value.

To estimate the content of lightweight aggregate, it can be assumed that the total volume of aggregate will usually be from 1.0 to 1.2 m³ per

Table 19.12: Air content of air-entrained and non-air-entrained concrete as given by ACI 211.2–81

Level of exposure	Recommended average *total* air content, per cent, for maximum size of aggregate		
	10 mm ($\frac{3}{8}$ in.)	12.5 ($\frac{1}{2}$ in.)	20 mm ($\frac{3}{4}$ in.)
	(a) Air-entrained concrete		
Mild	4.5	4.0	4.0
Moderate	6.0	5.5	5.0
Extreme	7.5	7.0	6.0
	(b) Non air-entrained concrete		
Approximate amount of entrapped air	3	2.5	2.0

Table 19.13: Approximate relation between cement content and strength of lightweight and semi-lightweight aggregate concrete according to ACI 211.2–81

Compressive strength of standard cylinders		Cement content, kg/m^3 (lb/yd^3)	
		All lightweight	Semi-lightweight
MPa	psi	aggregate	aggregate
17	2500	210–310 (350–520)	150–270 (250–460)
21	3000	240–325 (400–550)	190–310 (320–520)
28	4000	300–385 (500–650)	250–355 (420–600)
34	5000	355–445 (600–750)	300–415 (500–700)
41	6000	415–505 (700–850)	355–475 (600–800)

cubic metre (27 to 32 ft^3 per cubic yard) of concrete, as measured on a dry-loose basis. The proportion of fine aggregate is usually between 40 and 50 per cent. For closer estimates, it may again be useful to consult the aggregate supplier.

Knowing the dry-loose densities of fine and coarse aggregate, we calculate their masses on a dry basis, and then a trial mix is made using sufficient water to produce the required slump. This water consists of both added water and that absorbed by the aggregate. After measuring the density (unit weight) of the fresh concrete, the yield can be estimated so that the batch quantities can be calculated.

If the cement content of the trial mix turns out to be different from that specified but the other properties (such as air content, workability and cohesiveness) are satisfactory, the cement content should be adjusted. We assume that the volume of aggregate (dry-loose) should be increased by 0.0006 m^3 (0.01 ft^3) for each 1 kg (1 lb) decrease in cement content, and *vice versa*. This 'rule' applies only to small adjustments to the cement content so that the resulting small changes in the aggregate content will not significantly change the added water requirement.

As a first approximation, it may be assumed that the *total* water also remains unaffected by the adjustments. To allow for the total moisture contents of the aggregate, the dry quantities are simply multiplied by the appropriate total moisture contents, and this increase in the mass of the (wet) aggregates is deducted from the total water requirement to obtain the added water.

A more accurate method of adjusting the mix proportions is to use a so-called pycnometer *specific gravity factor, S,* which is defined as the ratio of the mass of the aggregate at mixing to the effective volume displaced by the aggregate (i.e. the volume of aggregate and its moisture). The mass of the aggregate thus includes any moisture, absorbed or free, at the time of placing the aggregate in the mixer. The pycnometer specific gravity factor differs from the bulk specific gravity (SSD) because it includes the free moisture (see page 54).

394

The value of S is given by (see page 51):

$$S = \frac{A'}{A' - (B - C)}$$

where A' = mass of the aggregate tested (moist or dry),

B = mass of pycnometer with the aggregate and then filled with water (usually after 10 min of sample immersion), and

C = mass of pycnometer filled with water.

By this means, the specific gravity factors for both fine and coarse aggregate can be obtained for different total moisture contents (for example, see Fig. 19.5).

The above, more accurate, procedure for adjusting the mix proportions of lightweight and semi-lightweight aggregate concretes is based on the effective volume approach. For example, if a trial mix has satisfactory properties of workability and cohesiveness but the strength is too low, an increase in the cement content is necessary. The total water requirement and coarse aggregate content are assumed to remain unchanged but the fine aggregate content will have to be reduced. Knowing the specific gravity of cement and the specific gravity factor of coarse aggregate on a dry basis, the volume of fine aggregate can be estimated by deducting the sum of the volumes of cement, coarse aggregate, water and air from the total volume of concrete. Since we know the specific gravity factor (dry) of fine aggregate, the mass of this ingredient can be calculated. For

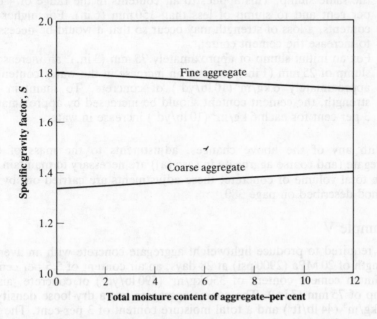

Fig. 19.5: Example of the relation between the pycnometer specific gravity factor S and the total moisture content of lightweight aggregate

semi-lightweight concrete, the mass of the normal weight fine aggregate is obtained by using its bulk specific gravity (SSD).

The above are dry quantities. However, for proportioning, we require the quantities on a wet basis, and the increases in the masses of lightweight fine and coarse aggregates are obtained simply by multiplying the dry masses by their respective total moisture contents; in the case of normal weight fine aggregate, the moisture content is used. Subsequently, the volume of added water is obtained by deducting the sum of the volumes of wet aggregates, cement and air from the total volume of concrete; the volume of added water multiplied by its density (in kg/m^3 or lb/ft^3) gives the mass of added water.

A trial mix should now be made. Density (unit weight), air content and slump should be measured. We should also check the finishing properties of the mix and verify that it does not segregate.

Batch quantities can then be calculated from the yield. When adjustments are required to fine aggregate, air content and slump, the following 'rules of thumb' are recommended:

(a) An increase in each percentage point of the fine to total aggregate ratio requires an increase in water content of approximately $2\,kg/m^3$ ($3\,lb/yd^3$) of concrete. To maintain the strength, the cement content should be increased by approximately 1 per cent for each $2\,kg/m^3$ ($3\,lb/yd^3$) increase in water content.

(b) An increase of 1 per cent in air content requires a decrease in water content of approximately $3\,kg/m^3$ ($5\,lb/yd^3$) of concrete to maintain the same slump. This applies to air contents in the range of 4 to 6 per cent and to slump of less than 150 mm (6 in). For higher air contents, a loss of strength may occur so that it would be necessary to increase the cement content.

(c) For an initial slump of approximately 75 mm (3 in.), an increase in slump of 25 mm (1 in.) requires an increase in the water content of approximately $6\,kg/m^3$ ($10\,lb/yd^3$) of concrete. To maintain the strength, the cement content should be increased by approximately 3 per cent for each $6\,kg/m^3$ ($10\,lb/yd^3$) increase in water.

With any of the above changes, adjustments to the mass of fine aggregate (and coarse aggregate in case (a)) are necessary to maintain the same total volume of concrete; these adjustments are carried out by the method described on page 380.

Example V

It is required to produce lightweight aggregate concrete with an average strength of 20 MPa (2900 psi) at 28 days, an air content of 5.5 per cent, a minimum cement content of $350\,kg/m^3$ ($590\,lb/yd^3$) of concrete, and a slump of 75 mm (3 in.). The coarse aggregate has a dry-loose density of $720\,kg/m^3$ ($44\,lb/ft^3$) and a total moisture content of 3 per cent. The fine aggregate has a dry-loose density of $900\,kg/m^3$ ($56\,lb/ft^3$) and a total moisture content of 7 per cent.

The procedure for choosing the mix proportions is as follows.

Assuming that the total volume of the two aggregates, measured on a dry-loose basis, is $1.2\,m^3$ per cubic metre ($32\,ft^3$ per cubic yard) of concrete, and that the coarse and fine aggregate volumes are equal, the trial mix quantities on a dry basis for $0.02\,m^3$ ($0.026\,yd^3$) of concrete are as follows:

cement:	$350 \times 0.02 =$	7.00 kg	(15.40 lb)
fine aggregate:	$0.60 \times 900 \times 0.02 =$	10.80 kg	(23.76 lb)
coarse aggregate:	$0.60 \times 720 \times 0.02 =$	8.64 kg	(19.01 lb)
water:		5.00 kg	(11.00 lb)
Total:		31.44 kg	(69.17 lb).

The mass of water is that found to produce a slump of 75 mm (3 in.) for the air-entrained trial mix, and consists of the added water and the water absorbed by the aggregates.

The measured density of fresh concrete is found to be $1510\,kg/m^3$ ($92.5\,lb/ft^3$). Hence, the yield is $\dfrac{31.44}{1510} = 0.0208\,m^3$ ($0.736\,ft^3$). The quantities on a dry basis in kg per cubic metre (lb per cubic yard) of concrete are as follows:

cement:	$7.00 \times \dfrac{1}{0.0208} =$	336	(567)
fine aggregate:	$10.80 \times \dfrac{1}{0.0208} =$	519	(874)
coarse aggregate:	$8.64 \times \dfrac{1}{0.0208} =$	415	(700)
water:	$5.00 \times \dfrac{1}{0.0208} =$	246	(405)
Total:		1510	(2546).

Now, the cement content in the trial mix is $14\,kg/m^3$ ($24\,lb/yd^3$) less than that specified. Since the other properties of the mix, including strength, are satisfactory and only a small change in the cement content is required, we can decrease the volume of aggregate on a dry-loose basis by $0.0006\,m^3$ for each 1 kg increase in the cement content. Allocating the reduction equally to fine and coarse aggregates, the quantities of these become:

fine aggregate: $519 - \frac{1}{2}(14 \times 0.0006 \times 900) = 515\,kg/m^3$
                 ($868\,lb/yd^3$) of concrete.

coarse aggregate: $415 - \frac{1}{2}(14 \times 0.0006 \times 720) = 412\,kg/m^3$
                 ($694\,lb/yd^3$) of concrete.

For these small adjustments, the required amount of added water will not be changed appreciably.

To allow for the total moisture content of the aggregate, the quantities of fine and coarse aggregate have to be increased and the water content decreased by the same amount. Hence, the final adjusted quantities on a damp basis in kg per cubic metre (lb per cubic yard) of concrete are as follows:

cement:	$336 + 14 =$	350	(590)
fine aggregate:	$515 \times 1.07 =$	551	(928)
coarse aggregate:	$412 \times 1.03 =$	424	(715)
added water:	$240 - [(551 - 515) + (424 - 412)] =$	192	(323)
Total:		1517	(2556)

Example VI

The same example will be used but let us suppose that the average strength of 20 MPa (2900 psi) at 28 days was not reached using the required minimum cement content of 350 kg per cubic metre (590 lb per cubic yard) of concrete.

To achieve the required strength, we assume that an increase in cement content of 50 kg per cubic metre (84 lb per cubic yard) of concrete is necessary. To adjust the above mix so as to have a cement content of 400 kg/m³ (674 lb/yd³) of concrete, a more accurate method is necessary. Now, we assume that the total water requirement and the coarse aggregate content are unchanged. Since the increased amount of cement provides a fine material, the quantity of fine aggregate should be decreased. To estimate this quantity, we use the pycnometer specific gravity factors, defined on page 394. We need to determine these factors, both on a dry basis and at the total moisture content as shown, for example, in Fig. 19.5.

Let us assume that the specific gravity factors have been found to be 1.78 and 1.35 on a dry basis, and 1.75 and 1.36 on the wet basis, for the fine and coarse aggregate, respectively.

On a dry basis, the volume of concrete with a cement content of 350 kg/m³ (590 lb/yd³) of concrete is

$$\underset{\text{(cement)}}{\frac{350}{3.15 \times 1000}} + \underset{\substack{\text{(fine} \\ \text{aggregate)}}}{\frac{515}{1.78 \times 1000}} + \underset{\substack{\text{(coarse} \\ \text{aggregate)}}}{\frac{412}{1.35 \times 1000}} + \underset{\text{(water)}}{\frac{240}{1000}} + \underset{\text{(air)}}{0.055} = 1.000 \, \text{m}^3 \, (1.308 \, \text{yd}^3).$$

For a mix with a cement content of 400 kg/m³ (674 lb/yd³) of concrete, the sum of the volumes of cement, coarse aggregate, water and air on a dry basis is

$$\frac{400}{1000 \times 3.15} + \frac{412}{1000 \times 1.35} + \frac{240}{1000} + 0.055 = 0.727 \, \text{m}^3 \, (0.951 \, \text{yd}^3).$$

Hence, the volume of fine aggregate is $1.00 - 0.727 = 0.273 \, \text{m}^3$ (0.357 yd³).

398

The mass of fine aggregate on a dry basis is now

$0.273 \times 1000 \times 1.78 = 486 \, \text{kg/m}^3 \, (819 \, \text{lb/yd}^3)$ of concrete.

To estimate the quantities on a wet basis, we allow for the total moisture contents of the aggregates as before so that, in kg per cubic metre (lb per cubic yard) of concrete, they become:

fine aggregate: $486 \times 1.07 = 520 \, (876)$
coarse aggregate: $412 \times 1.03 = 424 \, (715)$.

Now, instead of simply deducting from the total water, the sum of the increase in the mass of the aggregates (from the dry to the wet basis), the added water is obtained more accurately deducting the sum of the volumes of cement, wet aggregate and air from the total volume of concrete, viz.:

$$1.000 - \underset{\text{(cement)}}{\frac{400}{1000 \times 3.15}} - \underset{\substack{\text{(fine} \\ \text{aggregate)}}}{\frac{520}{1000 \times 1.75}} - \underset{\substack{\text{(coarse} \\ \text{aggregate)}}}{\frac{424}{1000 \times 1.36}} - \underset{\text{(air)}}{0.055} = 0.209 \, \text{m}^3 \, (0.273 \, \text{yd}^3).$$

(concrete)

(Note the use of wet specific gravity factors for the aggregates.)

Therefore, the mass of added water is

$1000 \times 0.209 = 209 \, \text{kg/m}^3 \, (352 \, \text{lb/yd}^3)$ of concrete.

Summarizing, the quantities on a wet basis in kg per cubic metre (lb per cubic yard) of concrete are as follows:

cement:	400	(674)
fine aggregate:	520	(876)
coarse aggregate:	424	(715)
added water:	209	(352)
Total:	1553	(2617).

A second trial mix is now carried out to establish whether the concrete is satisfactory, or whether adjustment to the batch quantities are necessary.

Bibliography

19.1 ACI COMMITTEE 211.1–89, Standard practice for selecting proportions for normal, heavyweight, and mass concrete, Part 1, *ACI Manual of Concrete Practice*, 1990.

19.2 ACI 211.2–81, Standard practice for selecting proportions for structural lightweight concrete, Part 1, *ACI Manual of Concrete Practice*, 1990.

19.3   ACI COMMITTEE 318–89, Building code requirements for reinforced concrete, Part 3, *ACI Manual of Concrete Practice*, 1990.

19.4   E. E. BERRY and V. M. MALHOTRA, *Fly ash in Concrete* (Canada Centre for Mineral and Energy Technology, Nov. 1984).

19.5   A. M. NEVILLE, *Properties of Concrete* (London, Longman, 1981).

19.6   S. POPOVICS, *Concrete-Making Materials* (Washington/London, Hemisphere Publishing Corporation, 1979).

19.7   D. C. TEYCHENNÉ, J. C. NICHOLLS, R. E. FRANKLIN and D. W. HOBBS, *Design of Normal Concrete Mixes*, pp. 42 (Building Research Establishment, Department of the Environment, London, HMSO, 1988).

19.8   R. J. TORRENT, A. ALVAREDO and E. POYARD, Combined aggregates: a computer-based method to fit a desired grading, *Materials and Construction*, Vol. 17, No. 98, pp. 139–44, 1984.

Problems

19.1   What is meant by a designed mix?

19.2   What is meant by a prescribed mix?

19.3   Comment on the relation between the maximum aggregate size and section of a concrete number.

19.4   What is meant by free water/cement ratio?

19.5   How do you allow for the moisture content of aggregate in calculating the batch quantities?

19.6   Briefly describe the modified replacement method for the design of concrete mixes with PFA (fly ash).

19.7   What are the main factors in designing concrete for durability?

19.8   Describe how the cost of labour is influenced by the workability of the mix.

19.9   What is the approximate relation between the entrapped air content and maximum aggregate size?

19.10 Compare the strengths and workabilities of concretes of the same mix proportions but one made with rounded coarse aggregate, the other with crushed aggregate.

19.11 You have a mix which is satisfactory. If you add an air-entraining admixture to this mix, what will be the consequences?

19.12 Explain what is meant by partial replacement of Portland cement by PFA.

19.13 What mix properties are specified for concrete which is to be exposed to freezing and thawing?

19.14 What are the arguments for specifying the water/cement ratio for concrete to be exposed to freezing and thawing?

19.15 What are the disadvantages of using too rich a mix?

19.16 Is there such a thing as an ideal grading of fine and coarse aggregates? Discuss.

19.17 Describe the mass method and the volume method for estimating the fine aggregate content per unit volume of concrete, using the ACI method.

19.18 What is the purpose of trial mixes when using a designed mix?

19.19 When designing a lightweight aggregate concrete mix, why is it difficult to determine the free water/cement ratio?

19.20 In the American method of mix design, what is the difference between the cement content–strength method and the mass method?

19.21 Explain what is meant by the pycnometer specific gravity factor.

19.22 The gradings of one fine aggregate and two coarse aggregates are as follows:

Sieve size		Cumulative percentage passing for:		
mm or μm	in. or No.	Fine aggregate	19.0–4.75 mm ($\frac{3}{4} - \frac{3}{16}$ in.)	38.1–19.0 mm ($1\frac{1}{2} - \frac{3}{4}$ in.)
38.1	$1\frac{1}{2}$	100	100	100
19.0	$\frac{3}{4}$	100	99	13
9.5	$\frac{3}{8}$	100	33	8
4.75	$\frac{3}{16}$	99	5	2
2.36	8	76	0	0
1.18	16	58		
600	30	40		
300	50	12		
150	100	2		

It is required to combine the three aggregates so that 24 per cent of the total aggregate passes the 4.75 mm ($\frac{3}{16}$ in.) sieve, and 50 per cent passes the 19.0 mm ($\frac{3}{4}$ in.) sieve. Calculate the grading of the combined aggregate.

Answer: For each of the sieve sizes listed, the cumulative percentage passing is: 100, 50, 34, 24, 17, 13, 9, 3, 0.5.

19.23 Use the American method to design a concrete mix which is required to have a specified mean strength of 30 MPa (4400 psi) at 28 days. The presence of reinforcement requires a slump of 75 mm (3 in.) and a maximum size of aggregate of 10 mm ($\frac{1}{2}$ in.). The aggregates are of normal weight and their gradings conform to the appropriate standard with a fineness modulus of 2.8. Assume: negligible absorption and moisture content; bulk density of coarse aggregate 1600 kg/m³; and extreme exposure conditions.

Answer: Using the first estimate of density and the mass method, the batch quantities in kg/m³ (lb/yd³) are:

cement:	435 (733); ordinary Portland (Type I) cement
fine aggregate:	819 (1380)
coarse aggregate:	736 (1240)
added water:	200 (337)
air content:	7.5 per cent

20

Special concretes

There exist many types of concrete which have been developed for special purposes. In general, the cement-based matrix is modified in some way so as to improve particular properties. Some of these concretes are very recent additions to the concrete scene, and this chapter briefly reviews the present situation so that the reader will become familiar with these newer materials as well as obtain some basic knowledge of their technology.

Polymer–concrete composites

Before discussing the various types of polymer–concrete composites, it may be appropriate to define some chemical terms. A *monomer* is an organic molecule which is capable of combining chemically with similar or different molecules to form a high molecular-weight material, known as polymer. A *polymer* consists of numerous monomers which are linked together in a chain-like structure, and the chemical process which causes these linkages is called *polymerization*. Polymers are classified as either thermoplastics or thermosets. Thermoplastics have long, linear and parallel chains which are not cross-linked, and these polymers exhibit reversibility on heating and cooling. On the other hand, thermosets have randomly oriented chains which are cross-linked, and these polymers do not exhibit reversibility with changes of temperature. Thermoplastics may be converted to thermosets by the use of cross-linking agents.

In general, polymers are chemically inert materials having higher tensile and compressive strengths than conventional concrete. However, polymers have a lower modulus of elasticity and a higher creep, and may be degraded by heat oxidizing agents, ultra violet light, chemicals and micro-organisms; also, certain organic solvents may cause stress cracking. Many of these disadvantages can be overcome by choosing a suitable polymer and by adding substances to the polymer, e.g. antioxidants to suppress oxidation and light stabilizers to reduce ultra violet degradation.

Polymers are used to produce three types of polymer–concrete composites: polymer impregnated concrete (PIC), polymer concrete (PC) and polymer Portland cement concrete (PPCC).

To make *polymer impregnated concrete,* conventional Portland cement concrete is dried and subsequently saturated with a liquid monomer, e.g. methyl methacrylate (MMA) and styrene (S). Polymerization is achieved by gamma radiation or by thermal-catalytic means. With either method, free radicals are generated to form the polymer. For example, using the above monomers, the thermoplastics polymethyl methacrylate (PMMA) and polystyrene (PS) are formed. Greater impregnation is achieved by evacuating the concrete after drying at a temperature of 150 °C (about 300 °F), followed by impregnation with monomer under pressure. For large members, the rates of heating and cooling should be controlled to prevent cracking of the concrete.

Compared with pre-treated concrete, the polymerized product has much higher compressive, tensile and impact strengths, a higher modulus of elasticity and exhibits lower creep and lower drying shrinkage; Table 20.1 gives some typical properties. Moreover, the treated concrete exhibits a higher resistance to freezing and thawing, to abrasion and to chemical attack than the untreated concrete. All of these improvements are due to the lower porosity and permeability of the polymer impregnated concrete, but the extent of change in properties depends on the water/cement ratio, the depth of impregnation, the efficiency of polymerization, the degree of continuity of the polymer phase, and on the mechanical properties of the polymer. On the negative side the coefficient of thermal expansion is higher for polymer impregnated concrete and the mechanical properties on exposure to fire are more seriously affected than in untreated concrete.

The main drawback of polymer impregnated concrete is its high cost but partial impregnation of concrete members may be economically viable. For example, the shear capacity of reinforced concrete beams, without shear reinforcement, can be increased by about 60 per cent and resistance to anchorage stresses is increased. Thus, impregnation at the ends of precast floor units will improve their load-carrying capacity and partial impregnation of bridge decks will increase their flexural strength, reduce deflection and improve water-tightness and surface durability.

Polymer concrete is formed by polymerizing a monomer mixed with aggregate at ambient temperature, using promoter-catalyst systems or curing agents. Early polymer concretes were made with polyester and epoxy resin systems but, at present, the monomer system is based on methyl methacrylate and styrene. When silane is added to the monomer system, it acts as a coupling agent, and the interfacial bond between polymer and aggregate and, therefore, the strength of the composite are improved.

The aggregate to be used in polymer concrete should have a low moisture content and should be graded so as to produce good workability with a minimum amount of monomer or resin; additives in the form of Portland cement or silica flour will improve workability. The fresh polymer concrete can be placed and compacted by vibration in a manner similar to conventional concrete, but solvents are required to clean equipment when epoxies and polyesters are used. Other monomer systems present no cleaning problems but some are volatile and evapor-

404

Table 20.1: Typical mechanical properties of PIC*

Monomer	Polymer loading, per cent of mass	Strength, MPa (psi)			Modulus of elasticity, GPA (10^6 psi)
		Compressive	Tensile	Flexural	
Unimpregnated	0	35 (5100)	2 (290)	4 (600)	19 (2.7)
MMA	4.6–6.7	142 (20 600)	11 (1600)	18 (2600)	44 (6.4)
MMA + 10 per cent TMPTMA	5.5–7.6	151 (21 900)	11 (1600)	15 (2200)	43 (6.2)
Styrene	4.2–6.0	99 (14 400)	8 (1200)	16 (2300)	44 (6.4)
Acrylonitrile	3.2–6.0	99 (14 400)	7 (1000)	10 (1400)	41 (5.9)
Chlorostyrene	4.9–6.9	113 (16 400)	8 (1200)	17 (2500)	39 (5.6)
10 per cent polyester + 90 per cent styrene	6.3–7.4	144 (20 900)	11 (1600)	23 (3300)	46 (6.7)
Vinyl chloride[1]	3.0–5.0	72 (10 400)	5 (700)	– –	29 (4.2)
Vinlidene chloride[1]	1.5–2.8	47 (6800)	3 (400)	– –	21 (3.0)
t-butyl styrene[1]	5.3–6.0	127 (18 400)	10 (1400)	– –	45 (6.5)
60 per cent styrene + 40 per cent MPTMA[1]	5.9–7.3	120 (17 400)	6 (900)	– –	44 (6.4)

* Concrete dried at 105 °C overnight; radiation polymerization.
[1] Dried at 150 °C overnight.
From: J. T. DIKEAU, Development in use of polymer concrete and polymer impregnated concrete, pp. 539–82 in V. M. MALHOTRA (Editor), Progress in Concrete Technology (Energy, Mines and Resources, Ottawa (June 1980)).

405

Table 20.2: Typical mechanical properties of polymer concretes

Polymer–Monomer	Polymer/aggregate ratio	Density		Compressive strength		Tensile strength		Flexural strength		Modulus of elasticity	
		kg/m³	lb/ft³	MPa	psi	MPa	psi	MPa	psi	GPa	10⁶ psi
Polyester	1:10	2400	150	117	17 000	13	1900	37	5400	32	46
Polyester	1:9	2330	146	69	10 000	–	–	17	2500	28	41
Polyester-styrene	1:4	–	–	82	11 900	–	–	–	–	–	–
Epoxy + 40 per cent dibutyl phthalate	1:1*	1650	103	50	7 300	130	18 900	–	–	2	3
Epoxy + polyaminoamide	1:9	2280	143	65	9 400	–	–	23	3300	32	46
Epoxy-polyamide	1:9	2000	125	95	13 800	–	–	33	4800	–	–
Epoxy-furan	1:1*	1700	106	65	9 400	7	1000	0.1	14	–	–
NMA-TMPTMA	1:15	2400	150	137	19 900	10	1400	22	3200	35	51

* Polymer mortar.

From: J. T. DIKEAU, Development in use of polymer concrete and polymer impregnated concrete, pp. 539–82 *in* V. M. MALHOTRA (Editor), *Progress in Concrete Technology* (Energy, Mines and Resources, Ottawa (June 1980)).

ate quickly to produce potentially explosive mixtures; in these instances, specialized non-sparking and explosion-proof equipment is necessary.

Some mechanical properties of polymer concretes are given in Table 20.2.

Typical uses of polymer concrete are in quick repairs for high-volume traffic highways, in the manufacture of precast wall panels reinforced with sheet fibreglass, in floor blocks, and in the production of thin-walled, fibreglass reinforced pipes for water and sewage.

Polymer Portland cement concrete is made by adding to the fresh concrete either a polymer in the form of an aqueous solution or a monomer which is polymerized *in situ*. Rubber latexes, acrylics and vinyl acetates are typical materials which are used together with an anti-foaming agent to minimize entrapped air; air-entraining agents should never be used. Optimum properties are obtained by moist curing from 1 to 3 days, followed by dry curing.

Compared with conventional concrete, the major benefits of polymer Portland cement concrete are its improved durability and better adhesion characteristics. Resistance to freezing and thawing, to abrasion and to impact loading is high but creep is higher than in conventional concrete. Typical uses are overlays for bridge decks and in prefabricated masonry curtain wall panels.

Sulphur–concrete composites

In recent years, there has been an overproduction of sulphur, especially in Canada, a situation which has led to its use as a low-cost construction material, viz. sulphur concrete and sulphur-infiltrated concrete.

Sulphur concrete consists of sulphur, and fine and coarse aggregate, but contains no water or cement. The powdered sulphur and aggregate are mixed in a conventional mixer equipped with a heater so as rapidly to raise the temperature of the mix to 140 °C (284 °F). At that temperature, the ingredients form a uniform mixture which can be cast into moulds. An alternative procedure is slowly to pre-heat the coarse aggregate to about 180 °C (356 °F), and then feed it into a tilting-drum mixer. A sufficient amount of sulphur is added to coat the coarse aggregate, then fine aggregate is added, followed by the remaining sulphur and a workability agent (such as silica flour). On casting, the moulds or forms are overfilled to allow for the contraction of sulphur on cooling, after which the surplus concrete is removed by sawing.

The mix proportions (by mass) for optimum strength and workability are, typically, 20 per cent sulphur, 32 per cent fine aggregate, 48 per cent coarse aggregate and 5 per cent silica flour. The grading of the aggregate should be selected so as to give a minimum void content.

Table 20.3 lists some typical properties of sulphur concrete.

Compared with Portland cement concrete, sulphur concrete gains strength very rapidly and attains about 90 per cent of its ultimate strength

Table 20.3: Physical and mechanical properties of sulphur concrete

Property	Range
Compressive strength, MPa (psi)	28–70 (4000–10 200)
Modulus of rupture, MPa (psi)	3–10 (400–1400)
Direct tensile strength, MPa (psi)	3–8 (400–1200)
Modulus of elasticity, GPa (10^6 psi)	20–45 (2.9–6.5)
Coefficient of thermal expansion, 10^{-6} per °C (10^{-6} per °F)	8–35 (4–19)
Thermal conductivity, J/m²s °C/m (BTU/ft²h °F/ft)	0.4–2.0 (0.23–1.56)
Water absorption, per cent	0–1.5

From: V. M. MALHOTRA, Sulphur concrete and sulphur infiltrated concrete, pp. 583–638 *in* V. M. MALHOTRA (Editor), *Progress in Concrete Technology* (Energy, Mines and Resources, Ottawa (June, 1980)).

in 6 to 8 hours under normal temperature and humidity conditions. The high early-age strength makes sulphur concrete suitable for use in precast units for outdoor applications and its good chemical durability makes it appropriate for industrial plant use. The disadvantages of sulphur concrete are its brittleness, high creep and the corrosive effect on reinforcing steel under wet or humid conditions. Also, sulphur concrete has a low melting point of 119 °C (240 °F) with a consequential loss of strength, it is vulnerable to combustion, with toxic gases being produced, and it has a poor resistance to freezing and thawing. These disadvantages make sulphur concrete unsuitable for most structural applications.

Sulphur-infiltrated concrete is made in a similar way to polymer impregnated concrete but, since sulphur is considerably cheaper than monomers and the impregnation technique is more simple, there is an obvious cost benefit. The water/cement ratio of the concrete to be infiltrated should be high, between 0.7 and 0.8, so that there is no need for external pressure to drive in the sulphur.

After moist curing for 1 day, followed by drying at about 125 °C (257 °F), the concrete is immersed in molten sulphur for a period of time which depends on the type and size of the member. For lean concrete, the procedure is similar but the immersion is in molten sulphur under vacuum for 2 hours before soaking for 30 minutes without vacuum. For low water/cement ratio mixes, external pressure is applied after the evacuation stage to achieve a greater penetration of sulphur.

Compared with conventional concrete having compressive and flexural strengths of 20 and 4 MPa (2900 and 600 psi), respectively, sulphur infiltration increases these values to about 90 and 18 MPa (13 000 and 1200 psi), respectively. The resistance of sulphate infiltrated concrete to attack by sulphates, acids and freezing and thawing is good but, generally, alkaline solutions leach out the sulphur.

Typical applications are precast products, especially where chemical attack can occur. However, like sulphur concrete, sulphur-infiltrated concrete cannot be used where there is a risk of temperature above 100 °C (212 °F).

Fibre reinforced concrete

According to ACI 544.1R–82, fibre reinforced concrete is defined as concrete made with hydraulic cement, containing fine or fine and coarse aggregate, and discontinuous discrete fibres. The fibres can be made from natural material (e.g. asbestos, sisal, cellulose) or are a manufactured product such as glass, steel, carbon, and polymer (e.g. polypropylene, kevlar).

In earlier chapters, we have referred to the somewhat brittle nature of both hydrated cement paste and concrete. The purposes of reinforcing the cement-based matrix with fibres are to increase the tensile strength by delaying the growth of cracks, and to increase the toughness[1] by transmitting stress across a cracked section so that much larger deformation is possible beyond the peak stress than without fibre reinforcement. Figure 20.1(a) demonstrates the enhanced strength and toughness of fibre reinforced concrete in flexure, while Fig. 20.1(b) illustrates the enhanced toughness in compression, the compressive strength not being affected. Fibre reinforcement improves the impact strength and fatigue strength, and also reduces shrinkage.

The quantity of fibres used is small, typically 1 to 5 per cent by volume, and to render them effective as reinforcement the tensile strength, elongation at failure and modulus of elasticity of the fibres need to be substantially higher than the corresponding properties of the matrix. Table 20.4 shows typical values. Moreover, the fibres should exhibit very low creep; otherwise, stress relaxation will occur. Poisson's ratio should be similar to that of the matrix to avoid induced lateral stress; any large lateral stress may affect the interfacial bond which must have a shear strength large enough to allow the transfer of axial stress from the matrix to the fibres.

Some other significant characteristics of the fibres are: aspect ratio (i.e. ratio of length to mean diameter), shape and surface texture, length, and structure. The fibre can withstand a maximum stress σ_f, which depends on the aspect ratio (L/d), viz.:

$$\sigma_f = \tau\left(\frac{L}{d}\right) \tag{20.1}$$

where τ = interfacial bond strength,
d = mean diameter of fibre, and
L = length of fibre $(L < L_c)$.

---

[1] The total energy absorbed prior to total separation of the specimen.

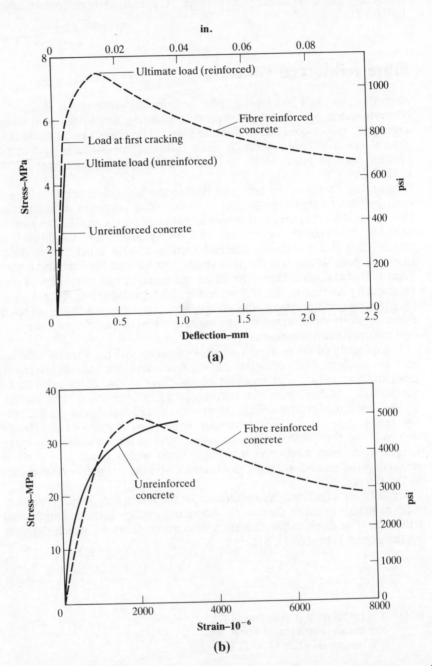

Fig. 20.1: Typical stress-deformation behaviour of fibre reinforced concrete and unreinforced concrete: (a) in flexure of a beam (b) in compression of a cylinder

410

Table 20.4: Typical properties of fibres

Type of fibre	Specific gravity	Tensile strength MPa	psi	Modulus of elasticity GPa	10^6 psi	Elongation at failure, per cent	Poisson's ratio
Crysotile asbestos	2.55	3 to 4.5	435 to 650	164	23.8	3	0.30
Alkali-resistant glass	2.71	2.0 to 2.8	290 to 410	80	11.6	2.0 to 3.0	0.22
Fibrillated polypropylene	0.91	0.65	95	8	1.2	8	0.29 to 0.46
Steel	7.84	1.0 to 3.2	145 to 465	200	29.0	3.0 to 4.0	0.30
Carbon	1.74 to 1.99	1.4 to 3.2	200 to 465	25u to 450	36.2 to 65.3	0.4 to 1.0	0.2 to 0.4
Kevlar	1.45	3.6	520	65 to 130	94.3 to 18.8	2.0 to 4.0	0.32

From: C. D. JOHNSTON, Fibre reinforced concrete, pp. 451–504 in V. M. MALHOTRA (Editor), *Progress in Concrete Technology* (Energy, Mines and Resources, Ottawa (June 1980)).

We can define L_c as the critical length of fibre such that, if $L < L_c$, the fibre will pull out of the matrix due to failure of bond, but if $L > L_c$ then the fibre itself will fail in tension. The length of the fibre should be greater than the maximum size of aggregate particles.

According to Eq. (20.1), the higher the interfacial bond strength the higher the maximum stress in the fibre. Interfacial bond strength is improved by fibres having a deformed or roughened surface, enlarged or hooked ends, and by being crimped. For example, the use of multifilament or fibrillated polypropylene fibres in an open or twined form gives a good interfacial bond, and overcomes the poor adhesion of plastic to cement paste.

Clearly, the orientation of the fibre relative to the plane of a crack in concrete influences the reinforcing capacity of the fibre. The maximum benefit occurs when the fibre is unidirectional and parallel to the applied tensile stress, and the fibres are of less benefit when randomly oriented in three dimensions. This statement is illustrated in Fig. 20.2, which also shows that higher fibre contents lead to a higher strength.

The ultimate strength of the fibre reinforced composite is related to the properties of the matrix and of the fibre as follows:

$$S_c = AS_m(1 - V_f) + BV_f\left(\frac{L}{d}\right) \tag{20.2}$$

where S_c and S_m are the ultimate strengths of the composite and of the matrix, respectively, V_f = volume fraction of fibres, and A = a constant.

The coefficient B of Eq. (20.2) depends on the interfacial bond strength and the orientation of the fibres.

It is useful to note that the modulus of elasticity of a fibre reinforced composite can be estimated from the individual properties of the matrix and of the fibre, using the two-phase composite approach of Chapter 1.

It is important that the fibres be undamaged in the process of incorporation into the matrix; otherwise, the reinforcing effect will be smaller or even absent. Whereas some fibres are robust, others are rather delicate, which means that the manufacturing process has to vary accordingly. For example, to make *asbestos cement*, the fibres, cement and water are mixed to form a slurry suspension, which is gently agitated whilst a thin layer of fresh composite is drawn by off by a moving belt.

To make thin sheets of *glass fibre reinforced cement* (GRC), fibres in the form of a roving (i.e. in a multi-fibrous form) are fed continuously into a compressed air gun which cuts them to the required length and sprays them together with the cement slurry and onto the form. Other methods also exist. The glass has to be resistant to chemical attack by the alkalis in cement.

For steel, plastic or glass fibre reinforcement, standard mixing techniques can be used but the order of feeding the fibres varies according to the size of the batch and to the type of fibre: the mixing operation should provide a uniform dispersion of the fibres and should prevent segregation or balling of the fibres. In this respect, the fibre aspect ratio, size of

412

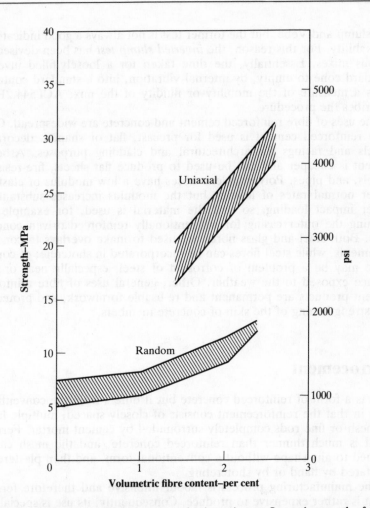

Fig. 20.2: Influence of fibre content and orientation on flexural strength of steel fibre reinforced matrices
(Based on: C. D. JOHNSTON, Fibre reinforced concrete, pp. 451–504 *in* V. M. MALHOTRA, *Progress in Concrete Technology* (Energy, Mines and Resources, Ottawa, (June, 1980).)

coarse aggregate, percentage volume of fibres, rate of addition, and whether the fibres are collated (bundled) or not, are all relevant.

Compared with conventional concrete mixes, fibre reinforced concrete generally has a higher cement content, a higher fine aggregate content and a smaller size of coarse aggregate. For a particular type of fibre, the mix proportions are best determined by trial mixes and the fibre and mix are adjusted as necessary to meet the requirements of workability, strength and durability.

The workability of fibre reinforced mixes decreases as the fibre content increases and as the aspect ratio increases. The usual tests are employed,

413

viz. slump and Vebe, but the former test is not always a good indicator of workability. For this reason, the *inverted slump test* has been devised for fibrous mixes. Essentially, the time taken for a loosely-filled inverted standard cone to empty, by internal vibration, into a standard container gives a measure of the mobility or fluidity of the mix; ACI 544.2R–78 describes the procedure.

The uses of fibre reinforced cement and concrete are widespread. Glass fibre reinforced cement is used for precast, flat or shaped, decorative panels and facings for architectural and cladding purposes. Asbestos cement is cheaper and can be used to produce flat sheets, fire-resistant panels, and pipes. Polypropylene fibres have a low modulus of elasticity under normal rates of loading but the modulus increases substantially under impact loading, so that the material is used, for example, for forming the outer casing for conventionally reinforced driven concrete piles. Both steel and glass fibres are used to make overlays to concrete pavements, while steel fibres can be incorporated in shotcrete; of course, there may be a problem of corrosion of steel, especially near or at a surface exposed to the weather. Other, general uses of fibre reinforced cement products are permanent and re-usable formwork, and protection and strengthening of the skin of concrete members.

Ferrocement

This is a form of reinforced concrete but it differs from the conventional type in that the reinforcement consists of closely spaced, multiple layers of mesh or fine rods completely surrounded by cement mortar. Ferrocement is much thinner than reinforced concrete, and the mesh can be formed to any shape without a conventional form, and then plastered or mortared by hand or by shotcreting.

The manufacturing process is labour intensive and therefore ferrocement is rather expensive to produce. Consequently, its use is specialized: boats, mobile homes, swimming pools, silos, water tanks and curved roofs. These uses are a reflection of a high tensile strength to mass ratio, and of a superior resistance to cracking compared with reinforced concrete.

Generally, the mortar matrix consists of Portland cement or Portland-pozzolan cement and a well-graded sand and, possibly, some small-sized gravel, depending upon the type and size of mesh. A sand/cement ratio of 1.5 to 2.5 by mass and a water/cement ratio of 0.35 to 0.55 by mass will produce a satisfactory matrix, which occupies about 95 per cent of the total volume of ferrocement.

There is a wide variety of reinforcing mesh materials, ranging from chicken wire (woven or interlocking) to welded steel mesh, expanded metal lath, and punched or perforated sheets. It should be noted that the ACI 549R–62 definition of ferrocement includes non-metallic materials such as natural organic fibres, and glass fibres assembled into a two-dimensional mesh.

Roller compacted concrete

Roller compacted concrete (RCC) or lean-rolled concrete, is a dry (no-slump) concrete which has been consolidated by vibrating rollers. Typical applications are in the construction of dams, rapid placement of single layer paving for highways and runways, and multiple layer placements for foundations.

For effective consolidation, roller compacted concrete must be dry enough to support the mass of the vibrating equipment but wet enough to allow the cement paste to be evenly distributed throughout the mass during the mixing and consolidation processes. To ensure adequate bonding of the roller compacted concrete to hardened concrete at the bottom of the lift or at cold joints, segregation must be prevented and a high plasticity bedding mix must be used at the start of placement. Roller compacted concrete can also be used for continuous placements without cold joints.

To minimize the heat of hydration in massive concrete placements, and therefore to minimize cracking, the cement content should be low and the largest possible size of aggregate which will not cause segregation should be used. The use of pozzolans is often more economical than of cement alone, besides having the advantage of a lower heat of hydration. In fact, high PFA (fly ash) contents of 60 to 80 per cent of the volume of cementitious material have been used successfully for the construction of dams.

Very-high strength concrete

Conventional concrete can be made to have strengths of up to 100 MPa (14 500 psi) at 28 days, using good quality and well-graded aggregates. The fundamental parameter is low porosity which is achieved by cement contents in excess of $500 \, kg/m^3$ (about $840 \, lb/yd^3$) of concrete, low water/cement ratios (<0.35), and by adequate compaction and curing. To achieve a normal workable mix, a superplasticizing admixture is necessary.

Very-high strength concrete may be defined as between 60 and 100 MPa (8700 and 14 500 psi). Since the modulus of elasticity does not increase at the same rate as strength, utilization of higher working stresses in high-strength concrete leads to higher strains or deformations than in normal-strength concrete. Also, the higher the strength the more brittle the concrete.

The advantages of very-high strength concrete are that column sections can be reduced in size or, for the same cross-section, the amount of steel reinforcement can be reduced. In tall buildings, there is an economic advantage because of an increased floor area for rental. In bridges, the use of high strengths can reduce the number of beams. For a given level of prestress in prestressed concrete, the loss of prestress due to creep is smaller than in a lower strength member. However, disadvantages appear to be a relatively low shear strength and increased creep and shrinkage due to a lower aggregate content.

415

A related topic is the recently developed special high-strength cements. One such material is known as a *macrodefect-free (MDF)* cement which is made from aluminous cement, mixed with an organic plasticizer and water to form thin products having similar properties to those of a fairly strong plastic. Compressive strengths of 100 to 300 MPa (14 500 × 43 500 psi), flexural strengths of 30 to 45 (4400 to 6500 psi) and moduli of elasticity of 35 to 50 GPa (5.1×10^6 to 7.2×10^6 psi) are typical.

Another product is known as *DSP (densified systems containing homogeneously arranged ultrafine particles)*, in which silica fume is a component. Silica fume is a waste product from the ferro-silicon manufacturing process, and reacts with the lime liberated during the hydration of Portland cement. Very low porosities are achieved due to the much smaller particle size of the silica fume, compared with Portland cement, and to the low water/cementitious material ratio of the mix; a superplasticizer is used to obtain a suitable workability.

Clearly, developments in the whole field of special concretes continue.

Bibliography

20.1 ACI COMMITTEE 207.5R–90, Roller compacted mass concrete, *ACI Manual of Concrete Practice*, Part 1: Materials and General Properties of Concrete, 1990.

20.2 ACI COMMITTEE 544.1R–82 (Reapproved 1986), State-of-the-art report on fiber reinforced concrete, *ACI Manual of Concrete Practice*, Part 5: Masonry, Precast Concrete, Special Processes, 1990.

20.3 ACI COMMITTEE 544.2R–89, Measurement of properties of fiber reinforced concrete, *ACI Manual of Concrete Practice*, Part 5: Masonry, Precast Concrete, Special Processes, 1990.

20.4 ACI COMMITTEE 548.1R–86, Guide for the use of Polymers in concrete, *ACI Manual of Concrete Practice*, Part 5: Masonry, Precast Concrete, Special Processes, pp. 7, 1990.

20.5 ACI COMMITTEE 549.R–88, State-of-the-art report on ferrocement, Part 5: Masonry, Precast Concrete, Special Processes, *ACI Manual of Concrete Practice*, 1990.

20.6 J. L. CLARKE and C. D. POMEROY, Concrete opportunities for the structural engineer, *The Structural Engineer*, Vol. 63A, No. 2, pp. 42–53 (Feb. 1985).

20.7 J. T. DIKEAU, Development and use of polymer concrete and polymer impregnated concrete, pp. 539–82 in V. M. MALHOTRA, *Progress in Concrete Technology*, (Energy, Mines and Resources, Ottawa (June 1980)).

20.8 M. R. H. DUNSTAN, Development of high fly ash content

concrete, *Proc. Inst. Civ. Engrs,* Part 1, Vol. 74, pp. 495–513 (Aug. 1983).

20.9 L. HJORTH, Development and application of high density cement-based materials, *Phil. Trans. Royal. Soc.,* A310, pp. 167–73, 1983.

20.10 C. D. JOHNSTON, Fibre reinforced concrete, pp. 451–504 *in* V. M. MALHOTRA, *Progress in Concrete Technology,* (Energy, Mines and Resources, Ottawa (June 1980)).

20.11 K. KENDALL, A. J. HOWARD, and J. D. PIRCHALL, The relation between porosity, microstructure and strength, and the approach to advanced cement-based materials, *Phil. Trans. Royal Soc.,* A310, pp. 139–53, 1983.

20.12 V. M. MALHOTRA (Editor), *Progress in Concrete Technology,* pp. 796 (Energy, Mines and Resources, Ottawa, (June 1980)).

20.13 V. M. MALHOTRA, Sulphur concrete and sulphur infiltrated concrete: properties, applications and limitations, pp. 583–638 *in* V. M. MALHOTRA (Editor), *Progress in Concrete Technology,* (Energy, Mines and Resources, Ottawa (June 1980)).

20.14 S. P. SHAH, Ferrocement: A new construction material, pp. 505-38 *in* V. M. MALHOTRA (Editor), *Progress in Concrete Technology* (Energy, Mines and Resources, Ottawa, (June 1980)).

Problems

20.1 Explain what is meant by polymer impregnated concrete.
20.2 Explain what is meant by polymer concrete.
20.3 Under what conditions would you recommend the use of polymer impregnated concrete?
20.4 Describe the effects of incorporating fibres in concrete.
20.5 Discuss the significance of the modulus of elasticity of fibres incorporated in concrete.
20.6 Discuss the uses of glass fibre reinforced cement.
20.7 What is polymerization?
20.8 What is meant by roller compacted concrete?
20.9 What is the difference between thermosets and thermoplastics?
20.10 Explain what is meant by polymer Portland cement concrete?
20.11 What is the difference between sulphur concrete and sulphur-infiltrated concrete?
20.12 State the advantages and disadvantages of sulphur concrete.
20.13 What is the inverted slump test?
20.14 Explain what is meant by ferrocement.
20.15 In what situation would you use a PFA (fly ash) content of 70 per cent by volume of cementitious material for making concrete?
20.16 Give three types of very high strength concrete.

21

An overview

A book of this type does not usually include an epilogue but, in this case, it may be useful to look back on what we try to achieve when designing and placing a concrete mix and to compare this ideal with experience.

It is only fair to say at the outset that the gap between the ideal and the practice is often wide. The two are closest to one another on large civil engineering works where many thousands of cubic metres of concrete are placed. In this type of construction, a great deal of effort is devoted to obtaining the best possible concrete for a given purpose.

It may be worthwhile to give, in some detail, the approach to concrete-making on a large project because this would provide a basis for a check-list from which items appropriate to a given project can be selected. The importance of the specification for concrete cannot be overestimated. This should recognize the several types of concrete which may be needed: perhaps one for general structural use, possibly another for prestressed concrete, a concrete with special durability characteristics, and these may include resistance to freezing and thawing, to specific chemical attack, or to abrasion. We are not encouraging a proliferation of concrete mixes (because this complicates site operation and increases the danger of mistakes) but only a recognition of the fact that the mix should be tailored to the technical needs.

Different criteria for the mix will lead to different bases in specification. For instance, strength would be assured by the water/cement ratio; chemical durability by the type of cement and the cement content or the water/cement ratio and possibly by the type of aggregate; resistance to freezing and thawing by the water/cement ratio and the content of entrained-air; placing in a part of the structure where reinforcement is heavily congested by specifying the workability or the use of a particular admixture; and so on.

The specification must, however, be written on the basis of knowledge of what is available or possible. For instance, specifying a type of cement which would have to be imported from afar may be an unduly expensive proposition, and the specification writer would do better to find another solution, such as the use of a high cement content so as to prevent penetration of the attacking fluid or even the use of a protective

418

membrane. We are not listing here the various possible solutions, but merely encouraging consideration of a wide range of means of achieving the desired end.

On a large project, it is unlikely that the aggregate would be brought in over a long distance. It is therefore useful for the specification writer to be familiar with the properties of the aggregates actually available and to design the mix around these properties. What is relevant may be the shape of the aggregate, possibly the content of flaky particles, or the fact that it leads to rather low-strength concrete.

In some cases, some very specific properties of aggregate may be of interest. For instance, in prestressed concrete pressure vessels for nuclear reactors, creep of concrete is of importance, and it may be necessary to test the aggregate for its creep properties before the design can be finalized.

The production of the concrete mix, by which we mean the accuracy and precision of batching and the uniformity of mixing, is clearly better in a large and automated plant than when a portable mixer is used and the batching is done manually. The reason for this is not only the technical quality of the batching and mixing system but also the standard of training of personnel. With a static plant, be it a central plant on site, be it a ready-mix plant, the personnel are likely to be permanent concrete specialists rather than those whose duties happen to include concrete making. For similar reasons, laboratory-produced concrete, even though a small mixer is used, is likely to be of high quality. While this fact is of no value for the purposes of site production, it is worth bearing in mind: it would be wrong to assume that the high quality of laboratory-made concrete can be reproduced on site. Accordingly, the target-strength values for laboratory specimens should be higher, and this is recognized by standards.

Handling, transporting, placing and compacting concrete all depend on the quality of the work force and of supervision. Again, these are better when those involved acquire higher skills through experience than when they work at the given task only occasionally or when the supervisor is concerned with concrete only now and again. It does not follow, however, that, on a small job, good quality concrete cannot be obtained; it can, but conscious and determined effort is required. It is hoped that the reader is by now persuaded of the need for this effort and of the recompense in terms of a structure of good quality.

How is the good quality verified and maintained? On a large construction project, there may well be a site laboratory dedicated to this purpose. The laboratory may be operated by the contractor but more often by an independent testing house. Such a laboratory has to be of high standard because testing houses are themselves subject to control, and, with the laboratory being on the spot, site problems can be resolved quickly and advice can be obtained informally. Protracted correspondence about low-strength test results between the resident engineer and site agent, let alone between their respective head offices, loses much of its value if unsatisfactory concrete continues to be produced while all this is going on. The size of the problem, that is the quantity of concrete which is suspect, increases and the solution becomes more difficult and more expensive.

The preceding paragraphs provide the background against which to consider the concrete quality in a given construction project. When the amount of concrete to be placed is moderate, some of the procedures and precautions may be unnecessary or too expensive. In consequence, safety margins, for instance, in the mean strength or in cement content, should be increased. There is a clear balance between control, on the one hand, and target values and minima, on the other. It is impossible to spell out this balance in a general way, but it is important to bear it in mind in planning every concreting operation.

The individual chapters in this book give the details of the various technical considerations but, in order to achieve good concrete, experience is also necessary. Experience without the background of knowledge may lead to pitfalls; conversely, it is not possible to achieve good concrete in a structure on the basis of book knowledge alone, even if backed by laboratory experience. The successful engineer, that is one who achieves a structure of good quality, is one who combines a sound knowledge of concrete and concrete materials with good experience.

This is a positive note on which to end a book about a material that has been used in innumerable structures but perhaps a final note of warning may be permitted. What has been presented in the book represents good practice of the day based on up-to-date knowledge. But new materials are being developed and new uses for concrete are being found. Often, at first, these changes may be viewed simply as improvements or extensions of *status quo*. This may be so in many respects but, in others, there may lurk an unsuspected side effect, often of the kind that manifests itself only with the passage of time. An intelligent engineer will be on the look-out for the possibilities and will ask himself *all* the necessary questions about possible consequences of any innovation, even those remote from the immediate effect of the change. This was not done in the past when some types of cement and admixture were used or when more efficient methods of compaction allowed a reduction in cement content. The untoward consequences in these cases cost millions of pounds or dollars and led to the need for extensive repairs. We hope that the preceding exhortations will ensure that this will not happen in the future.

Problems

21.1 Why is it better to build structures of concrete than of steel?

21.2 In what respect is steel superior to concrete?

21.3 Why is it necessary to maintain concrete structures?

Relevant American Standards

References to these standards in the text may bear an earlier date but are unchanged in substance.

Cement

ASTM C 109–90	Test for Compressive Strength of Hydraulic Cement Mortars (Using 2 inch or 50 mm cube specimens).
ASTM C 114–88	Standard Methods for Chemical Analysis of Hydraulic Cement
ASTM C 115–91	Test for Fineness of Portland Cement by the Turbidimeter
ASTM C 150–89	Specification for Portland Cement
ASTM C 151–89	Test for Autoclave Expansion of Portland Cement
ASTM C 186–86	Test for Heat of Hydration of Hydraulic Cement
ASTM C 191–82	Test for Time of Setting of Hydraulic Cement by Vicat Needle
ASTM C 204–91a	Test for Fineness of Portland Cement by Air Permeability Apparatus
ASTM C 266–89	Test for Time of Setting of Hydraulic Cement by Gillmore Needles
ASTM C 348–86	Test for Flexural Strength of Hydraulic Cement Mortars
ASTM C 349–82 (Reapproved 1987)	Test for Compressive Strength of Hydraulic Cement Mortars (Using Portions of Prisms Broken in Flexure)
ASTM C 451–89	Test for Early Stiffening of Portland Cement (Paste Method)
ASTM C 595–89	Specification for Blended Hydraulic Cements

ASTM C 618–91	Specification for Fly Ash and Raw or Calcined Natural Pozzolan for use as a Mineral Admixture in Portland Cement Concrete
ASTM C 845–90	Specification for Expansive Hydraulic Cement

Aggregate

ASTM C 29–90	Test for Unit Weight and Voids in Aggregate
ASTM C 33–90	Specification for Concrete Aggregates
ASTM C 40–84	Test for Organic Impurities in Sands for Concrete
ASTM C 70–79 (Reapproved 1985)	Test for Surface Moisture in Fine Aggregate
ASTM C 88–90	Test for Soundness of Aggregates by Use of Sodium Sulfate or Magnesium Sulfate
ASTM C 117–90	Test for Materials Finer than 75 μm (No 200) Sieve in Mineral Aggregates by Washing
ASTM C 123–83	Test for Lightweight Pieces in Aggregate
ASTM C 127–88	Test for Specific Gravity and Absorption of Coarse Aggregate
ASTM C 128–88	Test for Specific Gravity and Absorption of Fine Aggregates
ASTM C 131–88	Test for Resistance to Abrasion of Small Size Coarse Aggregate by Use of Los Angeles Machine
ASTM C 136–84a	Standard Method for Sieve Analysis of Fine and Coarse Aggregates
ASTM C 294–86 (Reapproved 1990)	Descriptive Nomenclature for Constituents of Natural Mineral Aggregates
ASTM C 330–89	Specification for Lightweight Aggregates for Structural Concrete
ASTM C 331–89	Specification for Lightweight Aggregates for Concrete Masonry Units
ASTM C 332–87	Specifications for lightweight Aggregates for Insulating Concrete

422

Admixtures

ASTM C 260–86	Specification for Air Entraining Admixtures for Concrete
ASTM C 494–90	Specification for Chemical Admixtures for Concrete
ASTM C 618–91	Specification for Fly Ash and Raw or Calcined Natural Pozzolan for Use as a Mineral Admixture in Portland Cement Concrete

Concrete

ASTM C 31–90a	Method of Making and Curing Concrete Test Specimens in the Field
ASTM C 33–90	Specification for Concrete Aggregates
ASTM C 39–86	Test for Compressive Strength of Cylindrical Concrete Specimens
ASTM C 42–90	Obtaining and Testing Drilled Cores and Sawn Beams of Concrete
ASTM C 78–84	Test for Flexural Strength of Concrete (Using Simple Beam with Third-Point Loading)
ASTM C 94–90	Specification for Ready-Mixed Concrete
ASTM C 138–81	Test for Unit Weight, Yield and Air Content (Gravimetric) of Concrete
ASTM C 143–90a	Test for Slump of Portland Cement Concrete
ASTM C 156–89	Test for Water Retention by Concrete Curing Materials
ASTM C 157–89	Test for Length Change of Hardened Cement Mortar and Concrete
ASTM C 171–69 (Reapproved 1986)	Specification for Sheet Materials for Curing Concrete
ASTM C 173–78	Test for Air Content of Freshly Mixed Concrete by the Volumetric Method

ASTM C 192–90a	Making and Curing Concrete Test Specimens in the Laboratory
ASTM C 215–85	Test for Fundamental Traverse, Longitudinal and Torsional Frequencies of Concrete Specimens
ASTM C 227–90	Test for Potential Alkali Reactivity of Cement–Aggregate Combinations (Mortar-Bar Method)
ASTM C 231–91	Test for Air Content of Freshly Mixed Concrete by the Pressure Method
ASTM C 232–87	Test for Bleeding of Concrete
ASTM C 234–91a	Test for Comparing Concretes on the Basis of the Bond Developed with Reinforcing Steel
ASTM C 260–86	Specifications for Air-Entraining Admixtures for Concrete
ASTM C 289–87	Test for Potential Reactivity of Aggregates (Chemical Method)
ASTM C 295–90	Petrographic Examination of Aggregates for Concrete
ASTM C 309–89	Specification for Liquid Membrane-Forming Compounds for Curing Concrete
ASTM C 360–82	Test for Ball Penetration in Fresh Portland Cement Concrete
ASTM C 403–90	Test for Time of Setting of Concrete Mixtures by Penetration Resistance
ASTM C 457–90	Recommended Practice for Microscopical Determination of Air-Void Content and Parameters of the Air-Void System in Hardened Concrete
ASTM C 469–87a	Test for Static Modulus of Elasticity and Poisson's Ratio of Concrete in Compression
ASTM C 470–87	Specification for Molds for Forming Concrete Test Cylinders Vertically
ASTM C 496–90	Test for Splitting Tensile Strength of Cylindrical Concrete Specimens
ASTM C 512–87	Test for Creep of Concrete in Compression

ASTM C 586–69 (Reapproved 1986)	Test for Potential Alkali Reactivity of Carbonate Rocks for Concrete Aggregates (Rock Cylinder Method)
ASTM C 597–83	Test for Pulse Velocity through Concrete
ASTM C 617–87	Capping Cylindrical Concrete Specimens
ASTM C 618–91	Specification for Fly Ash and Raw or Calcined Natural Pozzolan for Use as a Mineral Admixture in Portland Cement Concrete
ASTM C 666–90	Test for Resistance of Concrete to Rapid Freezing and Thawing
ASTM C 671–86	Test for Critical Dilation of Concrete Specimens Subjected to Freezing
ASTM C 672–91	Test for Scaling Resistance of Concrete Surfaces Exposed to Deicing Chemical
ASTM C 684–89	Making, Accelerated Curing, and Testing of Concrete Compression Test Specimens
ASTM C 779–89a	Test for Abrasion Resistance of Horizontal Concrete Surfaces
ASTM C 803–90	Test for Penetration Resistance of Hardened Concrete
ASTM C 805–85	Test for Rebound Number of Hardened Concrete
ASTM C 900–87	Test for Pullout Strength of Hardened Concrete
ASTM C 1084–87	Portland-Cement Content of Hardened Hydraulic-Cement Concrete

Relevant British Standards

Cement

B2 12: 1989	Portland Cements
BS 146: Part 2: 1973	Portland-Blastfurnace Cement
BS 915: Part 2: 1972 (1983)	High Alumina Cement
BS 1014: 1975 (1986)	Pigments for Portland Cement and Portland Cement Products
BS 1370: 1979	Low Heat Portland Cement
BS 3892: Part 1: 1982	Specification for Pulverised Fuel Ash for Use as Cementitious Component in Structural Concrete
BS 4027: 1980	Sulphate-Resisting Portland Cement
BS 4246: Part 2: 1974	Low Heat Portland Cement
BS 4248: 1974	Supersulphated Cement
BS 4550	Methods of Testing Cement
BS 4550: Part 0: 1978	General Introduction
BS 4550: Part 1: 1978	Sampling
BS 4450: Part 2: 1970	Chemical Tests
BS 4550: Part 3: 1978	Physical Tests
BS 4550: Part 4: 1978	Standard Coarse Aggregate for Concrete Cubes
BS 4550: Part 5: 1978	Standard Sand for Concrete Cubes
BS 4550: Part 6: 1978	Standard Sand for Mortar Cubes
BS 6588: 1985	Specification for Portland Pozzolana Cement
BS 6610: 1985	Specification for Pozzolanic Cement with Pulverised Fuel Ash as Pozzolana

Aggregate

BS 812	Testing Aggregates
BS 812: Part 1: 1975	Sampling, Shape, Size and Classification
BS 812: Part 2: 1975	Physical Properties
BS 812: Part 100: 1990	General Requirements for Apparatus and Calibration
BS 812: Part 4: 1976	Chemical properties
BS 812: Part 101: 1984	Guide for Sampling and Testing
BS 812: Part 102: 1984	Methods for Sampling
BS 812: Part 103: 1985	Methods for Determination of Particle Size Distribution
BS 812: Part 105	Methods for Determination of Particle Shape
BS 812: Part 105.1: 1985	Method for Determination of Flakiness Index
BS 812: Part 105.2: 1990	Elongation of Coarse Aggregate
BS 812: Part 106: 1985	Method for Determination of Shell Content in Coarse Aggregate
BS 812: Part 107 (Draft)	Method for Determination of Particle Density and Water Absorption
BS 812: Part 109: 1990	Methods for Determination of Moisture Content
BS 812: Part 110: 1990	Methods for Determination of Aggregate Crushing Value
BS 812: Part 111: 1990	Methods for Determination of Ten Percent Fines Value
BS 812: Part 112: 1990	Methods for Determination of Aggregate Impact Value
BS 812: Part 113 (Draft)	Method for Determination of Aggregate Abrasion Value
BS 812: Part 114: 1989	Method for Determination of Polished Stone Value
BS 812: Part 117: 1988	Method for Determination of Water-Soluble Chloride Salts
BS 812: Part 118: 1988	Method for Determination of the Sulphate Content
BS 812: Part 119: 1985	Method for Determination of Acid-Soluble Material in Fine Aggregate

427

BS 812: Part 120: 1989	Methods of Testing and Classifying Drying Shrinkage of Aggregates in Concrete
BS 812: Part 121: 1989	Method for Determination of Soundness
BS 812: Part 124: 1989	Method for Determination of Frost–heave
BS 882: 1983	Aggregates from Natural Sources for Concrete
BS 1047: 1983	Air-Cooled Blastfurnace Slag Coarse Aggregate for use in Construction
BS 3797: Part 2: 1976	Lightweight Aggregates for Concrete

Admixtures

BS 5075: Part 1: 1982	Specification for Accelerating Admixtures, Retarding Admixtures and Water Reducing Admixtures
BS 5075: Part 2: 1982	Specification for Air–Entraining Admixtures
BS 5075: Part 3: 1985	Specification for Superplasticizing Admixtures

Concrete

BS 1305: 1974	Batch Type Concrete Mixers
BS 1881	Testing Concrete
BS 1881: Part 5: 1970	Methods of Testing Hardened Concrete for other than Strength
BS 1881: Part 101: 1983	Methods of Sampling Fresh Concrete on Site
BS 1881: Part 102: 1983	Method for Determination of Slump
BS 1881: Part 103: 1983	Method for Determination of Compacting Factor
BS 1881: Part 104: 1983	Method for Determination of Vebe Time
BS 1881: Part 105: 1984	Methods for Determination of Flow of Fresh Concrete
BS 1881: Part 106: 1983	Methods for Determination of Air Content in Fresh Concrete
BS 1881: Part 107: 1983	Method for Determination of Density of Compacted Fresh Concrete
BS 1881: Part 108: 1983	Method for Making Test Cubes from Fresh Concrete

428

BS 1881: Part 109: 1983   Method for Making Test Beams from Fresh Concrete

BS 1881: Part 110: 1983   Method for Making Test Cylinders from Fresh Concrete

BS 1881: Part 111: 1983   Method of Normal Curing of Test Specimens (20 °C Method)

BS 1881: Part 112: 1983   Methods of Accelerated Curing of Test Cubes

BS 1881: Part 113: 1983   Method for Making and Curing No-Fines Test Cubes

BS 1881: Part 114: 1983   Methods for Determination of Density of Hardened Concrete

BS 1881: Part 115: 1986   Specification for Compression Testing Machines for Concrete

BS 1881: Part 116: 1983   Method for Determination of Compressive Strength of Concrete Cubes

BS 1881: Part 117: 1983   Method for Determination of Tensile Splitting Strength

BS 1881: Part 118: 1983   Method for Determination of Flexural Strength

BS 1881: Part 119: 1983   Method for Determination of Compressive Strength Using Portions of Beams Broken in Flexure (Equivalent Cube Method)

BS 1881: Part 120: 1983   Method for Determination of the Compressive Strength of Concrete Cores

BS 1881: Part 121: 1983   Method for Determination of Static Modulus of Elasticity in Compression

BS 1881: Part 122: 1983   Method for Determination of Water Absorption

BS 1881: Part 124: 1988   Methods of Analysis of Hardened Concrete

BS 1881: Part 125: 1986   Methods for Mixing and Sampling Fresh Concrete in the Laboratory

BS 1881: Part 127: 1990   Methods of Verifying Performance of a Cube Compression Testing Machine Using the Comparative Cube Test

BS 1881: Part 201: 1986   Guide to the Use of Non Destructive Methods of Test for Hardened Concrete

BS 1881: Part 202: 1986   Recommendations for Surface Hardness Testing by Rebound Hammer

BS 1881: Part 203: 1986	Measurement of the Velocity of Ultrasonic Pulses in Concrete
BS 1881: Part 204: 1988	Recommendations on the Use of Electromagnetic Covermeters
BS 1881: Part 205: 1986	Recommendations for Radiography of Concrete
BS 1881: Part 206: 1986	Recommendations for Determination of Strain in Concrete
BS 1881: Part 209: 1990	Recommendations for Measurement of Dynamic Modulus of Elasticity
BS 3148: 1980	Tests for Water for Making Concrete (Including Notes on the Suitability of the Water)
BS 3963: 1974 (1980)	Methods for Testing the Mixing Performance of Concrete Mixers
BS 5328	Concrete
BS 5328: Part 1: 1990	Guide to Specifying Concrete
BS 5328: Part 2: 1990	Methods for Specifying Concrete Mixes
BS 5328: Part 3: 1990	Specification for Procedures to be Used in Producing and Transporting Concrete
BS 5328: Part 4: 1990	Specification for Procedures to be Used in Sampling, Testing and Assessing Compliance of Concrete
BS 5497: Part 1: 1987	Guide for the Determination of Repeatability and Reproducibility for a Standard Test Method by Inter–laboratory Tests
BS 6100	Glossary of Terms for Concrete and Plaster
BS 6100: Part 6: Subsection 6.1: 1984	Binders
BS 6100: Part 6: Subsection 6.2: 1986	Concrete
BS 6100: Part 6: Subsection 6.3: 1984	Aggregates
BS 6100: Part 6: Subsection 6.4: 1986	Admixtures
BS 8110: Part 1: 1985	Structural Use of Concrete: Code of Practice for Design and Construction
BS 8110: Part 2: 1985	Structural Use of Concrete: Code of Practice for Special Circumstances

Index

431

437